D1239381

Valve Selection And Service Guide

John T. Mead

edited by
Matthew McCann

BNP Business News Publishing Company
Troy, Michigan
USA

Library of Congress Cataloging in Publication Data

Mead, John T.
Valve Selection and Service Guide.

1. Valves. I. McCann, Matthew. II. Title.
TJ223.V3M43 1986 621.8'4 85-25517
ISBN 0-912524-25-1

Printed in the United States of America

Contents

Disclaimer

Due to the infinite variations in valve applications and installations, this book can only be considered as a general guide. The author and publisher cannot be responsible for any misunderstanding, misuse or misapplications that may result in harm or loss of any kind, including material or personal injury.

Most valve manufacturers are fully staffed with trained application engineers who are more than willing to help you with specific problems. It is highly recommended that you discuss your valve applications with a manufacturers' representative, professional engineer or a consulting engineer.

Foreword

This guide is not intended to be a highly technical essay for engineers, but rather is designed as a guide for technicians, service personnel and those who must size, select, install or repair valves as parts of their daily tasks. Much of the material is aimed at the air conditioning, heating and refrigeration industry, but the basic valve characteristics and designs may be applied to numerous other fields.

Valve manufacturers have engineering personnel who can assist you in selecting the right material and valve design for a specific application. However, the recommendations they make will be based on the information you provide. If your information is sketchy or inadequate, an erroneous selection can result in a malfunction or failure of equipment involved, or worse, personal injury and/or considerable financial loss.

Some valves have been deliberately omitted in this treatment (specifically control valves and those valves used primarily in hydraulic and pneumatic applications). Not only are valves for those applications unique in their own right, they constitute an entirely separate species and warrant their own detailed studies. However, numerous other valve designs and configurations will be discussed within the course of this text, the objective being to increase the reader's appreciation and familiarity with valves and their proper applications.

1
Introduction

Valves are an important factor in American industrial production. They control thousands of processes that make many of the products we use and the foods we eat every day. Valves control our heating, air conditioning and refrigeration systems and provide comfortable, year-round environments in office buildings, apartments and homes. They assist in preservation of our foods. They control processes in oil refineries, power plants, ships at sea and mining operations. All these applications and many more require valves. They have become indispensible in almost every facet of modern technology, as seen in Figures 1-1, 1-2, 1-3, and 1-4.

Valves range in size from a fraction of an inch to 30 feet in height. They are fabricated from the familiar bronze and iron and from plastics, stainless steels and the exotic metals required by nuclear power plants and rocket engines.

Fluids they control include water, food products, gases, corrosive materials, chemicals, oil, toxic and radioactive substances, pharmaceuticals and cosmetics.

Pressures vary from thousands of pounds to high vacuums. Temperatures now range from over 1000 degrees Fahrenheit to cryogenic.

Because of the numerous uses, life cycles and

FIG. 1-1. Integrated screw compressor system uses numerous valves to control 235 tons of refrigeration. *Courtesy Frick Co.*

FIG. 1-2. Valves controlling fluid flow in a process industry. *Courtesy Valve Manufacturers Association.*

quality required, there are dozens of valve variations. All, however, are derived from a few basic configurations.

History

Rudimentary valve concepts can be traced back to antiquity. The Egyptians and Greeks used primitive methods to divert water flow for irrigation. Not until the Romans designed their aqueducts and other large scale water transport projects did the valve become more important and, in fact, a necessity.

From the beginning of the industrial revolution, with the development of the steam engine in the early 1700's, valves were required to regulate and contain the steam. As industries bloomed, in textiles, lumber, chemical, petroleum and the fabrication of war machinery, the valve played an ever-increasing role. Figure 1-5 outlines the pace of valve development.

As with many of the technical improvements through the years, the development of new materials brought a quick advancement in design and multiplicity of valves. The limits of valve capabilities were constantly being stretched and applied to more and more different applications. Valves capable of handling higher pressure, temperatures and a large variety of corrosive chemicals were developed. Good examples of material development enhancing valve uses is Teflon®, titanium, hastaloy and the family of PVC plastic materials. The ball valve came into its own because of Teflon®. Plastic-lined metal valves and all plastic valves and pipe with synthetic seating materials were developed 20 to 25 years ago and were a large boost to the petrochemical industry where corrosive chemicals were taking their toll on standard metal valves.

FIG. 1-3. Diaphragm valves in a food beverage plant. These valves can handle tomato puree, milk, beer, edible oils, orange juice, pickle brines, and wine. *Courtesy Flow Control Div., Rockwell International.*

FIG. 1-4. Sanitary valve manifold in a food processing plant. *Courtesy Ladish Co., Tri-Clover Div.*

What Fluids are regulated?

Valves regulate fluids in an almost infinite num-

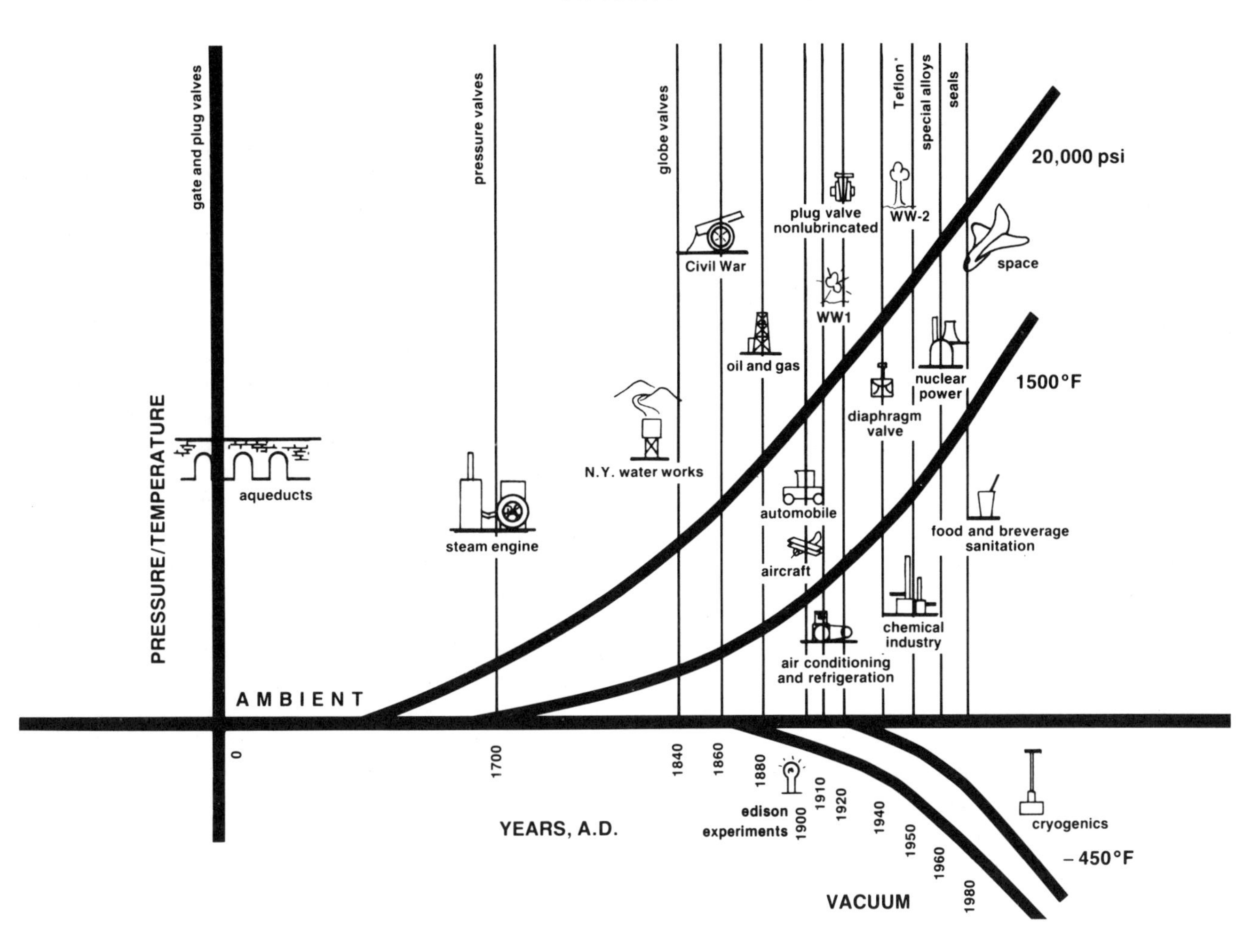

FIG. 1-5. Symbolic growth of the valve industry.

ber of applications and each has its own varying requirements. Mining has liquid slurries carrying solids and sand causing abrasive, erosive and shutoff problems. Refineries have high temperature applications where thermal and physical shock may cause structural failure of materials. The chemical industry is constantly handling corrosive acids, caustics and toxic materials. Rocket engines and airplanes handle cryogenic fluids, pressurized gases, vacuums, fuel and hydraulic oils. Many of our food products, from milk to peanut butter, have processes controlled by valves. Every ship contains hundreds of valves for handling sea water, steam and powerplant fluids, as well as the loading and unloading of bulk liquid and dry cargoes.

A recent report stated that many gas and oil reserves are at a depth of 30,000 feet with pressures of 20,000 psig and temperatures of 550°. Oil and gas at these depths contain corrosive contaminants, such as hydrogen sulfide and brines, that readily attack standard materials. New valve materials and designs will be required to handle these conditions.

Closer to home, apartments, condominiums, filling stations, stores and homes require valves to control air conditioning, water, fuel oil and gas for comfort and cooking. As many as 40 valves may be used in a single home. The automobile in the garage contains dozens of valves.

The advent of air conditioning, heat pumps and industrial refrigeration has required development of valves to control the flow of liquid and gaseous refrigerants from compressor discharge to suction lines. The hydronics industry has always used valves in balancing hot water flows and mixing fluids of different temperatures to control environments within buildings. The subsequent development of solar energy, heat recovery and basic energy conservation has extended the frontiers of valve uses.

Refrigerant circuits must handle gaseous as well as liquid refrigerants. Refrigerant flow is regulated to control capacities, condensing and evaporating pressures, refrigerant flows and direction. Various refrigerants attack different materials in varying degrees requiring careful material selection. One has to remember that all working refrigerant fluids are not the common R-12 or R-22 normally used, but include ammonia, bromides, air, hydrocarbons, etc.

Valve functions

Valve designs vary depending upon their service function and application. The most common valve functions can be categorized as follows:

Manual operation
- On-off service
- Sample/drain valves

Automatic operation
- Flow throttling
- Back flow prevention
- Pressure control
- Flow redirection
- Flow control

Valves may be operated by manual or powered controllers. There are overlapping functions in these styles; therefore, the proper selections must be made to meet each individual application. Manual and automatic operations may be interchanged, but they have been outlined and categorized here based upon their more frequent uses.

Automatically controlled valves usually throttle or prevent backflow, control pressure, temperatures, flow direction and flow volumes of fluids and gases. Most are automatically controlled by sensing pressures or temperatures and operate over a range of conditions.

Most on-off service and sampling valves are hand or remotely controlled.

About 50% of all valves used are for on-off service, 40% are for throttling and 10% for backflow prevention.

Economics

Companies spend millions of dollars each year in purchasing new and reconditioned valves. The chemical process industry alone may spend upwards of 20 to 30% of the piping costs of a new plant on valve procurement. Many of these plants cost hundreds of millions of dollars. Proper selection of valves is important because the cost can vary as much as 100%, depending upon material selected. Why pay for a stainless steel valve when a bronze valve will do?

Duty cycles, life vs. initial cost, severe duty, reliability, corrosive fluids within, and exterior corrosive environments must all be considered. Flashing vapors can cause pitting and impingement attack on materials, sand and gravel carried by liquids cause erosion while acids can literally eat up valve materials causing frequent maintenance. Because of all these potential problems, valve costs must be weighed against field failure which in turn means plant shutdowns or exposure of plant personnel to toxic chemicals or vapors with disastrous results. Industrial refrigeration equipment may be installed in chemical plants, deserts, ocean-going vessels, refineries, etc. Proper materials and finishes must be selected to withstand these environments.

The newly created energy conservation market has required a substantial input from the valve industry. From coal conversion to numerous solar programs, the valve industry is being challenged to develop new designs to meet these rigorous demands. Valves for energy-related projects will give an immediate boost to the valve industry and is expected to yield to a substantial growth in succeeding years. Energy-related industries, today, account for nearly 35% of the total industrial valve sales.

From the valve manufacturer's point of view, rising costs and foreign imports are the most immediate concerns. Like almost everything else since the energy crisis, the cost of valve manufacture has risen greatly.

FIG. 1-6. Valve bodies being fabricated in high volumes by automatic screw machines. *Courtesy Primore Sales, Inc.*

FIG. 1-7. Machine used to assemble and torque test components as a part of quality control. *Courtesy Hamilton-Pax, Inc.*

Valves require considerable energy input during their manufacture: vast amounts of heat are required for melting alloys in foundries, for forgings and for annealing ovens. Costs will continue to rise as energy becomes more and more expensive.

The acquisition of machine tools for valve manufacture requires considerable lead time, planning and money. Rigid codes and specifications by end users require rigidly-controlled manufacturing processes and testing. Because none of these changes is inexpensive, expansion and capital expenditures must be planned years in advance, see Figures 1-6, and 1-7.

American valve industry

About 600 firms manufacture valves in the U.S. Most, however, are categorized as speciality shops. Of the 600 firms, about 70 produce 80% of the valves. The table below shows the breakdown, by end-user industries, of domestic valve procurement as given by the Valve Manufacturer's Association.

• Petroleum industry	23%
• Oil and gas pipelines	12%
• Chemical, petrochemical industries	12%
• Waste and waste-water systems	11%
• Electric power plants	12%
• Construction	9%
• Others, including pulp, paper, food, metals, mining, air conditioning and refrigeration, and marine	21%

Valves for the air conditioning and refrigeration industry are mostly manufactured by companies specializing in the refrigeration field. The uniqueness of the refrigeration industry requires special valve styles and features that normally are not considered in the main line of valve designs required by other industries. Both styles will be examined since both are used in the air conditioning, heating, and refrigeration fields.

2
Types of Valves

Variations in valve designs are almost as numerous as the valve manufacturers themselves. When all these variations are reduced to the simplest common denominator, all valves have basic fluid handling functions such as on-off service, throttling, back flow prevention, pressure and flow control, flow diverting, and sampling.

Simple on-off valves place an object normal to the path of fluid flow, thereby stopping the flow. The gate, plug, ball, sliding, and U-valves are all on-off valves.

Throttling valves restrict fluid flow, from the fully closed to an infinite number of partially open positions, until the fully open position is reached. Globe, angle, Y-pattern, butterfly, and diaphragm valves are used for throttling.

Other flow control operations use variations of the on-off or throttling designs such as back flow prevention, pressure and flow control by globe-style regulators, and flow redirection by multiport designs. Sample and drain valves are special applications.

Patented variations, such as the sliding gate or eccentric rotating plug, are used in regulators and control valves for fine, close control.

The valve's design determines its head loss, that is, the pressure drop or resistance to flow. In low pressure applications, head loss is critical, consequently butterfly, gate, plug, and ball valves are used. Where pressure drop is of little importance, but flow control is desired, the globe valve is usually employed. In low pressure, flow control applications, the butterfly valve yields the best results.

Figure 2-1 compares head loss or pressure drop, as a function of valve size, for the various valve designs. It shows that the globe valve has, by far, a higher head loss than the straight-through, full-ported styles. Note that the head-loss scale is logarithmic.

Gate valves

Gate valves — the oldest of all designs — can be traced to antiquity. They are used to stop the flow of fluid and are never used for throttling! A partially open gate valve will chatter and wear excessively, through erosion, causing early failure and replacement.

The advantage of the gate valve is that, in the fully open position, fluid flow is totally unobstructed, Figure 2-2. This is important where low pressure drop is vital to system performance. The gate valve operates slowly because the stem must be turned a number of revolutions to open or close the valve. This very slowness prevents *fluid hammer* which is detrimental to piping systems.

Gate valve discs and seats are available in several configurations, depending on the required degree of positive shut-off, the type of fluid being controlled, and the reliability needed in a particular application.

Gate valves are mainly use in hydronic, petrochemical, and municipal water systems, marine applications, and the chemical industry. They are not suitable where volatile liquids, such as refrigerants, can easily vaporize and *migrate* to other parts of the system.

Gate valves have found new applications in the solar industry where pipe and pumping systems are selected for low energy consumption, and minimum pressure drop, while slowly moving various heat-transfer media.

Hydronic heating and cooling systems use gate valves to isolate portions of the system during servicing and to shut off hot and cold water lines leading to boilers and chillers. Many of the gate valves used by the hydronics industry are produced by companies that serve that industry exclusively.

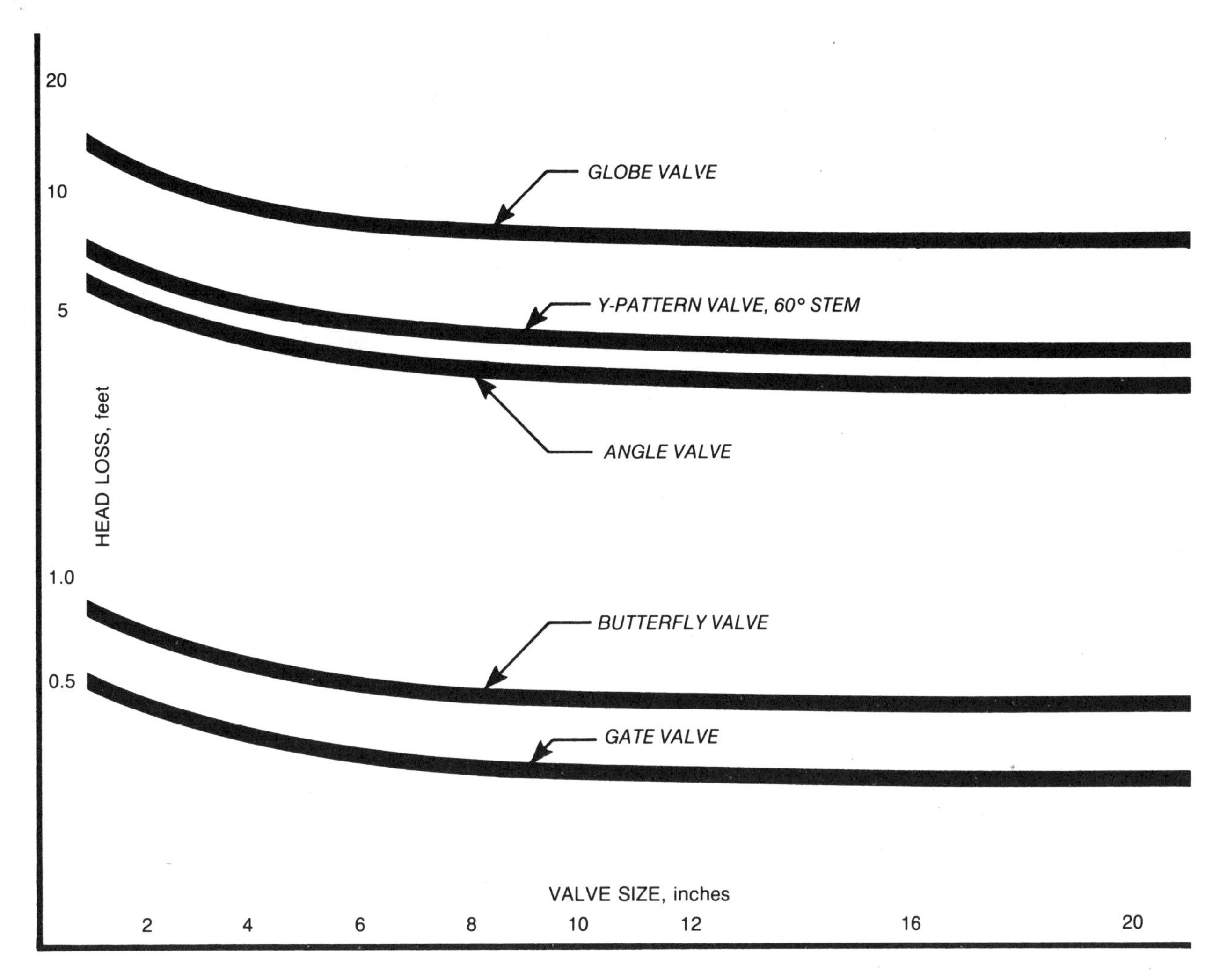

FIG. 2-1. Relative head loss vs. valve size. Fluid is water at constant velocity with turbulent flow.

Plug valves

The plug valve has a history and a design similar to the gate valve: it blocks fluid flow normal to the valve centerline. The nonlubricated type is a refinement of the simple cock, Figure 2-3. Like the gate valve, the plug valve has an unobstructed flow, minimizing turbulence and pressure drop. The plug valve has an additional advantage: it requires little headroom.

Only a 90° turn of the handle is required to open or close the valve and the number of wearing parts is far less than in a gate valve. Stem corrosion is minimal because there are no screw threads like those found in a gate valve.

Plug valves lend themselves to multiporting because of their basic construction. Plug valves with up to five ports have been fabricated.

The plug valve can have sealing problems. Sealing by metal-to-metal contact between the body and the plug has led to galling and the need for lubrication. The lubricant is pressure injected into the valve and maintains a leak-tight seal while easing rotation of the plug. The lubricant may, however, contaminate the fluid being handled. Therefore, the lubricant must be carefully selected. In some cases, the flowing fluid may act as a degreaser and wash out the lubricant, accelerating wear and causing galling of the plug and the body.

Nonlubricated valves have become more common in the last few years mainly because of material developments. Teflon® and other plastic materials allow the plug to turn in the body with little or no frictional resistance, eliminating the need for exterior lubrication. However the temperature and

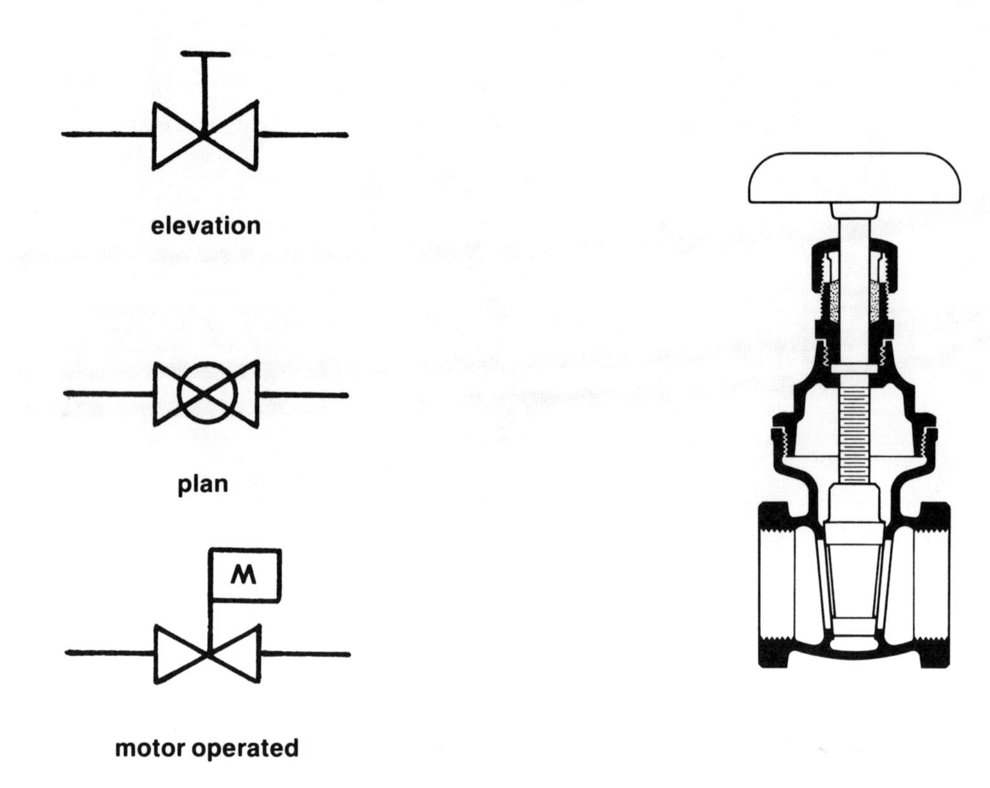

FIG. 2-2a. Gate valve, drafting symbol. *American National Standards Institute (ANSI).*

FIG. 2-2b. Gate valve, cross-sectional drawing. *Courtesy Valve Manufacturers Assoc.*

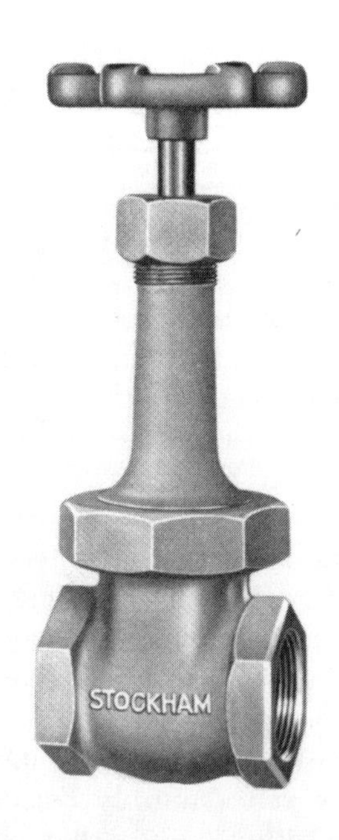

FIG. 2-2c. Gate valve, exterior. *Courtesy Stockholm Valves and Fittings.*

FIG. 2-2d. Gate valve, interior. *Courtesy DeZurik.*

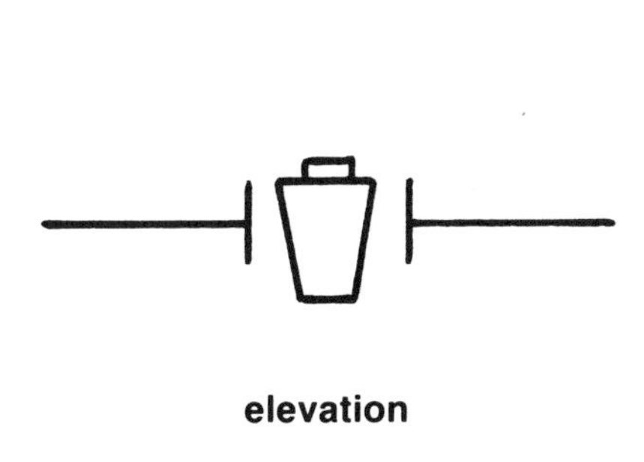

FIG. 2-3a. Plug valve, drafting symbol. *American National Standards Institute (ANSI).*

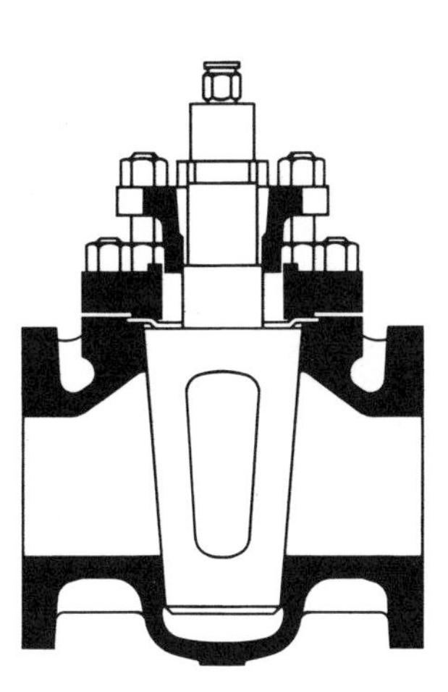

FIG. 2-3b. Plug valve, cross-sectional drawing. *Courtesy Valve Manufacturers Assoc.*

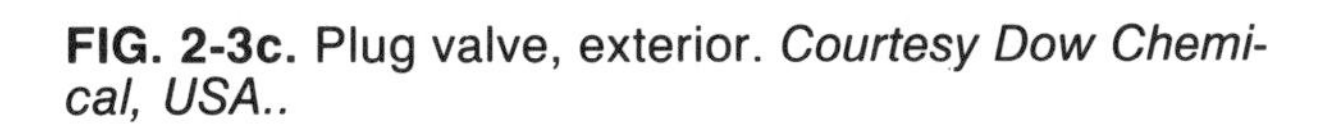

FIG. 2-3c. Plug valve, exterior. *Courtesy Dow Chemical, USA..*

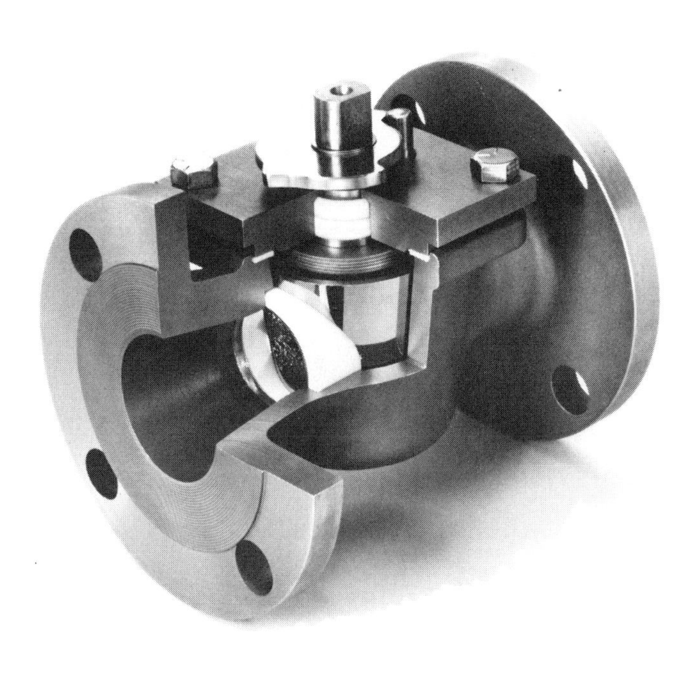

FIG. 2-3d. Plug valve, interior. *Courtesy DeZurik.*

FIG. 2-4a. Ball valve, drafting symbol. *American Society of Heating, Air-Conditioning and Refrigeration Engineers (ASHRAE).*

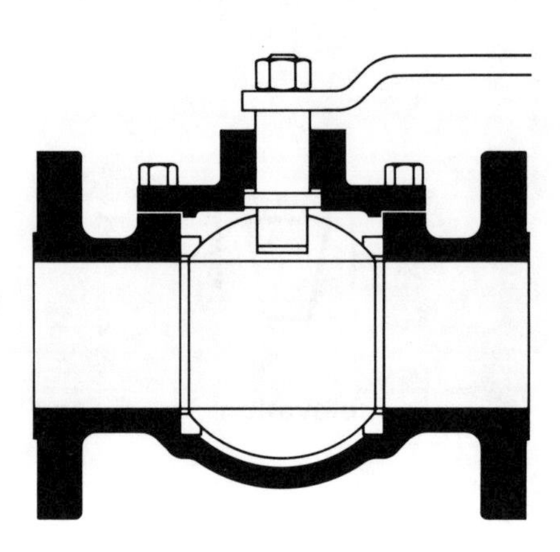

FIG. 2-4b. Ball valve, cross-sectional drawing. *Courtesy Valve Manufacturers Assoc.*

FIG. 2-4c. Ball valve, exterior. *Courtesy The Lunkenheimer Co., a Condec Co.*

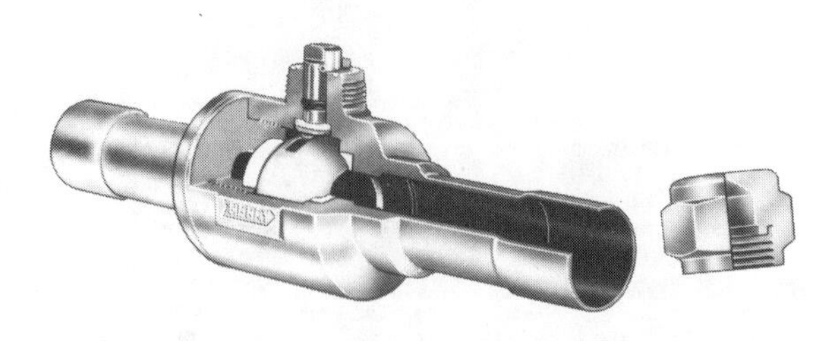

FIG. 2-4d. Ball valve, interior. *Courtesy Henry Valve Co.*

FIG. 2-5a. Butterfly valve, drafting symbol. *American Society of Heating, Air-Conditioning and Refrigeration Engineers (ASHRAE).*

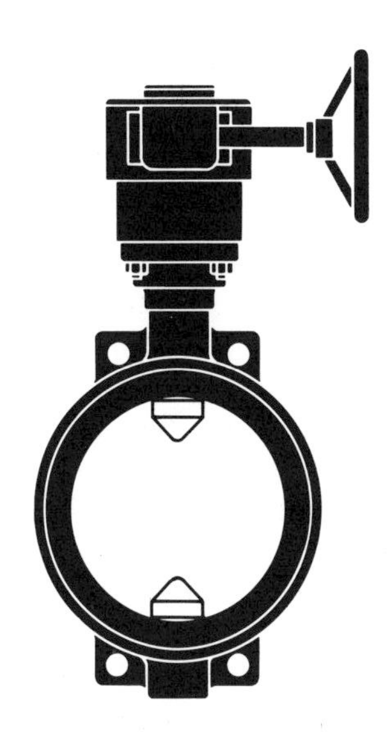

FIG. 2-5b. Butterfly valve, cross-sectional drawing. *Courtesy Valve Manufacturers Assoc.*

FIG. 2-5c. Butterfly valve, exterior. *Courtesy Flow Control Division, Rockwell International.*

FIG. 2-5d. Butterfly valve, interior. *Courtesy Jamesbury Corp.*

pressure ranges may be restricted, when compared to lubricated plug valves, because of the limited range of plastic materials.

Plug valves are used in hydraulic and pneumatic control systems where quick opening, closing, and diverting of flows is quite common. Like the gate valve, however, the plug valve has little use in air conditioning and refrigeration equipment. They are used in hydronic systems, as balancing cocks in terminals, to control pressure drop between zones.

Plug valves must be carefully selected if they have a multiport configuration. Because of the clearance required for the plug to rotate, some play is necessary and side movement of the plug is possible. In most cases, this movement is small but, if differential pressures are high, there may be leakage from one port to another.

Most plug valves have a tapered plug. Pressure, applied mechanically through the stem, will improve its seating over the valve's life.

Ball valves

The ball valve, Figure 2-4, like the plug valve, goes back a long way. However, like many ideas, it was never fully utilized until suitable materials were developed. One was, again, Teflon®. This plastic material is used for seats and seals and has allowed the ball valve to come into its own. Tight closure is achieved by squeezing the ball against the seat rings.

The ball valve, like the plug, has a low pressure drop, quick opening and closing, simple construction and easy maintenance. It is designed for full open or closed operation and is not recommended for throttling service, steam, slurries, or liquids with solid particles. The ball is so highly polished that any scratch on its surface will cause valve leakage.

Ball valves have recently found applications in refrigeration service where a sealed cap over the stem, in lieu of a handle, assures refrigerant containment. The cap must always be replaced, after the valve has been used, to prevent leaking.

Butterfly valves

Figure 2-5 shows a butterfly valve. When fully open, the disc of a butterfly valve is parallel to the flow. When closed, it is perpendicular to the flow, like the disc of a gate valve.

The butterfly disc is mounted on a shaft running through the center of the valve body. The fluid seal is maintained by O-rings, rubber liner inserts, or special, patented, pressure-loaded seal rings. Because nonmetallic components are used, this valve is limited by the temperature range of the elastomeric materials. Some fluids may attack the elastomeric materials and fluid selection is of critical importance.

Two basic body styles are available:

- Wafer type
- Double flange type

In the wafer type, the valve is held in place by two pipe flanges. The inner lining overlaps the body of the valve and creates a liquid-tight seal against the flanges.

The double flange type has flanges fabricated as part of the valve body which, in turn, are bolted to the pipe flanges.

The butterfly valve is suitable for handling large flows of gases and liquids at relatively low pressure drops. It is also used in slurry applications or with liquids that carry suspended particles.

Like the plug and ball valves, the butterfly valve is opened or closed with a 90° turn of the handle. Its added advantage is that it is an excellent throttling valve. Space and weight savings is another feature, as shown in Figure 2-6.

Refrigeration applications do not utilize the butterfly valve directly. However, large chilled water units use the valve to shut off the cooling water or brine systems. These cooling line mains, in apartments or factories, may be upwards of 12 inches in diameter. Butterfly valves can save weight and cost while providing very good service in these applications.

Globe valves

The globe valve, Figure 2-7, is a *recent* development, being only 120 years old.

Flow through the valve is a Z pattern that permits throttling control without excessive or localized wear of the seat and disc. The disc is forced against the fluid flow, rather than across it, thereby eliminating wear, erosion, and chatter that are so often found in gate valves when throttling. In the globe valve, the disc is in close proximity to the seat even when the valve is fully open. The time required to close the valve is less than that of a gate valve due to the shorter travel of the disc.

Because of the Z pattern, the pressure drop through a globe valve, even when fully open, is greater than with valves having straight through flow. Fluid systems using globe valves must allow for the higher pressure drop. It will be shown later that, in some cases, the additional pressure drop can

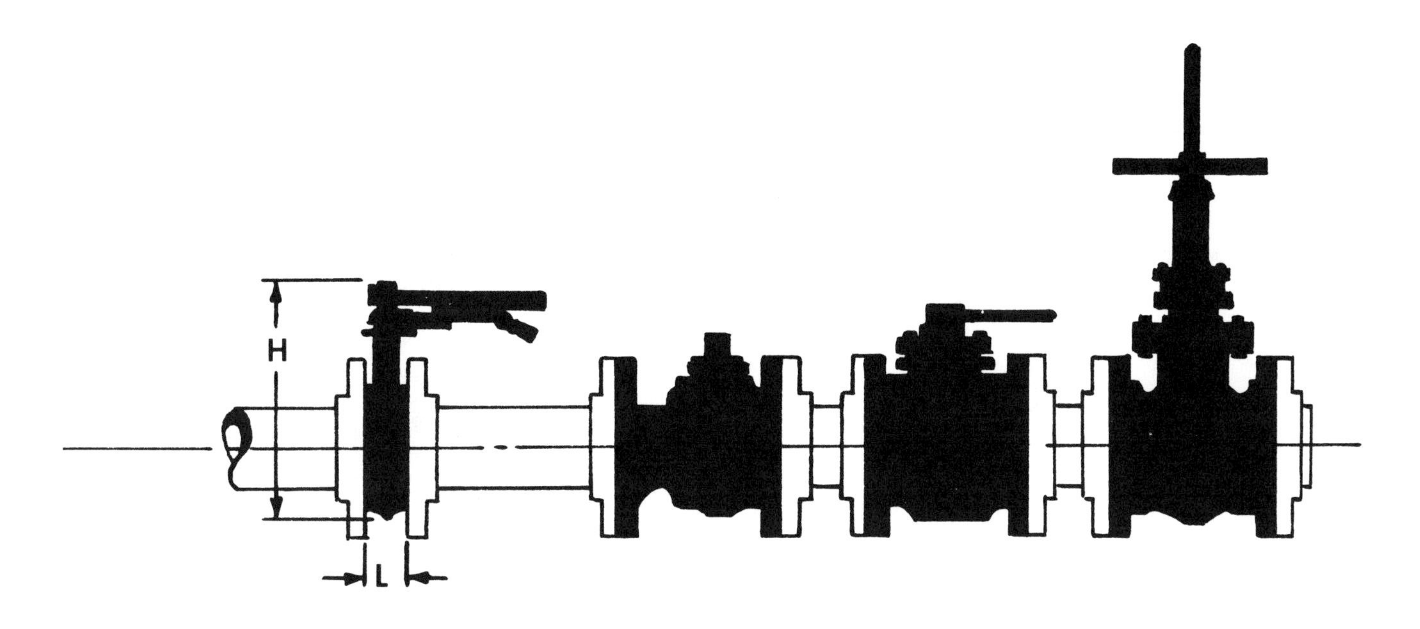

VALVE COMPARISONS BY RATIOS

	BUTTERFLY	*BALL*	*PLUG*	*GATE*
Weight	1	3.0	4.0	5.8
Length	1	5.4	5.4	5.4
Height	1	.97	1.3	3.4
Torque	1	4.4	high	high

FIG. 2-6. Butterfly valves compared with (left to right) plug, globe, gate, and ball valves. *Courtesy Flow Control Division, Rockwell International.*

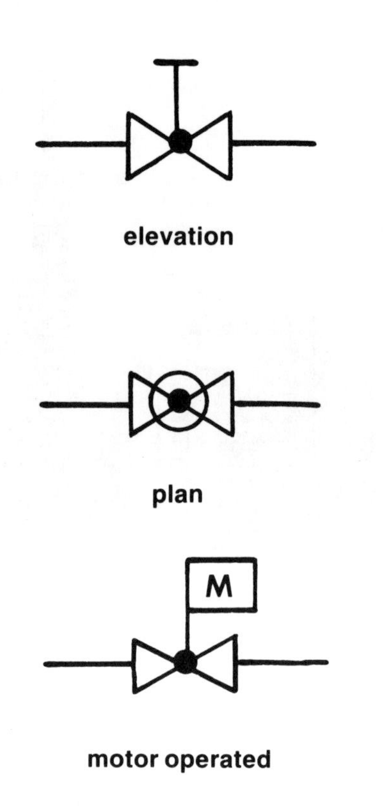

FIG. 2-7a. Globe valve, drafting symbols. *American National Standards Institute (ANSI).*

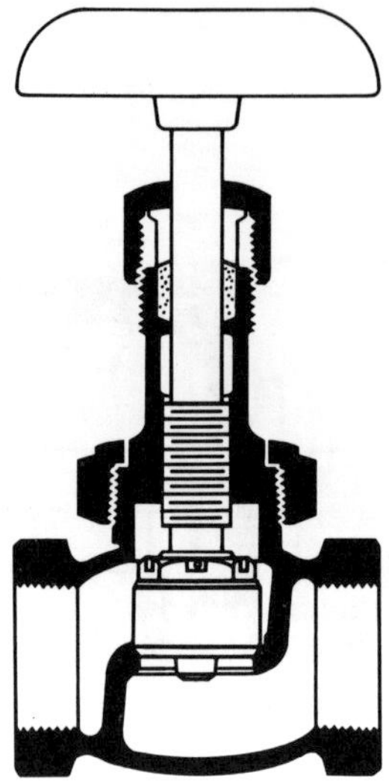

FIG. 2-7b. Globe valve, cross-sectional drawing. *Courtesy Valve Manufacturers Assoc.*

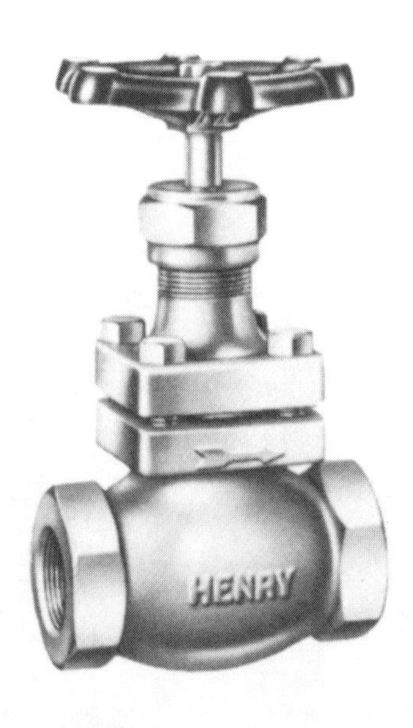

FIG. 2-7c. Globe valve, exterior. *Courtesy Henry Valve Co.*

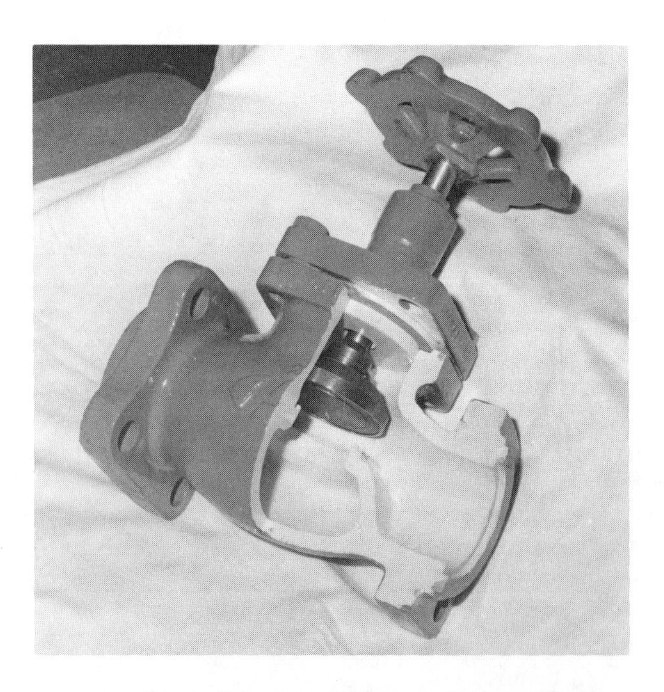

FIG. 2-7d. Globe valve, interior. *Courtesy Frick Co.*

be used to advantage and is needed for proper flow control.

Almost all air conditioning and refrigeration valves are globe valves. Most are the shut-off type and the globe design assures positive fluid containment. Figure 2-7 shows numerous valves used in refrigerant shut-off applications. These valves have additional sealing provisions in the cap and some are packless valves that do not use mechanical packings.

Flow and pressure control valves, as well as diverting valves, follow the globe pattern. These valves, also known as regulators, have two basic advantages:

- Positive shut off
- Ability to more evenly control flow and/or pressure over a large range of settings.

Angle valve

The angle valve, Figure 2-8, utilizes the globe valve seating principle while adding a 90° elbow to the piping. It is less resistant to flow than the standard globe valve and requires fewer fittings during installation, saving both material and labor. Pressure drop through the angle valve is about one-half that of the combination of a globe valve and a 90° elbow of the same size.

Y-pattern valve

The Y-pattern valve, Figure 2-9, approaches the flow characteristics of the gate or plug valves, but has the throttling performance of the globe valve. The pressure drop through the valve is considerably less than for a conventional globe valve.

Sliding valve

The sliding valve is a special version of the gate valve that is closed by the sliding action of a pivoted lever arm, Figure 2-10. This action forces away any foreign material carried by the fluid and, at the same time, cleans and grinds the seating surface for tight shut-offs. The pressure of the fluid behind the disc presses against the seat to create a tight seal. This valve is used in many slurry operations where the fluid may carry large quantities of foreign material.

Stationary boilers, used for heating water or producing steam, use this valve for *blow down* because it will shut down tightly time after time.

U-valve

The U-valve, Figure 2-11, is similar to the gate valve but, instead of using a metal wedge as a disc, it employs a conical rubber plug with a U-shaped cross section. Because the rubber plug is resilient, it conforms to the body cavity and creates a seal. Domestic, light commercial, and industrial water applications use this design. Its use is limited due to the material used in the plug.

Diaphragm valve

Figure 2-12 shows a diaphragm valve. Shut-off and throttling are effected by flexing the diaphragm against a mating seat, weir, that is directly opposite the diaphragm. The diaphragm covers the entire bonnet area of the valve, thereby isolating the operating mechanism and stem from the fluid. No stem packing is required.

This design is suited to handling viscous solutions and slurries. However, temperature limitations of the diaphragm membrane limit its application.

Many food-related piping systems employ diaphragm valves because the diaphragm isolates the stem from the fluid and sanitation is more easily maintained.

Another diaphragm valve does not utilize the raised bridge, weir, in the body and, therefore, has less pressure drop when wide open. However, the flexing distance, required for the diaphragm to seat tightly, is considerably greater.

The raised bridge, shown in Figure 2-12, reduces closing time and, most importantly, diaphragm flexing which yields a longer service life and allows a wider material selection than the straight through design.

Multiport valves

Multiport valves are available in plug, ball, or shear styles, with various flow patterns. Figure 2-13 shows the ball valve type. They are also available in full port, reduced port diameter, and venturi designs with three-, four-, and five-way configurations.

The multiport valve can be quite helpful in industrial piping where various processes flow through *common piping* systems. This strategy eliminates the expense of many valves and reduces the amount of piping and separate fittings. Installation is faster and simpler, and there is greater reliability.

Multiport valves are used in heating and air conditioning systems as mixing, diverting, and tempering valves, Figure 2-13e and 2-13f. The terms *mixing* and *diverting* refer to the direction of fluid flow through the valve. The difference between the valves is the internal construction of the valve

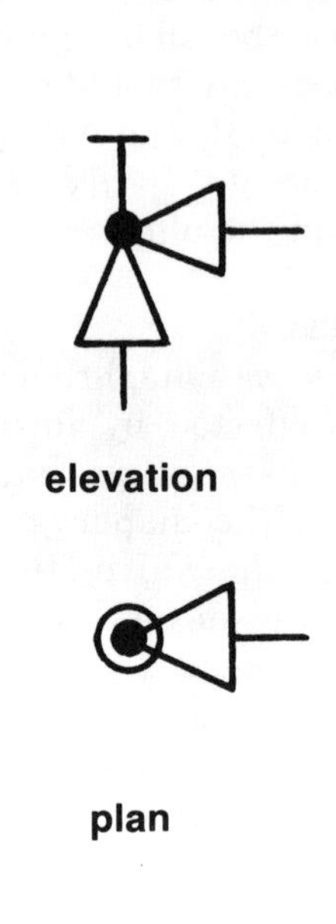

FIG. **2-8a.** Angle valve, drafting symbol. *American National Standards Institute (ANSI).*

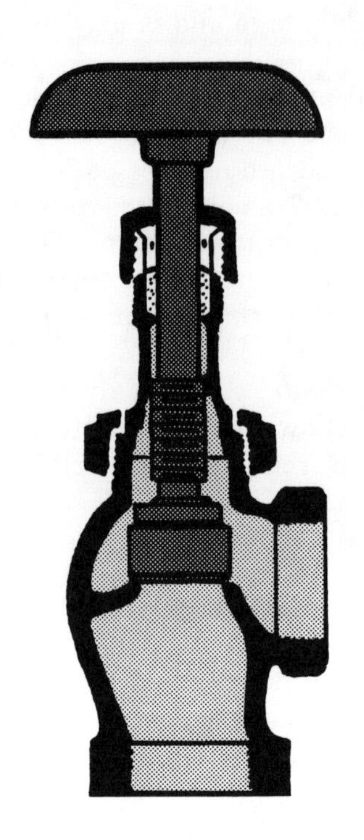

FIG. 2-8b. Angle valve, cross-sectional drawing. *Courtesy Valve Manufacturers Assoc.*

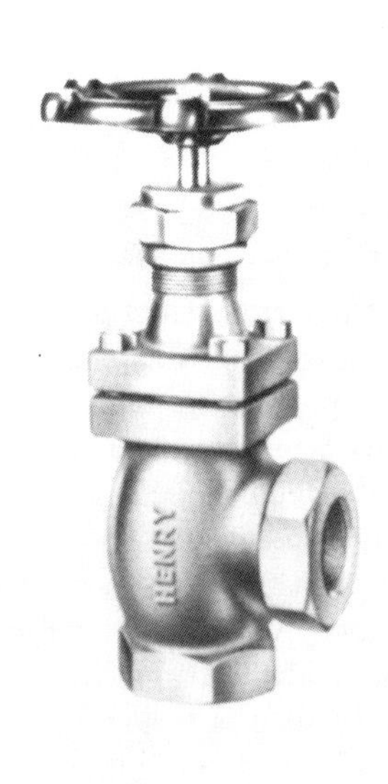

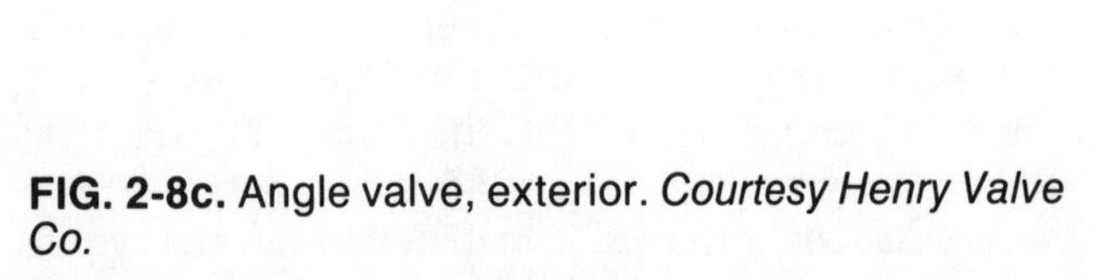

FIG. 2-8c. Angle valve, exterior. *Courtesy Henry Valve Co.*

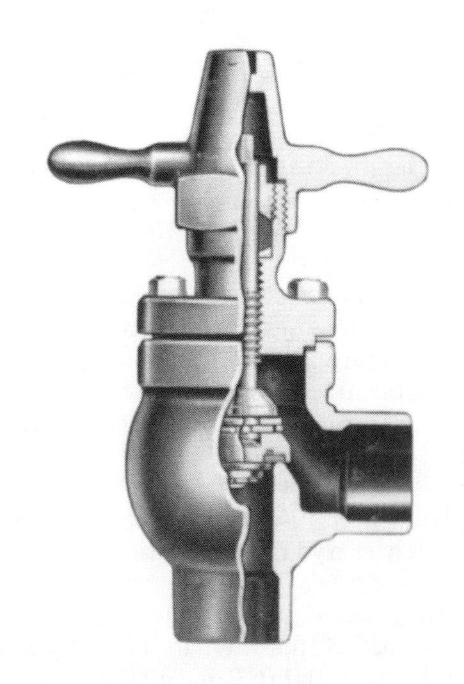

FIG. 2-8d. Angle valve, interior. *Courtesy Henry Valve Co.*

FIG. 2-9a. Y-pattern valve has no standarized drafting symbol.

FIG. 2-9b. Y-pattern valve, cross-sectional drawing. *Courtesy Valve Manufacturers Assoc.*

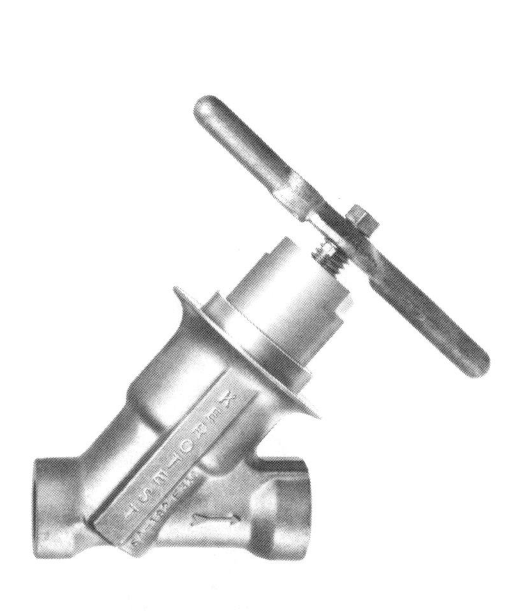

FIG. 2-9c. Y-pattern valve, exterior. *Courtesy Kerotest Mfg. Corp.*

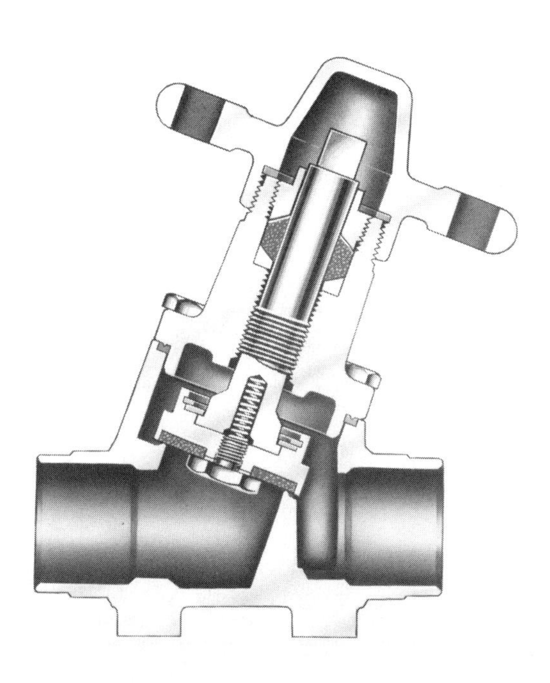

FIG. 2-9d. Y-pattern valve, interior. *Courtesy Superior Valve Co.*

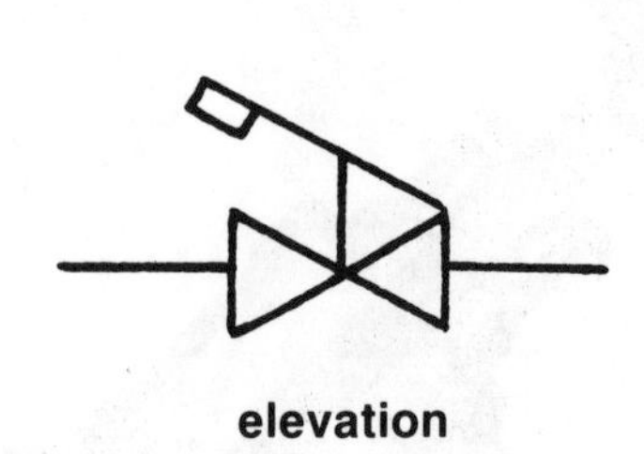

FIG. 2-10a. Sliding valve, drafting symbol. *American National Standards Institute (ANSI).*

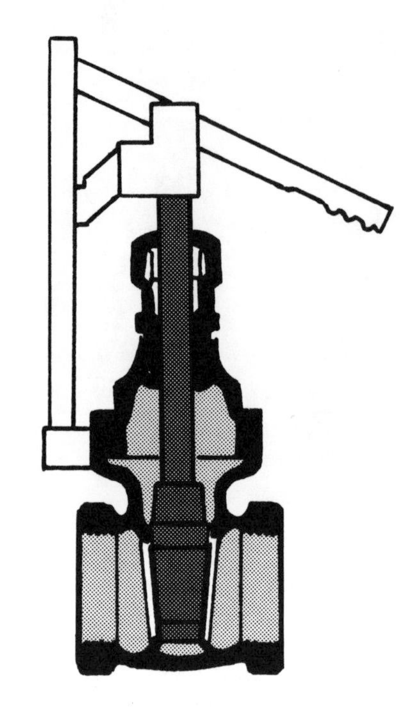

FIG. 2-10b. Sliding valve, cross-sectional drawing. *Courtesy Valve Manufacturers Assoc.*

FIG. 2-10c. Sliding valve, exterior. *Courtesy Stockham Valves and Fittings.*

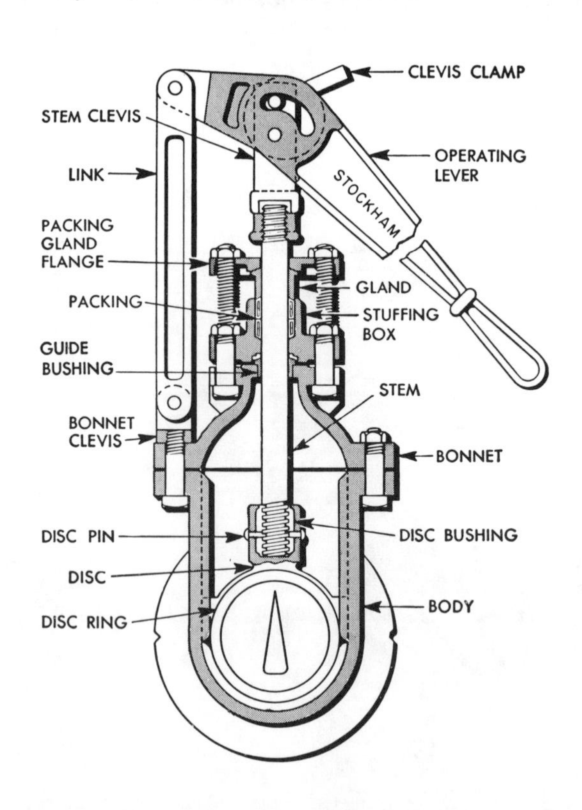

FIG. 2-10d. Sliding valve, interior. *Courtesy Stockham Valves and Fittings.*

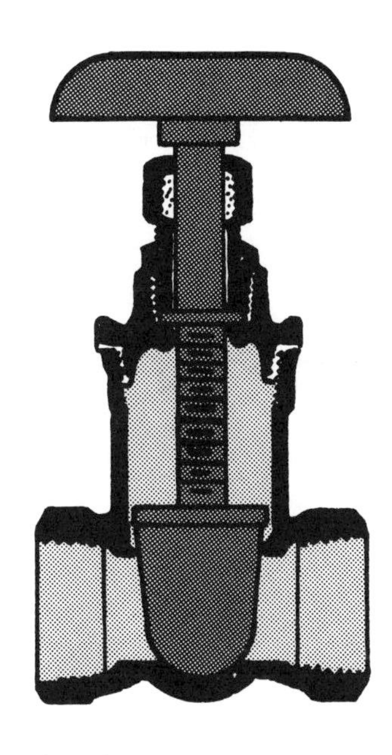

FIG. 2-11a. U-valve has no standardized drafting symbol.

FIG. 2-11b. U-valve, cross-sectional drawing.

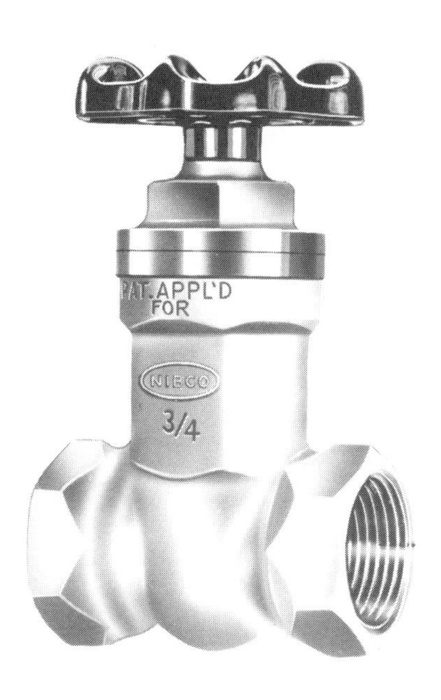

FIG. 2-11c. U-valve, exterior. *Courtesy NIBCO Inc.*

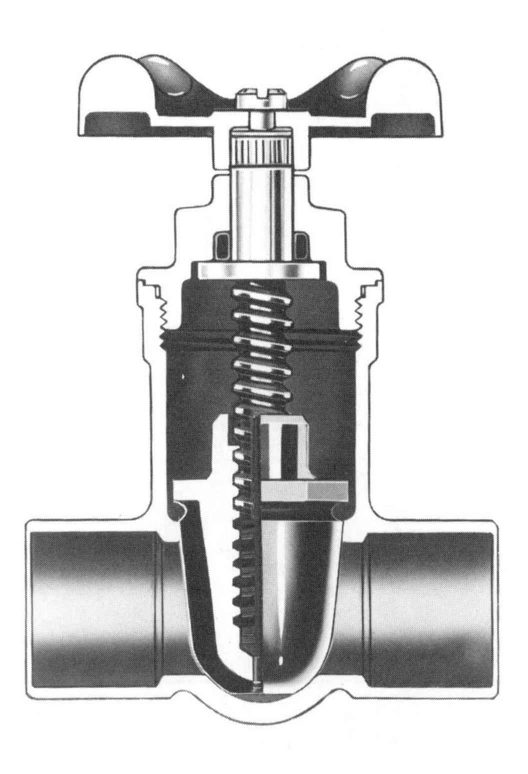

FIG. 2-11d. U-valve, interior. *Courtesy NIBCO Inc.*

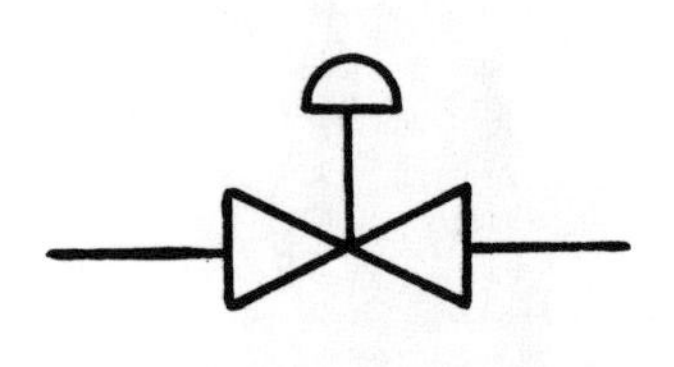

FIG. 2-12a. Diaphragm valve, drafting symbol. *American National Standards Institute (ANSI)*

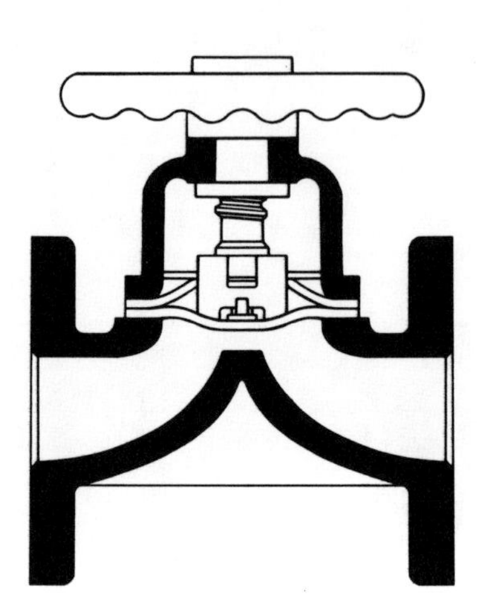

FIG. 2-12b. Diaphragm valve, cross-sectional drawing. *Courtesy Valve Manufacturers Assoc.*

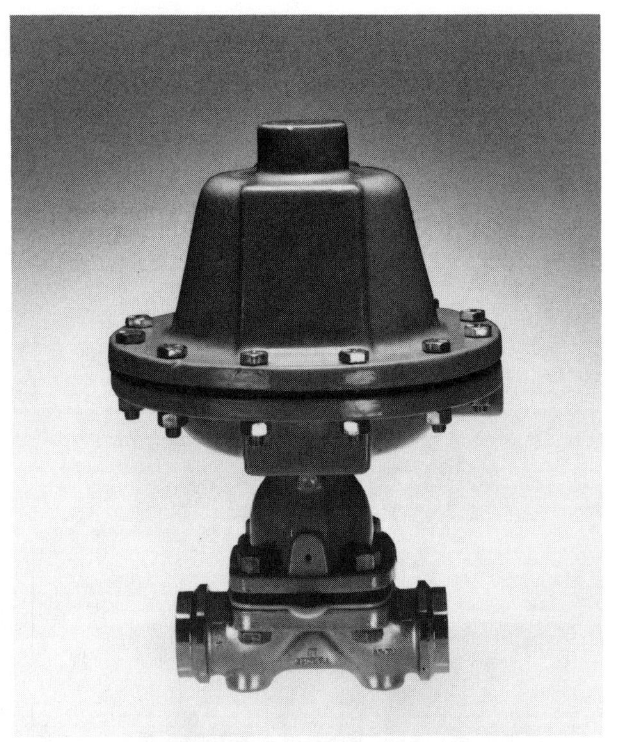

FIG. 2-12c. Diaphragm valve, exterior. *Courtesy Flow Control Division, Rockwell International.*

FIG. 2-12d. Diaphragm valve, interior. *Courtesy Flow Control Division, Rockwell International.*

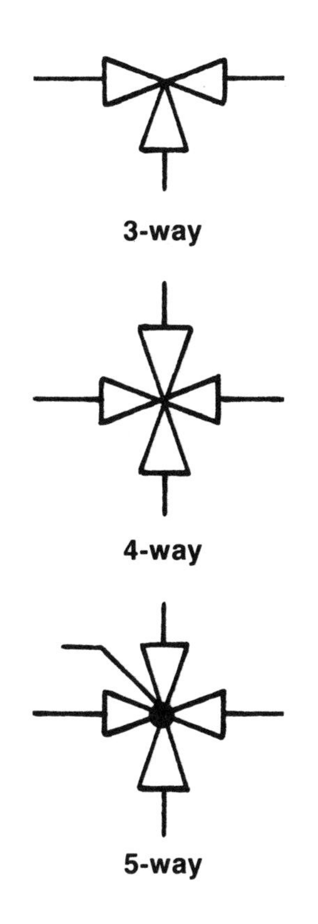

FIG. 2-13a. Multi-port valve, drafting symbols. *American National Standards Institute (ANSI).*

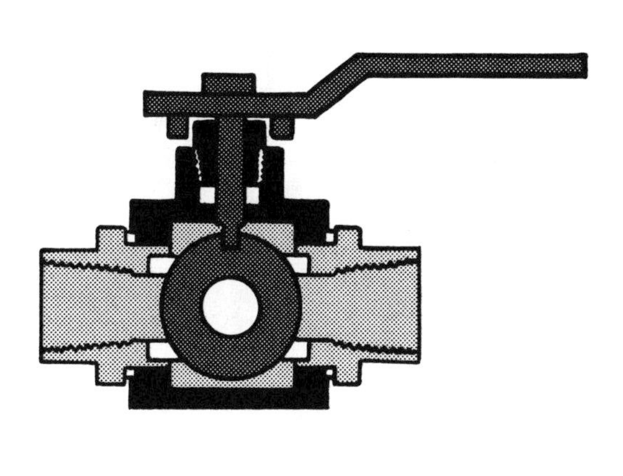

FIG. 2-13b. Multi-port valve, cross-sectional drawing.

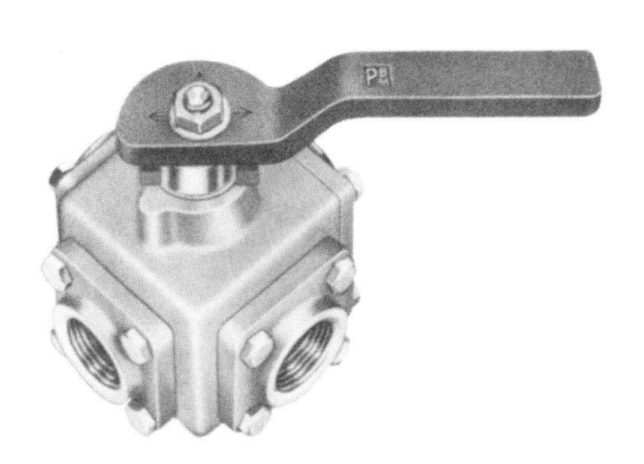

FIG. 2-13c. Multi-port valve, exterior. *Courtesy Pittsburg Brass Mfg. Co.*

FIG. 2-13d. Multi-port valve, interior. *Courtesy Pittsburg Brass Mfg. Co.*

FIG. 2-13e. Multi-port tempering valve mixes hot and cold water. *Courtesy Watts Regulator Co*

FIG. 2-13f. Multi-port three-way valve used for sequencing hot and cold water. *Courtesy Barber-Colman Co.*

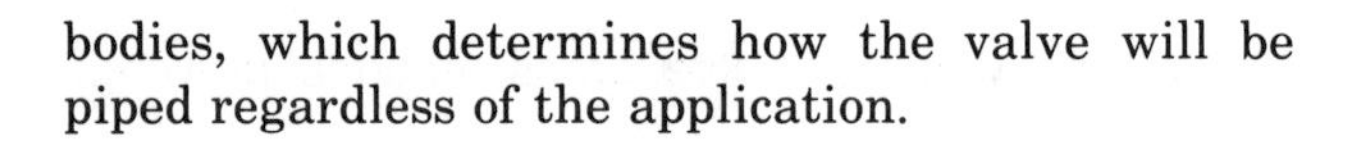
bodies, which determines how the valve will be piped regardless of the application.

Sanitary valves

Valves for sanitary use must be considered separately due to the stringent codes that they must meet. They are similar to other valves but, because sanitation is of major concern, ease in cleaning, quick disassembly, easy removal from the piping system, elimination of packings, and polished stainless steel construction make them unique. Four such valves are shown in Figure 2-14.

The plug valve is used primarily in the dairy industry. It can be cleaned in-place or readily dissembled for hand cleaning. These valves have no packings or seals. Sealing is by metal-to-metal contact of the tapered plug in the housing and a 90° turn of the plug opens or closes the valve. Lapped, close tolerance surfaces are a must and 316 stainless steel is most frequently used.

Butterfly and angle valves are also used with food process piping. The butterfly valve has an O-ring and an O-ring seating surface, both of which must be removable for cleaning.

The angle valve is usually operated by pneumatic controls from remote control stations.

A simple globe valve is also used. Fluid shut-off is accomplished by a tapered disc resting against a shoulder cut into the valve body. The disc can be either hard rubber or stainless steel. A variation of this design utilizes a long tapered plug for throttling flow in lieu of the blunt plug used in on-off service.

Diaphragm valves are used with food that contains particles or solids. The diaphragm design eliminates the packing or O-ring used to seal the area around the stem.

Relief valves

Relief valves protect steam power boilers, hot water heating boilers, pneumatic and hydraulic systems, and refrigeration and air conditioning equipment. They are rated and applied to conform to the ASME (American Society of Mechanical Engineers) code and tested using procedures defined by NB (National Board of Boiler and Pressure Vessel Inspectors). Relief valves marked NB meet the requirements of both of these governing bodies for the flow conditions listed on the identification plate, when the valve is fully open.

Figure 2-15 lists the the types of relief valves used by the air conditioning, heating, and refrigeration industries. Figure 2-16 shows the various relief

FIG. 2-14a. Sanitary plug valve. *Courtesy Ladish Co., Tri-Clover Div.*

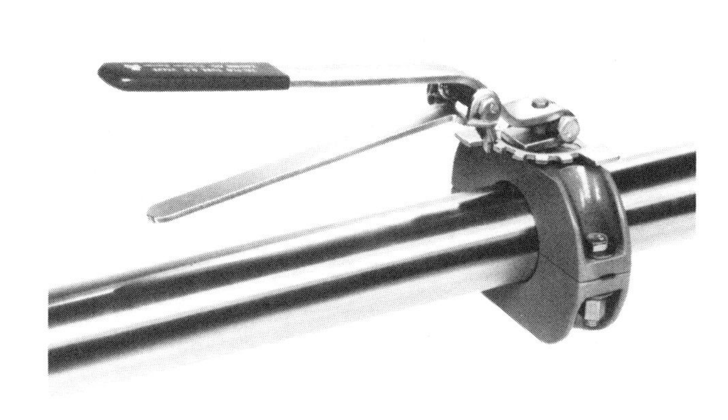

FIG. 2-14b. Sanitary butterfly valve. *Courtesy Ladish Co., Tri-Clover Div.*

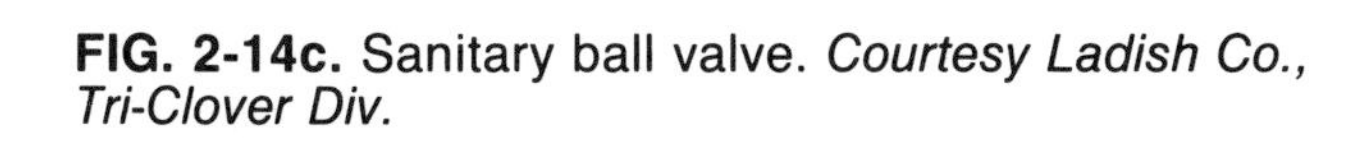

FIG. 2-14c. Sanitary ball valve. *Courtesy Ladish Co., Tri-Clover Div.*

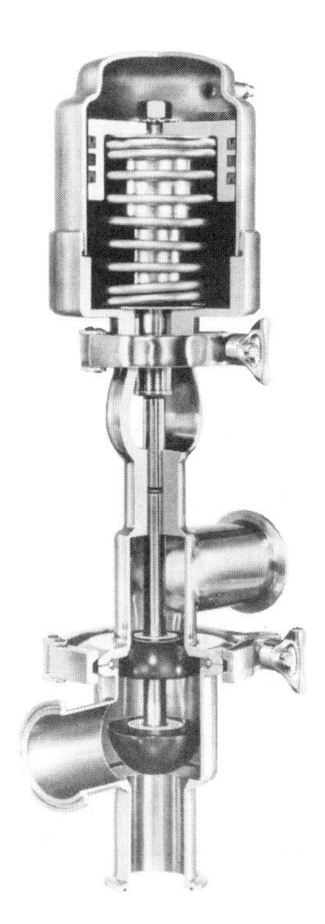

FIG. 2-14d. Sanitary angle-diverting valve. *Courtesy Ladish Co., Tri-Clover Div.*

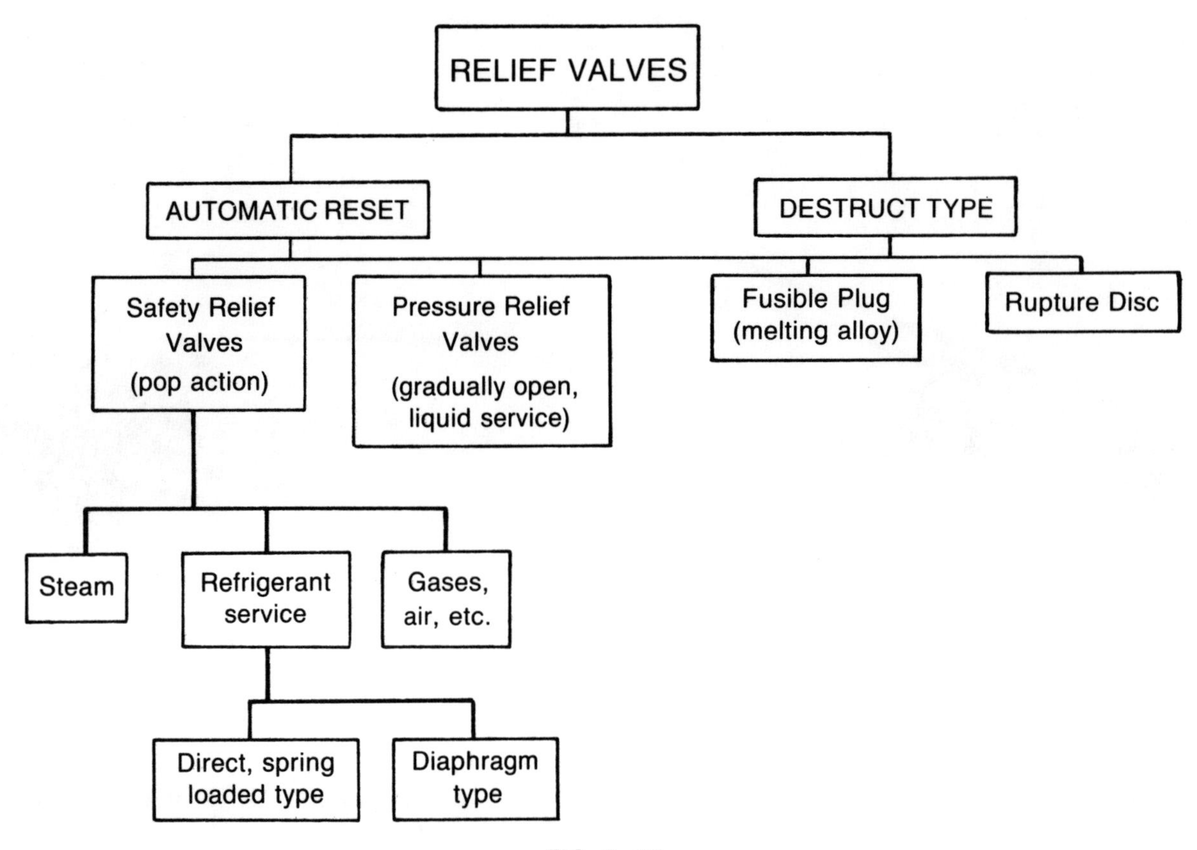

FIG. 2–15

valves available.

Safety relief valves used for steam, air, and gaseous service are *pop-action* valves. Pressure relief valves for hydronic heating and cooling systems open gradually, to relieve pressure, and are used for liquid service only.

Pressure relief and safety relief valves are designed to automatically relieve a vessel of excess pressure at a safe design limit. These relief devices are categorized as:

- Direct, spring-loaded
- Diaphragm
- Fusible plug
- Rupture disc

The first two relief devices reclose after actuating while the fusible plug and the rupture disc must be replaced after opening and before the system can be restarted.

Relief valves prevent the pressure in refrigerant receivers from rising above a safe limit. Excessive pressure, caused by control failure or excessive heat due to a fire, can rupture the receiver. The most important fact to remember is that, at the moment of rupture, the pressure in the vessel is suddenly reduced, allowing the the liquid refrigerant to expand into vapor with explosive force. The same is true of a steam boiler when a rupture allows the pressurized water to vaporize into explosive steam. The relief valve, when actuated, will release the refrigerant or steam at a controlled rate and maintain pressure within prescribed limits.

Many relief valves have a spring that exerts force on a disc or piston. As pressure nears the spring's setting, it will allow the piston to move slightly and permit vapor seepage until there is enough flow to *pop* the piston and provide full discharge.

All-metal relief valves are used with steam and corrosive fluids. The disc surfaces are machined to close tolerances and form a metal-to-metal seal.

In dual, pressure/temperature relief valves, the disc can be raised by either pressure or a temperature sensitive element. As the temperature rises, a wax fill material expands, driving a piston upward to unseat the disc.

Check valves

The purpose of the check valve, Figure 2-17, is to prevent back flow. Fluid pressure on the upstream

FIG. 2-16a. Relief valve, drafting symbol. *American National Standards Institute (ANSI).*

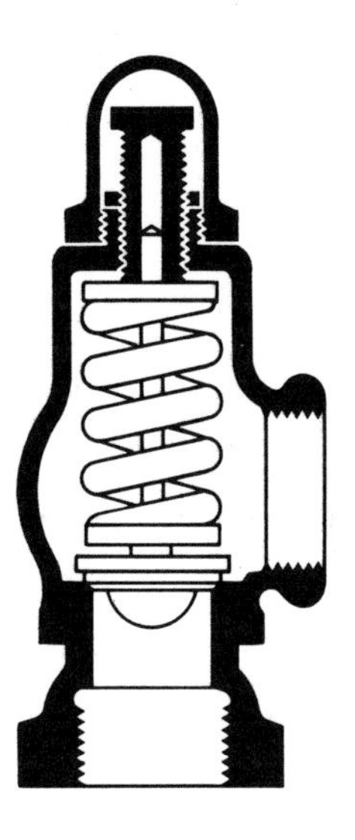

FIG. 2-16b. Relief valve, cross-sectional drawing. *Courtesy Valve Manufacturers Assoc.*

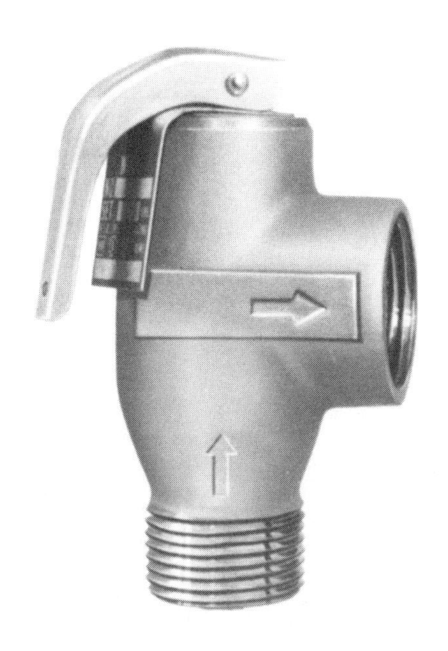

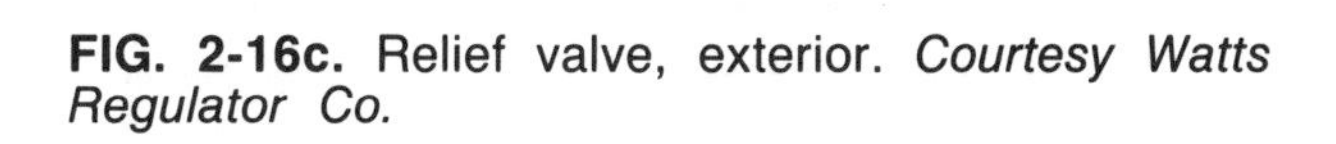

FIG. 2-16c. Relief valve, exterior. *Courtesy Watts Regulator Co.*

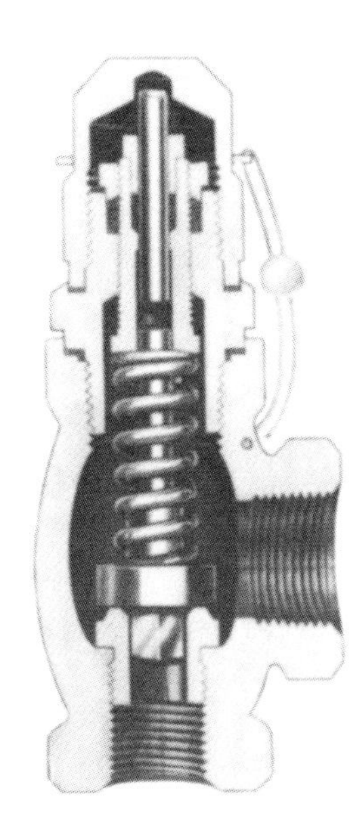

FIG. 2-16d. Relief valve, interior. *Courtesy Henry Valve Co.*

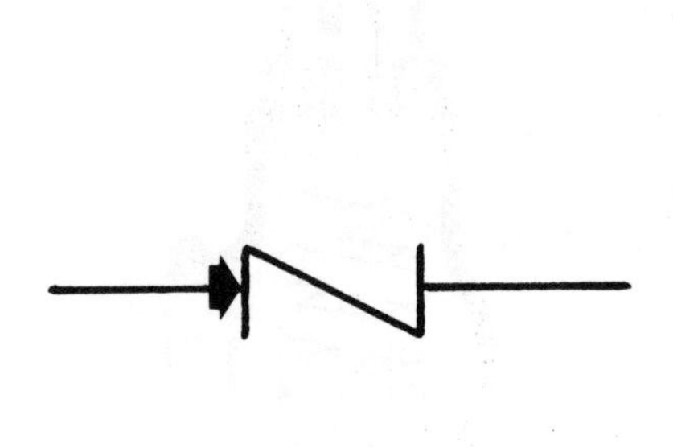

FIG. 2-17a. Check valve, drafting symbol. *American National Standards Institute (ANSI).*

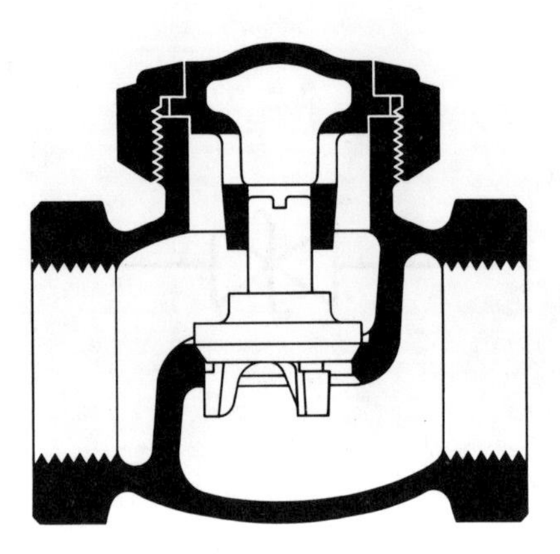

FIG. 2-17b. Check valve, cross-sectional drawing. *Courtesy Valve Manufacturers Assoc.*

FIG. 2-17c. Check valve, exterior. *Courtesy Henry Valve Co.*

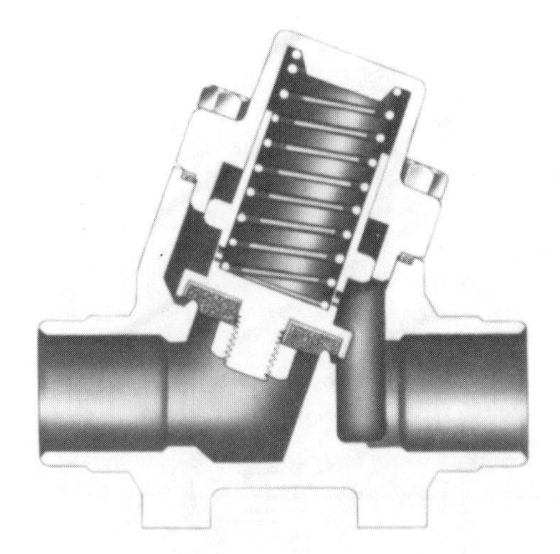

FIG. 2-17d. Check valve, interior. *Courtesy Superior Valve Co.*

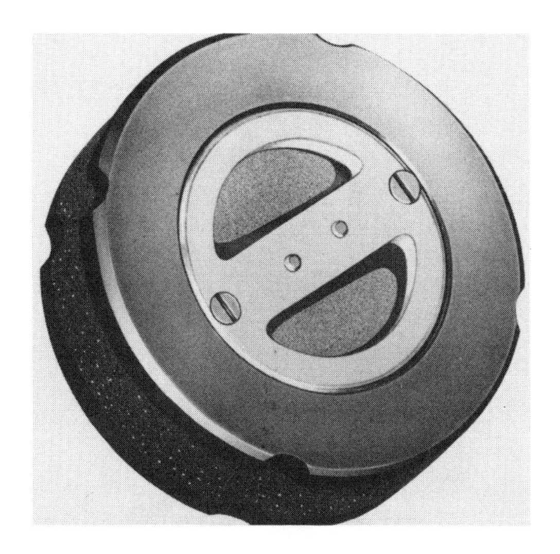

FIG. 2-17e. Nonslam, slow closing check valve for hydronic systems. *Courtesy ITT Fluid Handling Div.).*

FIG. 2-17f. Backflow preventer protects water supply from back siphonage or backflow. *Courtesy Watts Regulator Co.*

side opens the valve while increased pressure on the downstream side will close it to prevent back flow. The check valve, like the relief valve, is a self-actuated valve since it is operated by the fluid forces within the system.

Check valves come in three configurations:

• In-line type

This valve is usually used in smaller, pneumatic or hydraulic systems where checking is accomplished by a piston or ball that seats against a port.

• Swing type

Swing check valves have a disc that is hinged on a pin. The disc swings open in the direction of fluid flow and closes when flow is reversed.

• Piston type

The piston type is constructed like a globe valve. Proper flow will raise a disc that is guided by a shaft or cage above the piston. However, like the globe valve, this check valve creates a greater pressure drop because of the Z path of fluid flow. Higher pressure industrial steam systems and refrigeration systems use this style most frequently. When used with refrigeration, this valve is modified by spring loading the piston to assure a tight seal.

Hydronic systems can be exposed to severe fluid shock due to hydraulic forces within the system. When a moving fluid is suddenly halted, the resulting impact may measure thousands of pounds and actually distort internal valve components and cause failure. Nonslam, slow-closing check valves, Figure 2-17e, prevent this situation.

Specially designed check valves, Figure 2-17f, are required in many water systems to prevent back flow or back siphoning and to assure a safe potable water supply.

Regulators

Regulators, like check and relief valves, are self-actuated devices. This style of valve falls between the simple, hand operated valve and the more expensive and sophisticated valves normally used for process control.

The regulator controls pressures, flows, and temperatures, without an external power source in most cases. The fluid within the system acts on a spring-loaded diaphragm or piston, causing the disc to open or close by modulating the disc away from the seat.

Most designs are based on the globe valve pattern although sliding gates and other variations are available. Figure 2-18 shows regulators for steam, hydronic, and refrigerant applications.

Regulators are used in steam systems, to reduce

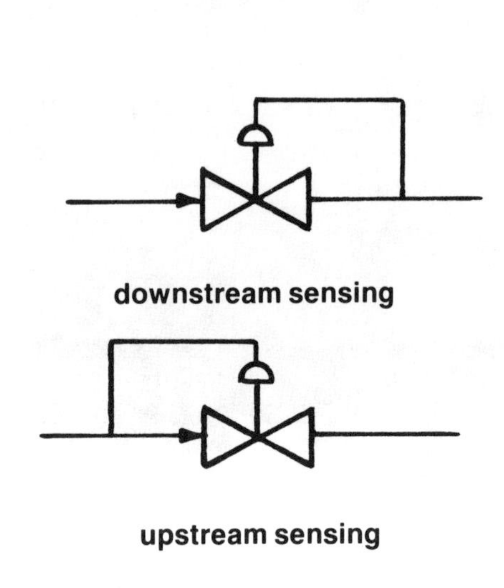

FIG. **2-18a.** Regulator, drafting symbol. *American National Standards Institute (ANSI).*

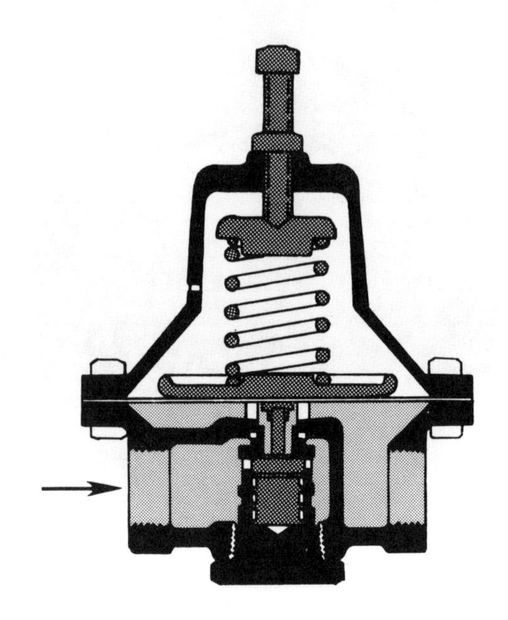

FIG. 2-18b. Regulator, cross-sectional drawing.

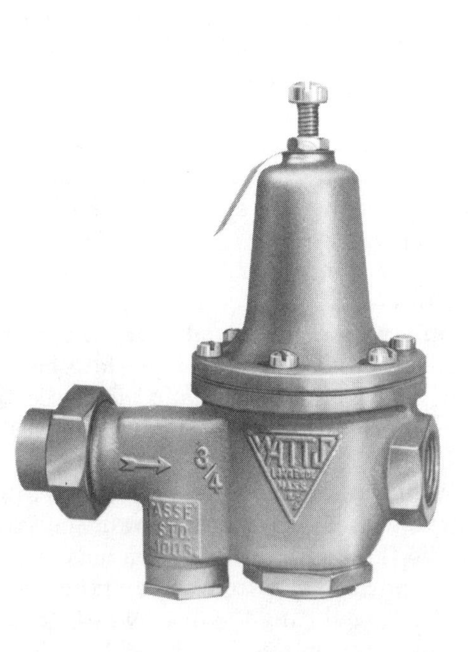

FIG. 2-18c. Regulator, exterior. *Courtesy Watts Regulator Co.*

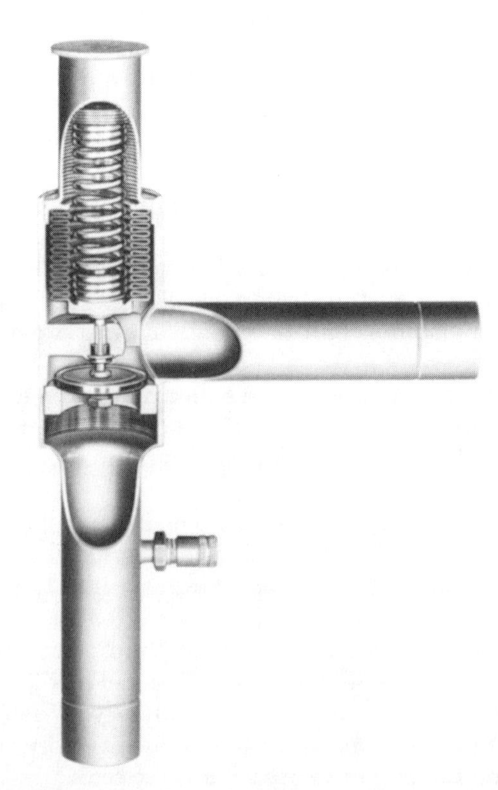

FIG. 2-18d. Regulator, interior. *Courtesy Sporlan Valve Co.*

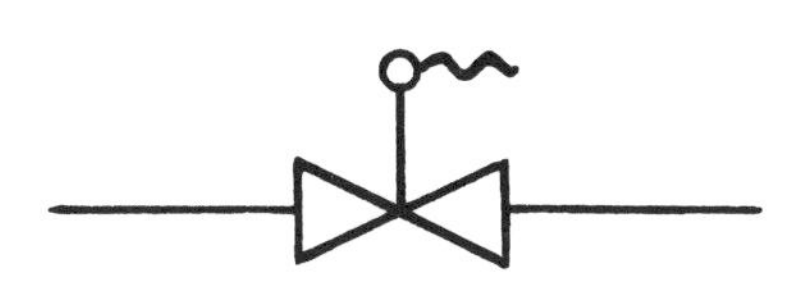

FIG. 2-19a. Solenoid valve, drafting symbol. *American National Standards Institute (ANSI).*

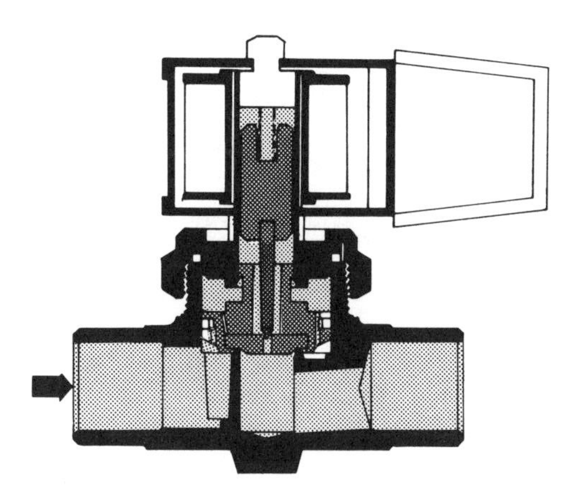

FIG. 2-19b. Solenoid valve, cross-sectional drawing.

FIG. 2-19c. Solenoid valve, exterior. *Courtesy Magnatrol Valve Corp.*

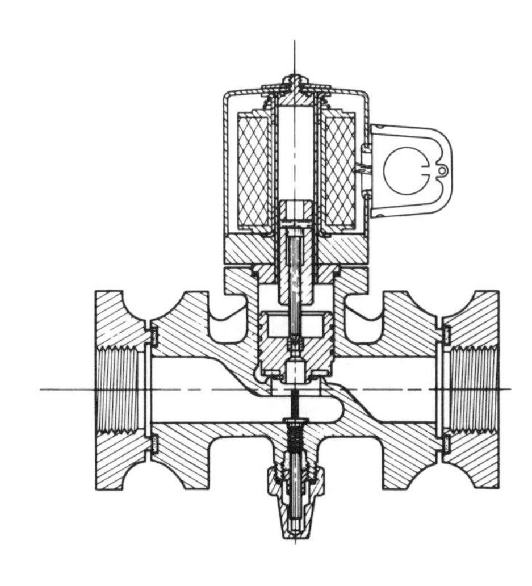

FIG. 2-19d. Solenoid valve, interior. *Courtesy Sporlan Valve Co.*

pressures for various heating or process requirements; in pneumatic systems, to control air pressures required by process controls or pneumatic controls; and in refrigeration systems, to control pressures or temperatures.

Direct, internal-pressure-sensing regulators are the most common. Temperature, flow, and combination pressure/temperature regulators are also used in industrial processes.

Several methods are used to control fluid or gaseous pressures; each affects the accuracy, cost, and output function of the regulator. They include direct-operated, pilot-operated, direct-operated externally-connected, and air-loaded regulators. The latter types approach control valve accuracy, using *typical* regulator features, without going the sophisticated control valve route.

Regulators fall into three categories:

- Pressure reducing
  Reduce pressure from upstream to downstream sections of the system.
- Back pressure (relief regulators)
  Maintain upstream pressure at a given maximum level.
- Differential pressure
  Maintain constant pressure differential between upstream and downstream sections of the system.

The self-contained, direct-operated regulator is the least costly, but its size is limited to up to 2 inches. Its accuracy ranges from 10 to 20% of maximum set point at full flow. In many cases, this is sufficient control for a process.

For greater accuracy, larger size regulators are usually pilot operated. That is, they have a smaller regulator or valve, either externally or internally mounted, to control the main valve. Accuracy of 1 to 5% is possible at greater flow rates.

Solenoid valves

Solenoid valves are electrically-operated on-off valves that handle a variety of fluids, some of which are hot and cold water, steam, oil, chemical solutions, refrigerants, and air.

Variations include two-way, three-way, and four-way versions. The disc may be a diaphragm, piston or pencil point plunger, Figure 2-19.

Normally open (NO) or normally closed (NC) are two designs commonly found in air conditioning and refrigeration. Normally open valves close when energized, normally closed valves open when energized. These valves may be direct-acting or pilot-operated, depending on the size of the disc and its maximum operating pressure differential (MOPD).

Key words

Gate valve: shut-off valve.

Plug valve: ¼ turn open or close, lubricated or nonlubricated.

Ball valve: ¼ turn open or close, Teflon® sealing rings.

Butterfly valve: ¼ turn open or close, infinite throttling positions.

Globe valve: Z flow pattern, throttling application, highest pressure drop, positive shut off.

Angle valve: globe valve with a 90° turn.

Y-pattern valve: globe valve with less pressure drop than a standard globe valve.

Sliding valve: gate valve with positive sealing.

U-valve: gate valve that replaces gate with conical rubber plug.

Diaphragm valve: flexible diaphragm against a seat.

Multiport valve: usually a ball valve with three, four or five pipe connections.

Sanitary valve: special valve, with few or no crevices within the fluid chamber, that can be quickly disassembled for cleaning.

Safety relief valve: pops open to relieve vessel pressure.

Pressure relief valve: Gradually opens to relieve pressure in liquid systems.

Check valve: prevents back flow.

Regulator: self-actuated by fluid in system.

Solenoid valve: electrically operated to either open or close.

3
Valve Components

The variety of valves and their applications is almost endless. Even though the configurations, sizes, and shapes of valves seem to vary and their exteriors all look different, the basic building blocks are the same. Like a puzzle, each valve can be assembled, Figure 3-1, from the following components:

- Body
- End connections
- Bonnet
- Seat and disc (trim)
- Stem
- Packing containment device
- Packing
- Operator

Body

The valve body holds all the parts of the valve together and supplies the connections to the piping system. The body may be a casting, forging, machined from bar stock or even molded from a variety of plastic compounds. The material used will be one of the parameters that define the temperature and pressure limitations of the valve.

Cast, forged, and machined valves have their own advantages and disadvantages. Casting is the most economical form of construction. Cast valves are poured into sand molds with cavities formed by patterns that are adaptable to a number of valve designs or a family of valves of similar size. With smaller size valves, dozens of units may be cast in one mold, thereby saving time and space during the molding operation. This flexibility also reduces the pattern inventory.

Common patterns can be used to cast valves of various alloys only if the shrinkage, while cooling, is the same for the all the alloys.

Valve body castings have a grain structure that is multidirectional. Since fluid pressures act in all directions, castings are ideal for containing fluids. Cast alloys also have better high temperature properties than comparable forgings.

Due to their porosity, conventional castings cannot be used to contain refrigerants. However, special alloys are used to eliminate this problem. Smaller size refrigerant valves are usually forgings or machined from bar stock and can be easily mass produced. Larger valves are castings because of the economics involved.

Cast valve bodies are used almost exclusively for chilled and hot water systems, solar and steam applications.

Foundries can vary the alloys for castings to suit the service requirements. Bronze valves are a good example. Bronze contains a number of metals, mainly copper, tin, zinc, lead, and trace elements. These metals can vary in percentage from alloy to alloy, yielding castings capable of withstanding a variety of pressures and temperatures. Alloys are identified by numbers assigned by the American Society for Testing Materials (ASTM).

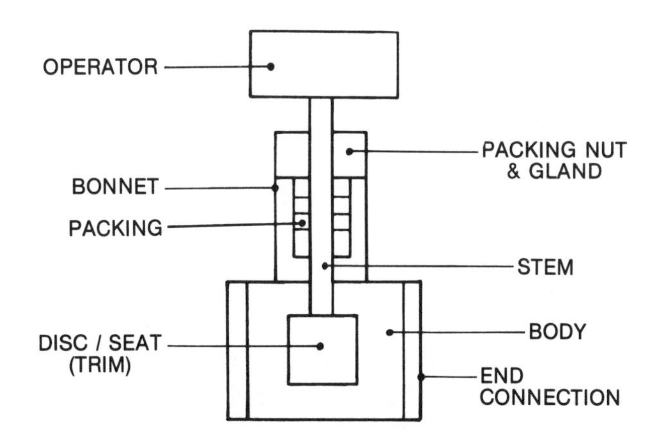

FIG. 3-1. Basic valve elements.

TABLE 3-1

Cast Composition Bronze ASTM B-62 Recommended for steam, water, oil, or gas up to 450°F.

CHEMICAL COMPOSITION (%)			PHYSICAL PROPERTIES	
	Minimum	Maximum		Minimum
Copper	84.00	86.00	Tensile strength	30,000 PSI
Tin	4.00	6.00		
Lead	4.00	6.00		
Zinc	4.00	6.00	Yield point	14,000 PSI
Nickel		1.00	Elongation in 2 in.	20%
Iron		0.30		
Phosphorus		0.05		

Cast Steam or Valve Bronze ASTM B-61 Recommended for steam, water, oil, or gas up to 550°F.

CHEMICAL COMPOSITION (%)			PHYSICAL PROPERTIES	
	Minimum	Maximum		Minimum
Copper	86.00	90.00	Tensile strength	34,000 PSI
Tin	5.50	6.50		
Lead	1.00	2.00		
Zinc	3.00	5.00	Yield point	16,000 PSI
Nickel		1.00	Elongation in 2 in.	22%
Iron		0.25		
Phosphorus		0.05		

A specification sheet, similar to Table 3-1, compares the materials used in two bronze alloys. The one, ASTM-B62 is a good choice for standard applications with normal temperatures and pressures. However, with a slight change in the alloy composition, ASTM-B61, which has higher strength and greater toughness, is created. By changing the copper content from 84% to 86% (minimum), tin from 4% to 5.5%, lead from 4% to 1%, and zinc from 4% to 3%, tensile stength increases from 30,000 psi to 34,000 psi and yield strength increases from 14,000 psi to 16,000 psi. ASTM-B61 is used for castings employed in higher pressure and temperature service.

Understanding ASTM standards and alloy designations is important. All valve materials will not react in the same way when in contact with corrosive fluids and gases. Many industrial fluids carry solids, abrasives, and dissolved gases, each attacking valve materials in its own way. In hydronic systems, there is enough variation in water quality from one place to another to influence valve life and care must be taken to select the proper valve material and water treatment chemicals.

The major drawback to castings is defects such as *blow holes* and sand inclusions in the metal. Metallurgical problems caused by uneven alloying, uneven cooling, or too rapid cooling will affect the structural integrity of the casting. Extra care must be taken with larger valves due to the greater mass of metal and the high cost of pouring such valves. Automation of foundry processes, from the initial mold and sand preparation, to pouring, cooling, and shakeout has eliminated many of the inconsistencies previously encountered.

A number of valve manufacturers are now operating their own foundries to assure better quality control and to maintain supervision over special processes, techniques, and proprietary procedures.

Many foundries are now required to perform nondestructive testing on castings prior to shipment. Under close quality control, as many as 50% of all castings may be rejected due to flaws. Testing includes radiographic inspection of sections up to seven inches thick. Magnetic particle, magnaglo, and dye penetrant testing are also used. Spectographic analysis of the molten metal yields a quick and accurate alloy analysis just minutes prior to pouring.

After the valve body is cast or forged, it is machined to close tolerances, to receive the various components that make up the valve assembly, and checked to see if it meets specifications. Specialty and larger sized valves are handled on a *one at a time* basis during machining while standard valves are chucked into automatic machines that may perform as many as six operations in sequence.

The basic valve body may have a number of accessories:

- Drains allow the valve to be cleared of corrosive fluids or materials that could be detrimental.
- Cleanouts, usually large enough to remove sludge, scale or foreign material, employ cover plates bolted to the body casting.
- Bypass ports are used to warm downstream piping prior to opening the valve. With larger size piping or high pressure systems, it may be necessary to equalize pressure on both sides of the valve, before it is opened, to prevent damage to the main valve or downstream piping due to shock from surging fluids.
- Taps, used to access the piping system, are shown in Figure 3-2. An important feature of air conditioning and refrigeration systems, the tap allows the application of a vacuum pump or refrigerant charging lines while servicing the system.
- Jacketed valves have cast cavities around the

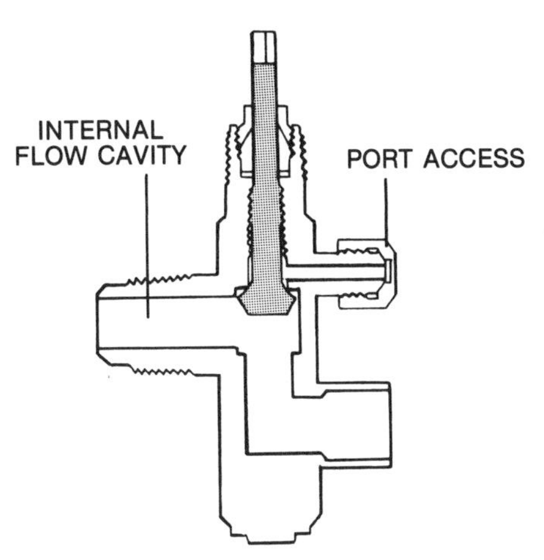

FIG. 3-2. Refrigerant-side access port is used to evacuate the system or to add refrigerant.

valve body or bolted-on jackets to contain heating media such as steam. Certain fluids may congeal or solidify, if not warmed, and block the valve port.

Forged valves are usually smaller sized, with simple internal shapes, and produced in the high quantities needed to amortize die costs. Forging is not suitable if internal passages are complicated or with some alloys. Many refrigerant valves are forged brass, but only in the smaller sizes.

Machined bar stock valves can be mass produced using just about any alloy.

Bodies for gate and globe valves are usually single castings with provision for end connections, bonnet assembly, and seat. Most seats are integral with the body but, where there is severe duty, the seat may be welded or screwed in place.

Some ball valves or regulators may have two-piece bodies that are screwed or bolted together. This construction simplifies the replacement of gaskets, seats, and discs. Other ball valves have swing-out bodies that allow replacement of the seat and disc without removing the valve from the line.

Valve ratings are stamped or cast on the body of the valve, indicating the allowable pressure and the type of fluid. Fluid codes are: S for steam, W for water, O for oil, G for gas. The body may also have an arrow indicating flow direction for proper control.

The pressure drop across a valve is the sum of the individual drops across the:

- Inlet and outlet fittings
- Internal flow passages
- Port area of the disc and seat

The interior of the valve body is designed with smooth curves to create a flow pattern that avoids turbulence. In on-off applications, pressure drop through the valve must be minimal to keep pumping horsepower requirements low. However, in throttling operations, pressure drop across the port area may be needed for good control.

Most of this discussion holds true, to a varying degree, for all types of valves. The manufacturing techniques will vary with the basic design, but all valves share common functions:

- Contain a fluid at the required temperature and pressure.
- Hold the body in the piping system.
- Allow for opening and closing the port.
- Control flow from shutoff to fully open.

End connections

End connections, that connect the valve body to the piping system, are cast, forged, or machined into the valve body during fabrication. The selection of the end connection depends upon the valve size, material, service conditions, pressure, temperature, mechanical stress, and field abuse.

Screw thread Threaded end connections, Figure 3-3a, are the most common and are generally reliable. They are easy to use in installing and removing a valve and require no special skills. The threads may be tapered pipe threads or straight machine threads as in flare or flareless fittings.

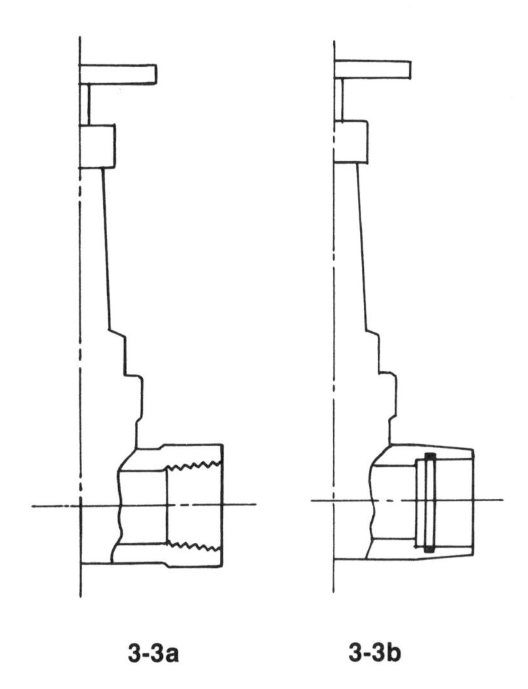

FIG. 3-3. End connections. **a.** Pipe thread. **b.** Socket braze.

Socket braze A valve designed for socket brazing has a socket or well into which tubing is inserted, Figure 3-3b. In some designs, a groove in the well holds a ring of silver solder. After cleaning and fluxing, the tube is inserted into the well and the joint is heated by a torch. At about 1200 °F, the silver solder will melt and seal the joint.

Other, smaller valves, used mostly in the plumbing and refrigeration industries, Figure 3-4a, have the silver solder applied by the mechanic. When heat is applied to the joint, capillary action will draw the solder into the joint to create a leak-proof assembly.

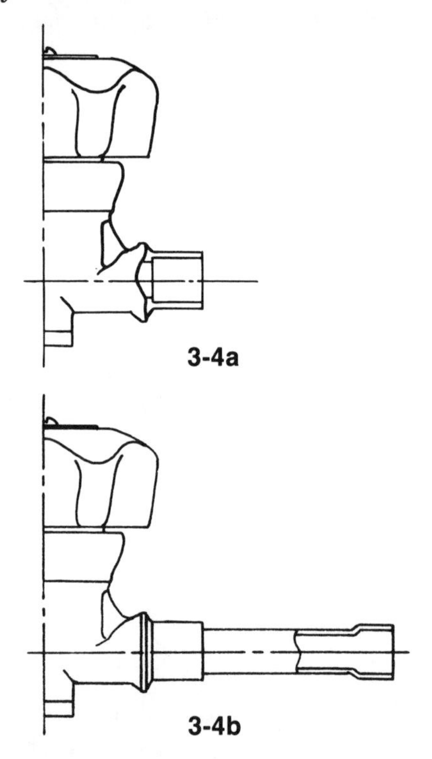

FIG. 3-4. End connections. **a.** Solder end. **b.** Extended solder end.

When soldering valves, critical internal components must be protected against overheating. Some manufacturers offer valves with extended end connections to increase the distance between the heated area and the valve components, Figure 3-4b.

Butt weld Welded joints, Figure 3-5a, are strong and are used in high temperature and pressure applications. However, this joint requires a welder with skills beyond those needed to work with threaded or brazed ends. Tube and/or pipe preparation is required and the cost of installation is high. Constricted areas, where the assembler cannot easily see or weld, may impede assembly and inspection.

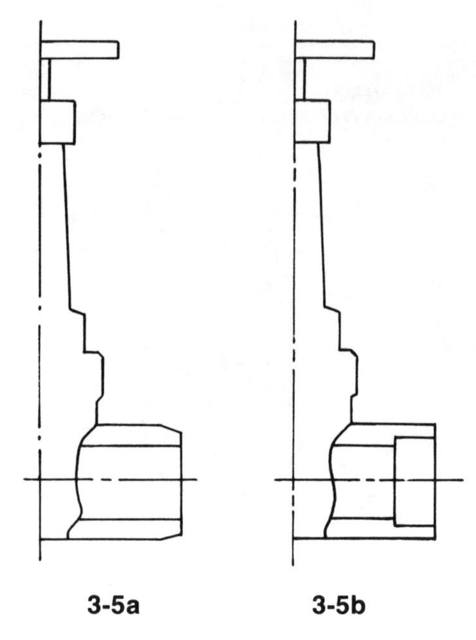

FIG. 3-5. End connections. **a.** Butt weld. **b.** Socket weld.

Socket weld Socket welds, Figure 3-5b, are similar to butt welds but provide a more positive seal. The valve has a machined socket that encases the pipe, similar to the well in a soldered joint.

Weld flange Welded flanges are suitable for larger, high pressure valves and can be unbolted if

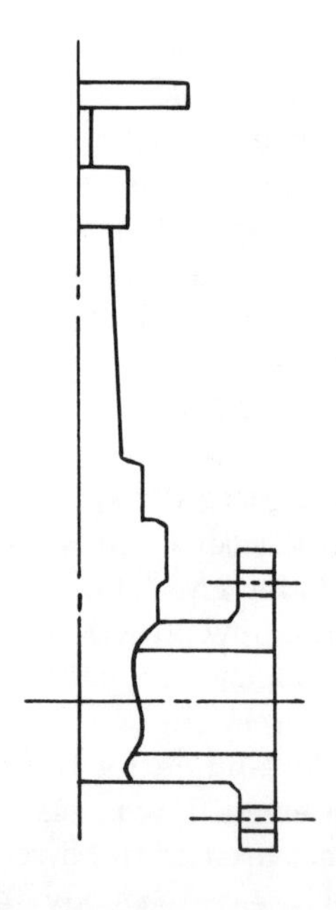

FIG. 3-6. End connection, flange.

need be, Figure 3-6. Flanges are extremely cumbersome and difficult to field-weld to the end of the piping, requiring a highly skilled welder. Aligning the assembly is time consuming and field inspection of the welds can be difficult.

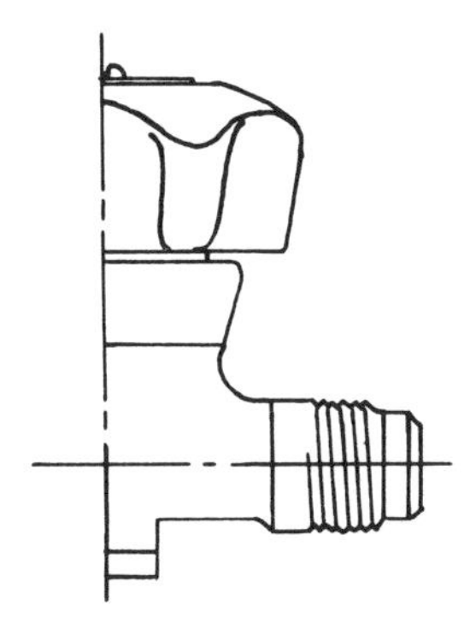

FIG. 3-7. End connection, 45° refrigeration flare.

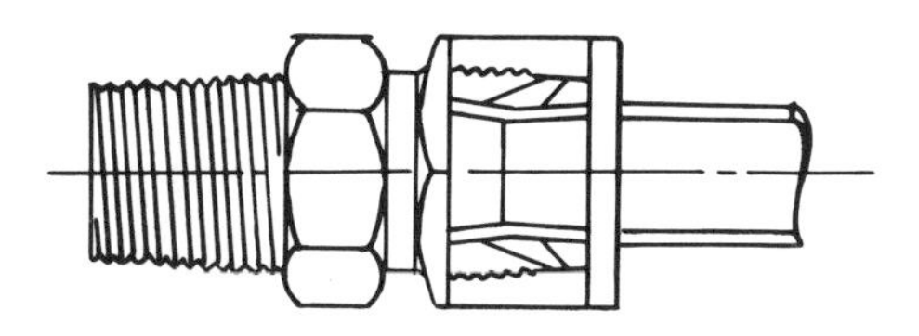

FIG. 3-8. End connection, patented flareless end.

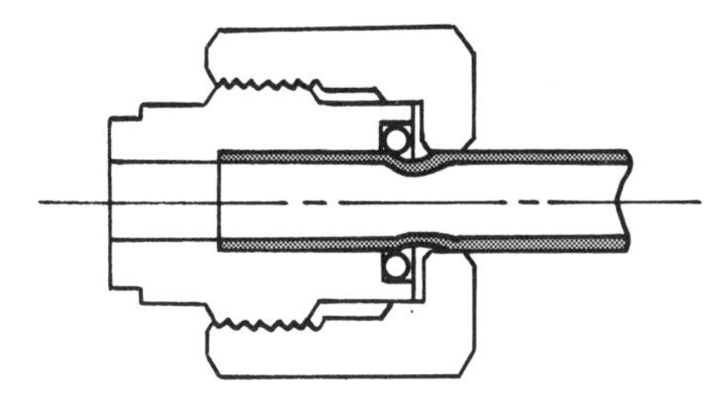

FIG. 3-9. Mec-Lock® patented end fitting. *Courtesy Primore Sales, Inc.*

Flare end The flared end is used in conjunction with a flare nut and is connected to a soft metal tube end that has been flared. Most air conditioning and refrigeration systems use a 45° flare with a single flare nut, Figure 3-7. A 37° flare is common with hydraulics and a sleeve is used with the flare nut to achieve a leak-proof joint. Many automotive applications use a double flare.

Flareless fittings, used in pneumatic and hydraulic systems and high vibration applications, can withstand thousands of pounds pressure. Patented,

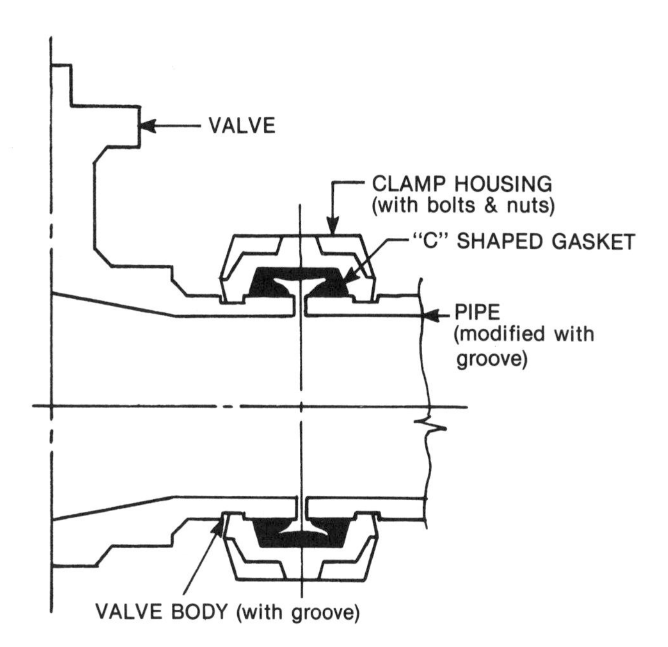

FIG. 3-10. Proprietary piping joint for heating/cooling water and steam systems. *Courtesy Victaulic Co. of America.*

proprietary designs are available for pneumatic, hydraulic, chemical, and vacuum applications, Figure 3-8.

Another patented design that is used with refrigeration systems is the Mec-Lock® which consists of coupling nut, crimp collar, and O-ring, Figure 3-9. The basic idea is to eliminate tube flaring and soldering.

Many heating and cooling applications use proprietary pipe couplings, Figure 3-10, in sizes up to 24 inches. This Victaulic system is comprised of a C-shaped gasket, a metallic gasket enclosure, and bolts and nuts to secure the joint. However, there must be a special groove on the valve end and the ajoining pipe before using this system. Overall, it is easier to assemble and less time consuming than welded or flanged joints.

Sanitary end connections Three basic end connections are available for sanitary usage:

- Bevel seat, Figure 3-11a
- Tri-clamp®, Figure 3-11b
- Welded, Figure 3-11c

The tri-clamp and bevel seat end connections are quickly and easily disassembled for cleaning. The welded connection, which requires a skilled craftsman, is expensive and time consuming. However, welded joints are required in many food plants to meet FDA requirements.

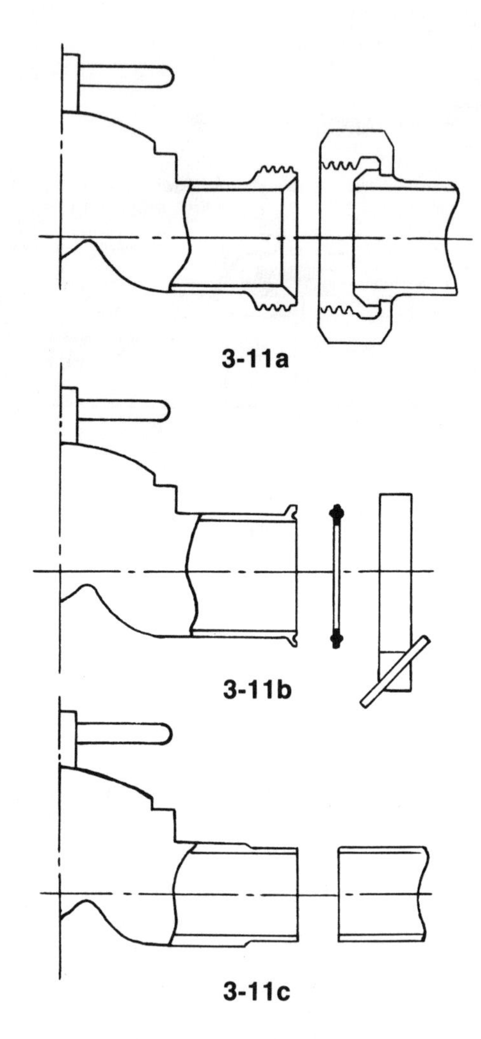

FIG. 3-11. Sanitary end connections. **a.** Bevel seat. **b.** Tri-clamp. **c.** Butt weld.

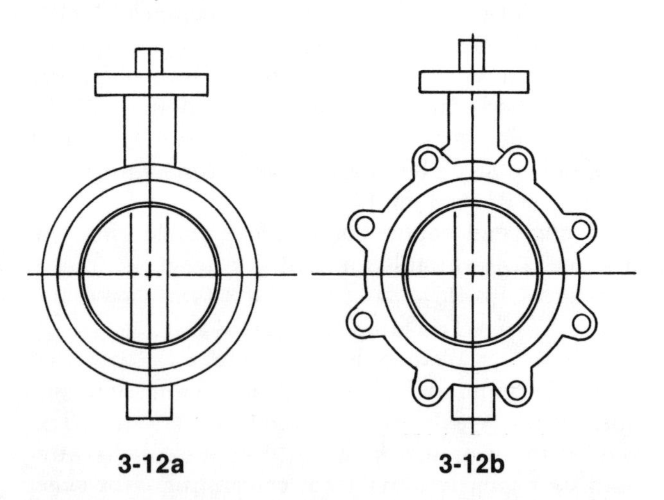

FIG. 3-12. Butterfly valves. **a.** Wafer type body. **b.** Lug wafer body.

There are some valves that have no end connections. The butterfly valve and some check valves are so thin that they are sandwiched into the piping system. The wafer type, Figure 3-12a, is held in place by bolts passing outside the periphery of the body. The lug type, Figure 3-12b, has standard pipe flanges, integral with the body, that match the flanges on the connecting pipe.

Bonnets

The bonnet is the top portion of the valve that contains the valve stem and packing. There are several types:

Screwed bonnet By far the most common type, screwed bonnets are seldom used for pressures over 150 psig, Figure 3-13a. Mating threads and surfaces must be leak-tight and accurately machined. After

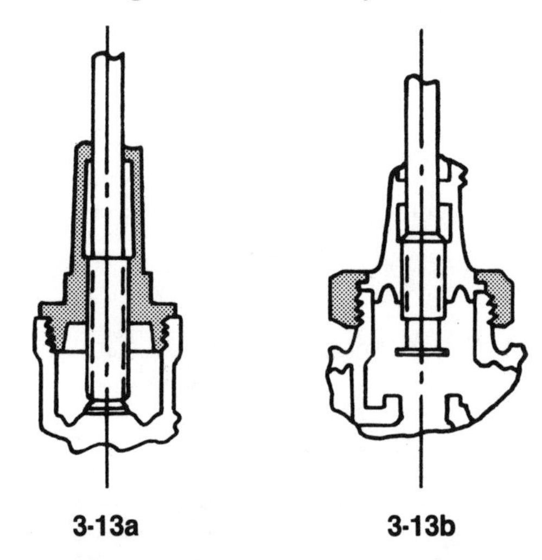

FIG 3-13. a. Screwed bonnet. **b.** Union bonnet.

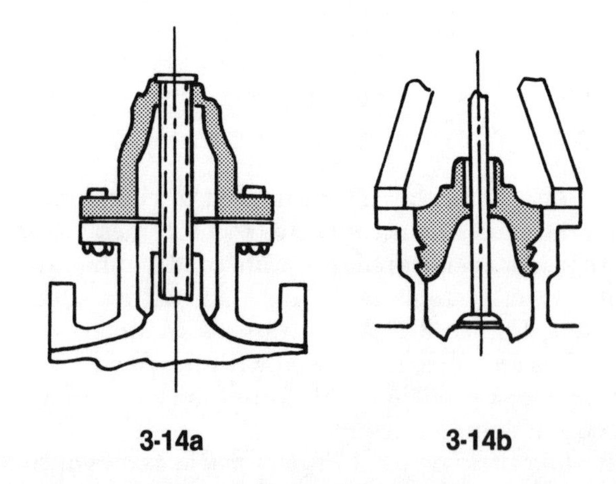

FIG. 3-14. a. Bolted bonnet. **b.** Pressure seal bonnet.

years of service, this style may not be easily disassembled due to accumulated corrosion.

Union bonnet The union bonnet, Figure 3-13b, has a separate, threaded ring that mates with external threads on the valve body. It is ideal for valves requiring frequent dismantling and for pressures exceeding 125 psig.

Bolted bonnet Commonly found in industrial applications, the bolted bonnet has mating flanges that are machined and drilled, Figure 3-14a. While this design can be adapted to any size valve, the flange thickness and the number of bolts will vary with the pressure rating.

Pressure seal The pressure seal bonnet, Figure 3-14b, is used in steam service and other high temperature, high pressure applications and is, by far, the most expensive and complicated. The bonnet joint is inside the valve body and the seal ring is wedge shaped. As pressure increases, the seal ring is wedged tighter and tighter between the body and the bonnet. This design is capable of 6000 psi and 700 °F and 4500 psi in the gate valve.

These bonnet configurations are most commonly found on gate and globe valves. The angle, Y-pattern, sliding, U-valve, diaphragm, and solenoid valves have bonnets similiar to the union bonnet although there are variations between manufacturers.

The plug, butterfly, ball, and multiport valves, along with regulators and relief valves, have unique bonnet configurations. The packing gland is usually enclosed in the valve body, creating a design with considerably less height than gate or globe valves.

Packless refrigerant valves have a shortened bonnet containing a diaphragm to hermetically seal and contain the refrigerant.

The body of bar stock valves is also the bonnet and contains all the components.

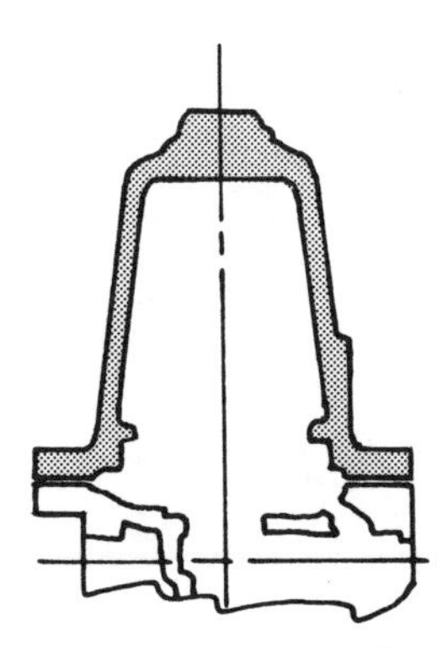

FIG. 3-15. Pressure regulator bonnet.

In some regulators, the bonnet contains the spring, adjusting stem, and diaphragm. In others, the body contains all the parts, including the bellows for hermetic sealing and the adjusting stem, Figure 3-15.

In cryogenic applications, the bonnet is of unusual length and fabricated from materials with low thermal conductivity. Since cryogenic fluids may be at temperatures that are hundreds of degrees below zero, a bonnet of normal length would subject the packing to extreme temperatures and sealing would be quite difficult. With a long bonnet, the packing is closer to ambient temperature and sealing is more reliable. In some designs, the bonnet is contained within a vacuum sleeve to minimize heat transfer.

Valve disc and seat (trim)

The port area, consisting of the disc and seat, is the heart of any valve and where the action takes place. Without a proper disc, and matching seat, tight shutoff cannot be accomplished nor can good throttling characteristics be achieved.

In less expensive valves, seats are machined into the body. In more critical applications, discs and seats are fabricated from different materials than the body and are usually more noble. This minimizes corrosion problems that could arise.

Since the disc and seat are the heart of the valve, choosing the proper materials is a must. Long, trouble-free service requires the proper combination of hardness, wear-resistance, and resistance to erosion, galling, seizing, and temperature. In severe cases, metal seats and/or discs may be replaced by more exotic materials like stellite and ceramic compositions. These materials are exceedingly hard and assure long life with little wear. Nonmetallic disc materials are used for tight sealing of hard-to-hold materials such as gasoline.

Since the trim is the heart of any valve, the disc and seat will *see* the pressures and fluid velocities in the system, as well as any contaminants. It is imperative that the trim materials be capable of handling the type of fluid and the pressures within the system in order to yield long service life, tight shutoff, and the required flow control.

Forces within the valve depend on the valve configuration. With flat discs, normal to flow, the disc will tend to bend the stem in the direction of flow. Gate valve discs will chatter and wear severely if used to throttle. With butterfly valves, the disc can be throttled even though subject to some of the same forces as a gate valve. Here, the disc is streamlined,

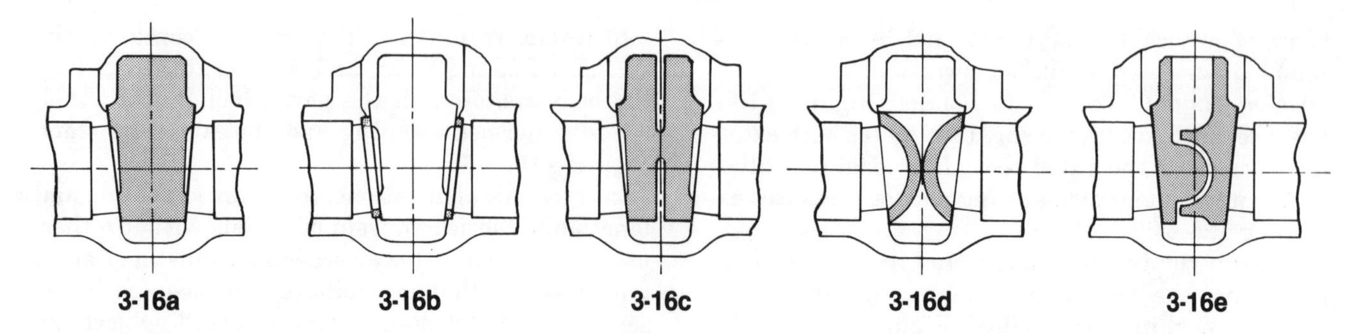

FIG. 3-16. Gate valve discs. **a.** Solid wedge. **b.** Solid wedge, removeable seats. **c.** Flexible wedge. **d.** Double wedge. **e.** Ball and socket.

to minimize turbulence, and a lower bearing is used to stabilize stem deflection.

Globe valve discs allow flow over the disc face parallel to the stem. This eliminates some of the bending forces on the stem.

The best way to describe the various disc/seat combinations is to outline them by valve type, that is gate, globe, and butterfly. Many of the other valves fall into one of the above three groupings.

Gate valve The gate valve has five basic configurations, as shown in Figure 3-16.

- Solid wedge disc This is the most popular, least expensive, and most rugged of the gate disc/seat arrangements. The disc has a wedge V design that fits into a tapered seat that is machined into the valve body, Figure 3-16a. Wedging, as the disc is lowered into the body, seals the mating surfaces and effects a tight shutoff. Basic disadvantages of the solid wedge are excessive wear, misalignment, and galling. If corrosion is negligible, this design can give years of useful service.
- Solid wedge, removeable seats Removeable seats, Figure 3-16b, employing alloy materials that are carefully selected to minimize corrosion between the body and the removeable seats, have greatly increased the useful life of gate valves.
- Flexible wedge disc This variation of the solid disc has two halves that flex independently of each other, Figure 3-16c. Independent flexing permits each half to seal tighly against its own seat even if body distortion has occured. Flexible wedges are widely used in high pressure and temperature service and in larger size valves.
- Double wedge The split wedge has even greater adaptablity than the flexible wedge disc since each half is free to seat independently of the other, Figure 3-16d. This design is used with sticky or viscous materials where complete shutoff can be a problem.
- Ball and socket This design incorporates a rotating disc, Figure 3-16e, that presents a new seating surface each time the valve is opened and closed. Extensive wear and misalignment can be tolerated without a loss of sealing ability.

Globe, angle, Y-pattern, check valves, regulators

The globe valve does not have the straight through

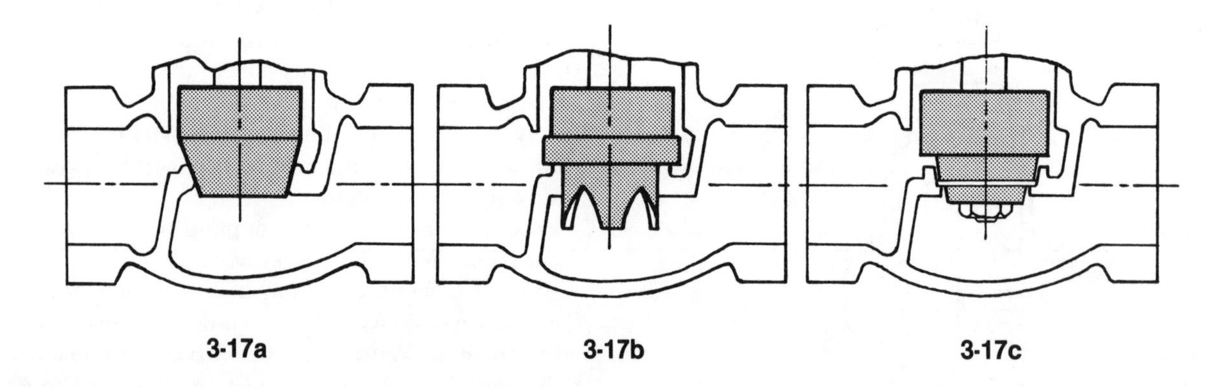

FIG. 3-17. Globe valve discs. **a.** Plug. **b.** V-port plug. **c.** Wafer.

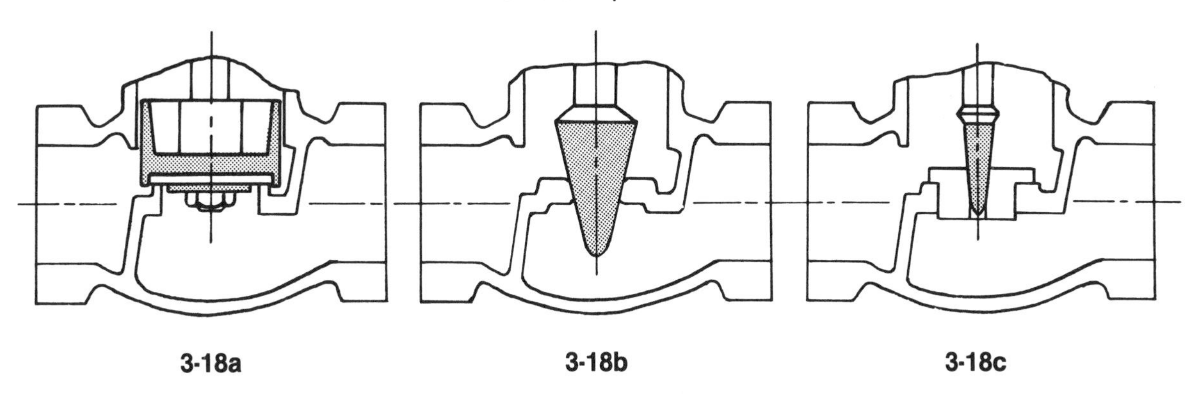

FIG. 3-18. Globe valve discs. **a.** Composition. **b.** Taper. **c.** Needle disc and removeable seat.

fluid flow of the gate valve, rather it is a Z pattern. The extra resistance of this pattern allows for closer regulation of the flow which can be enhanced by the shape of the disc and its relationship to the matching seat. In the case of regulators, the match-up between disc and seat is of prime importance.

Figures 3-17 and 3-18 show the basic seat and disc configurations available with globe valves.

- Plug disc The plug disc, Figure 3-17a, is the most widely used disc in globe valves. With its conical shape, wear is distributed by the shearing action of the mating surfaces. Due to the wide bearing surfaces, any minor damage to the surface, or buildup of dirt or scale, has little effect on sealing. This design is most effective in resisting erosion, due to close throttling, if constructed of proper materials.
- V-port plug This design, Figure 3-17b, is employed in many corrosive environments where deposits tend to accumulate on the seating surfaces. This disc will cut through deposits and create a tight metal-to-metal seal.
- Wafer disc This arrangement, Figure 3-17c, is used to achieve a better seal in more critical applications. A disc of nonmetallic material is sandwiched in the plug to assist metal-to-metal sealing by the addition of the safety sealing material.
- Composition disc Sealing by a composition disc, Figure 3-18a, is accomplished against a flat, raised seat, ring type, circular surface. The disc material is nonmetallic and effects a good, tight shutoff even in the presence of suspended solids or corrosion. It is not recommended for close regulating or throttling which can cut the disc material. The disc material must be compatible with the controlled fluid since many fluids or gases can chemically attack the composition and destroy it in short order. Chemical compatibility tables are shown in the Appendix.

  Some refrigerant valves utilize a nylon disc to assure a tight shutoff. However, extreme care, even disassembly, is required before the valve can be soldered to a line. Overheating can destroy the nylon disc, rendering the valve useless.
- Taper disc The taper disc, Figure 3-18b, is usually found in small size valves and is used in hydraulic and pneumatic systems where flow is used to do useful work.

  With a tapered disc, the size of the flow orifice changes very little as the valve is opened or closed. The stem threads have a finer pitch than ordinary valves, requiring a greater number of turns to alter the flow.
- Needle disc Refrigeration systems use the needle disc in hand expansion valves which may have removeable seats, Figure 3-18c. As long as the refrigeration load does not vary substantially, this valve controls the flow to a fine degree.

The globe valve is restricted to various throttling operations in industrial applications. Figure 3-19 shows three disc configurations and their flow control profiles. The percentage of opening is the distance the disc travels from fully closed to fully open or 0% to 100% respectively.

Poor flow control is a characteristic of the *quick opening* flat disc shown in Figure 3-19a. With only a 10% movement of the disc, flow is 40% of maximum.

Finer control is achieved by the conical disc shown in Figure 3-19b. A 10% or 20% or 30% movement of the disc increases flow by a corresponding 10% or 20% or 30%. The graph is a straight line, therefore, this disc creates a linear flow valve.

The third shape, Figure 3-19c, is the equal percentage disc. Here control is extremely fine since a 60% movement of the disc increases flow by only 20%.

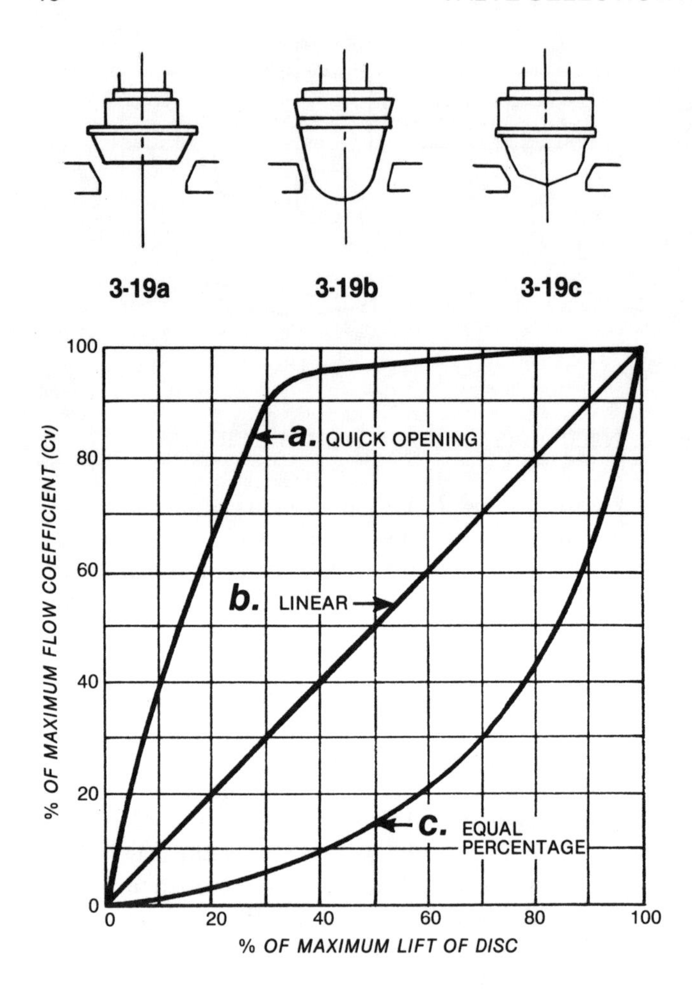

FIG. 3-19. Flow control characteristics. **a.** Flat disc. **b.** Conical disc. **c.** Equal percentage disc.

Cages Cages are cylindrical guides, with machined flow ports, that surround the disc in some globe valves. They maintain uniform distribution of flow around the disc and restrain side movement of the disc during high pressure drops.

Butterfly disc seat The butterfly arrangement is similar to the plug and ball valve styles because it takes a 90° turn of the stem to open and close the disc against its seat. The disc and seat arrangement, however, is completely unique. Figure 3-20 shows several proprietary designs.

Most butterfly valves seal by seating the disc against a rubber liner. This design is limited by the pressure, temperature, and chemical resistance of the liner material. At high pressures, the liner can bead and tear under normal use. High temperatures and chemical exposure can cause the rubber to swell and deteriorate. However, in many normal applications, this design is adequate and will give good service over long periods.

In many designs, the liner wraps around both faces of the body or extends beyond the body, Figure 3-20a. When the valve is bolted between two pipe flanges, the extended or overlapping liner forms a positive seal.

Teflon®, Figures 3-20b and 3-20c, has permitted increases in valve performance ratings and a general upgrading. It can provide a bubble tight shutoff over long periods of use and withstand exposure to more active chemicals.

Seats for butterfly valves are metal, rather than resilient, where temperatures range from 700° to 1000°F and pressures are up to 1500 psig. They are

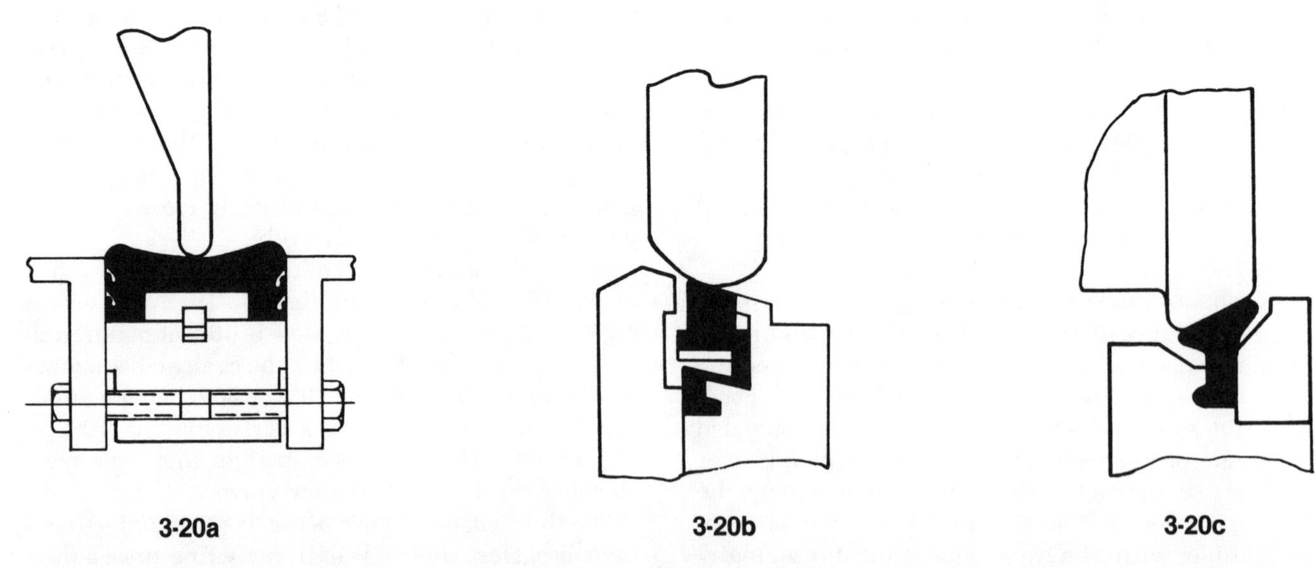

FIG 3-20. Butterfly disc seats. **a.** Resilient seat liner. **b.** Spring-loaded Teflon® seal. **c.** Teflon® flexing lip seal.

expensive, due to the sophisticated machining and finishes needed to provide a tight seal under these demanding conditions.

The disc of the butterfly valve rotates from perpendicular to the fluid flow (closed), through an infinite number of positions, until parallel to the flow (open). The disc remains submersed in the flowing fluid and can be considered to act like an airfoil or sail. Disc development, in the past few years, has employed wind tunnel tests to optimize the disc's shape and minimize drag, flutter, and torque.

The shaft of the standard butterfly valve passes through the seat while, in a high performance valve, the shaft is mounted off-center. This eccentric shaft permits 360° of uninterrupted surface for more reliable sealing and easier replacement of the seal since the disc does not have to be removed.

Offsetting the shaft reduces seat/disc interference. When opening, the disc lifts off the seat, reducing wear on the liner and lowering the operating torque.

Because the disc rotates, it requires bearing support at its top and bottom and care must be taken that the system pressures do not overload these bearings. A solid, minimum-deflection shaft and adequate bearings will assure long service life.

Plug, ball, and multiport valves The original 90°-turn valve was the plug valve. Its tapered disc was lapped into a tapered seat which is part of the body. A hole was bored or drilled through the disc or plug to allow fluid to pass through the valve. When the hole is parallel to the flow, the valve is open. When perpendicular (normal) to the flow, the valve is closed.

The plug valve disc was nonlubricated, that is, sealing was accomplished by metal-to-metal contact. This produced galling and sealing surface roughness, resulting in severe binding, extemely high operating torque, and leakage. The answer: lubricate the mating surfaces by cutting grooves in the disc face to retain grease. As long as the controlled fluids are compatible with the grease, there is no problem. This arrangement is still used in smaller valves.

Development of elastomers (rubber compounds) and Teflon® (TFE) has changed the characteristics and widened the usage of the plug valve. With new seating materials, the ball and multiport valves became practical. Figure 3-21 shows a nonlubricated plug valve, used in the food industry, that is excellent for sanitary applications. Galling and seizing have been minimized by fabricating the body and plug from different corrosion-resistant materials. A plug with TFE seats is shown in Figure 3-22.

The ball valve is shown in Figure 3-23. The disc is a highly polished ball suspended between two TFE seats. The secret of the ball valve's success is the seat design. Sealing is achieved by either line contact, specially shaped seats, or preloaded, patented seat designs.

The TFE seat gives or deflects when in contact with the ball. This spring effect will compensate for

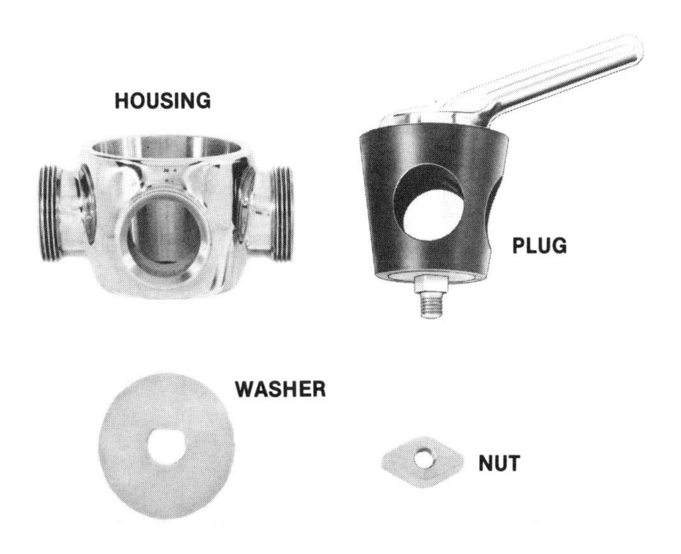

FIG. 3-21. Non-lubricated plug valve. *Courtesy Ladish Co., Tri-clover Div.*

FIG. 3-22. PTFE-coated plug and seat provide excellent sealing with minimum torque. *Courtesy Dow Chemical U.S.A.*

FIG. 3-23. Patented high performance ball valve with self-adjusting seats. *Courtesy Flow Control Div. Rockwell International.*

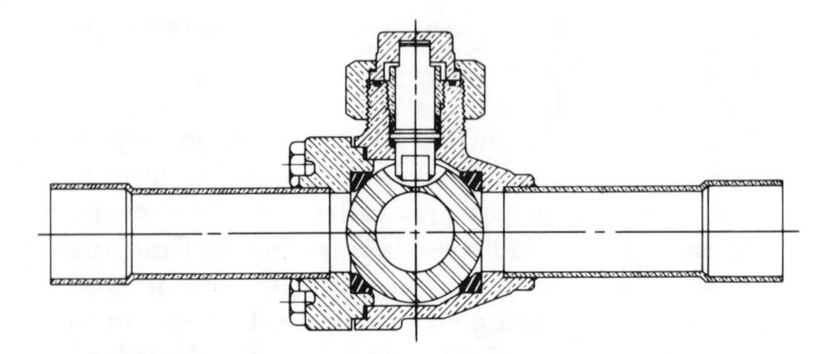

FIG. 3-24. Ball valve with Teflon® seats and triple stem packing. *Courtesy Superior Valve Co.*

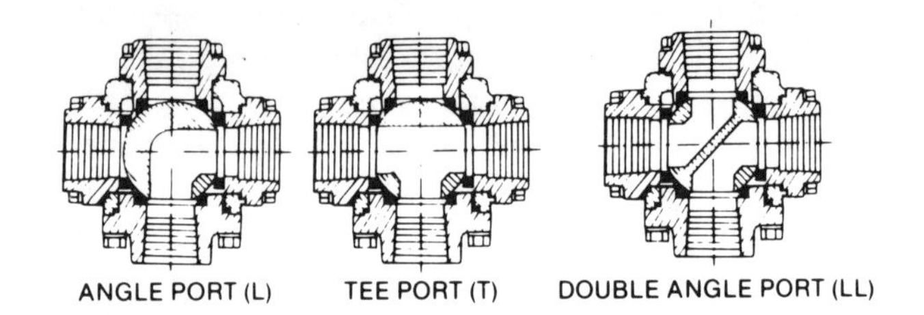

FIG. 3-25. Three styles of multiport ball valves. *Courtesy Pittsburg Brass Mfg. Co.*

wear and reduce operating torque to a minimum. The seat also wipes the ball clean each time the valve is actuated. In more expensive ball and multiport valves, the seat rings can be adjusted to compensate for wear.

Recently, ball valves have become available for direct use in refrigerant circuits, Figure 3-24.

The multiport valve is an extension of the two-way ball valve design. In this case, the ball is not simply drilled through. In the L configuration, two holes, 90° apart, meet in the center of the ball. In the T configuration, one of the ports in the L ball is drilled through. In the LL configuration, two L ports, 90° apart, are separated by a web of metal. These variations are shown in Figure 3-25.

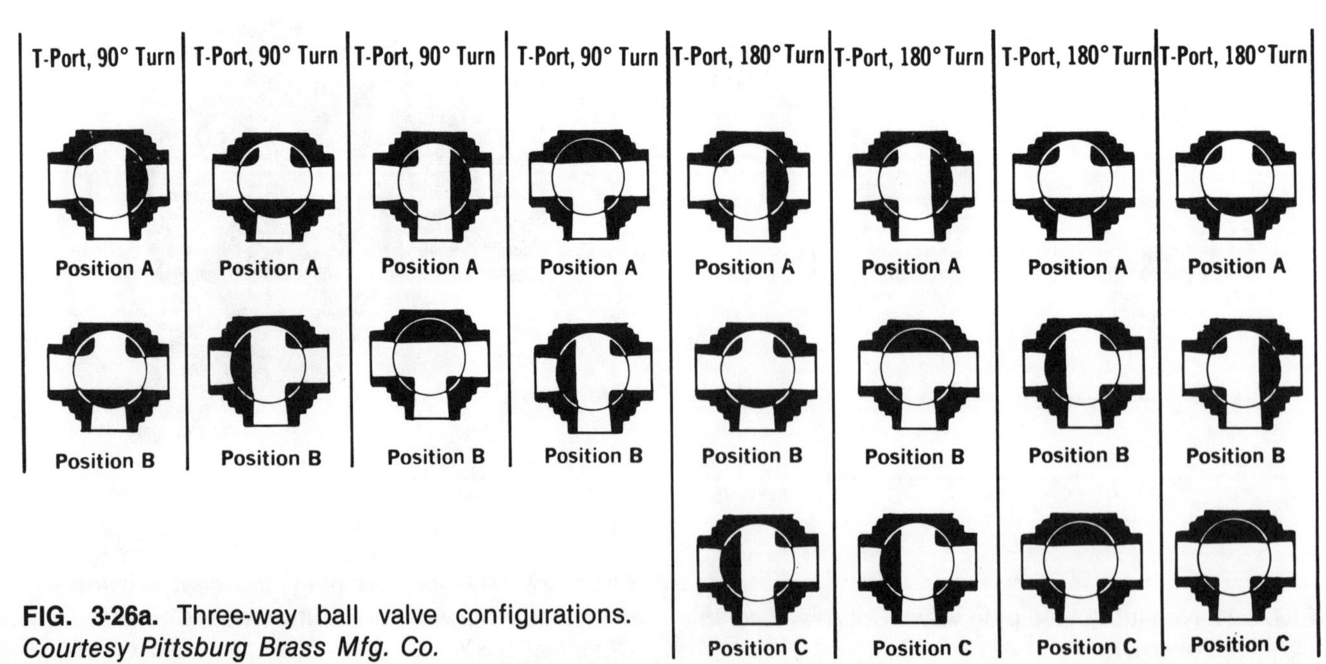

FIG. 3-26a. Three-way ball valve configurations. *Courtesy Pittsburg Brass Mfg. Co.*

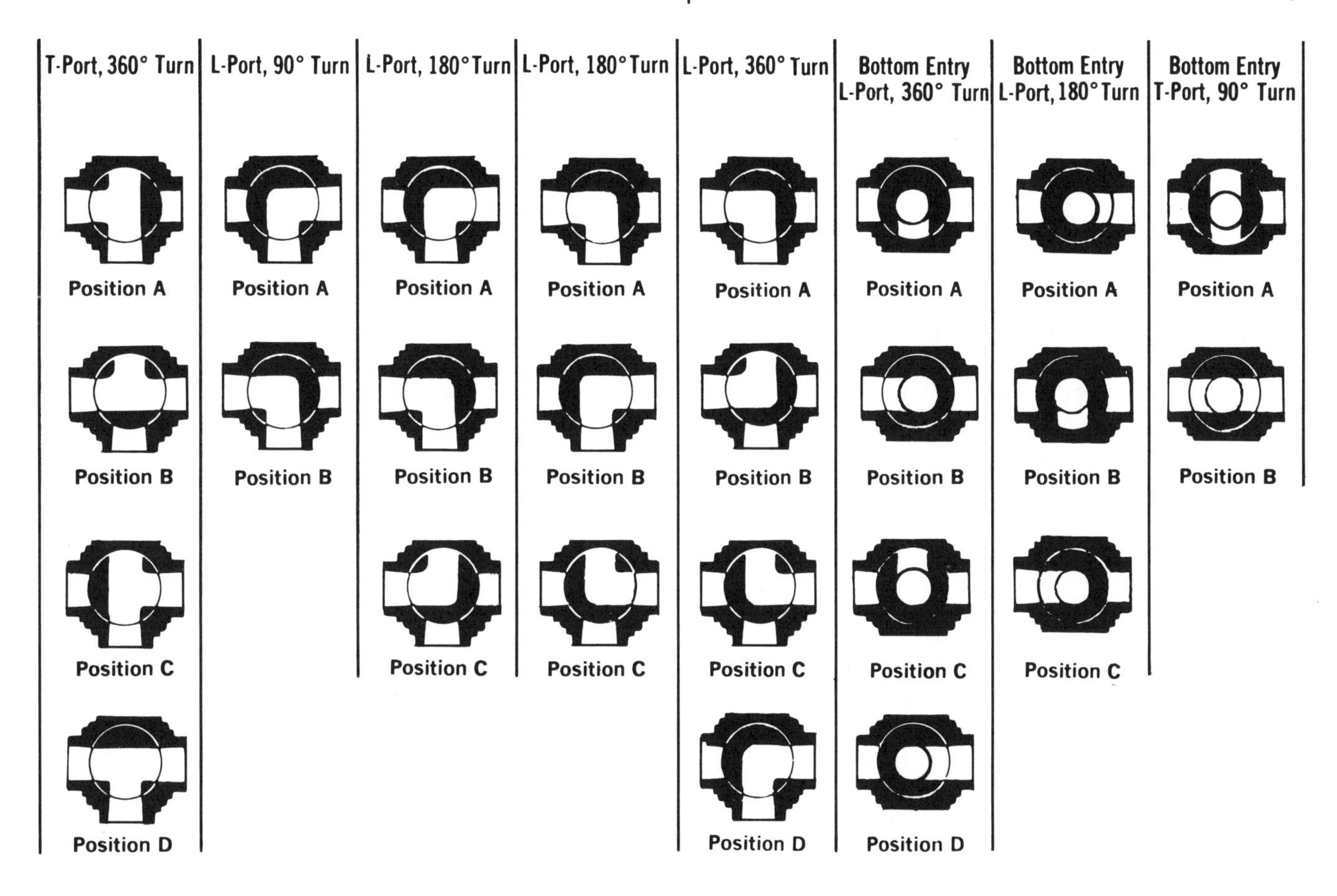

FIG. 3-26b. Three-way ball valve configurations. *Courtesy Pittsburg Brass Mfg. Co.*

4-Way Ball Valves

LL-Port, 90° Turn — Position A, Position B

L-Port, 180° Turn — Position A, Position B

L-Port, 360° Turn — Position A, Position B, Position C, Position D

T-Port, 180° Turn — Position A, Position B

Straight 2 Port 90° Turn — Position A, Position B

Bottom Entry L-Port, 360° Turn — Position A, Position B, Position C, Position

Bottom Entry T-Port, 90° Turn — Position A, Position B

5-Way Ball Valves

Bottom Entry L-Port, 360° Turn — Position A, Position B, Position C, Position D

Bottom Entry LL-Port, 180° Turn — Position A, Position B

Bottom Entry T-Port, 90° Turn — Position A, Position B

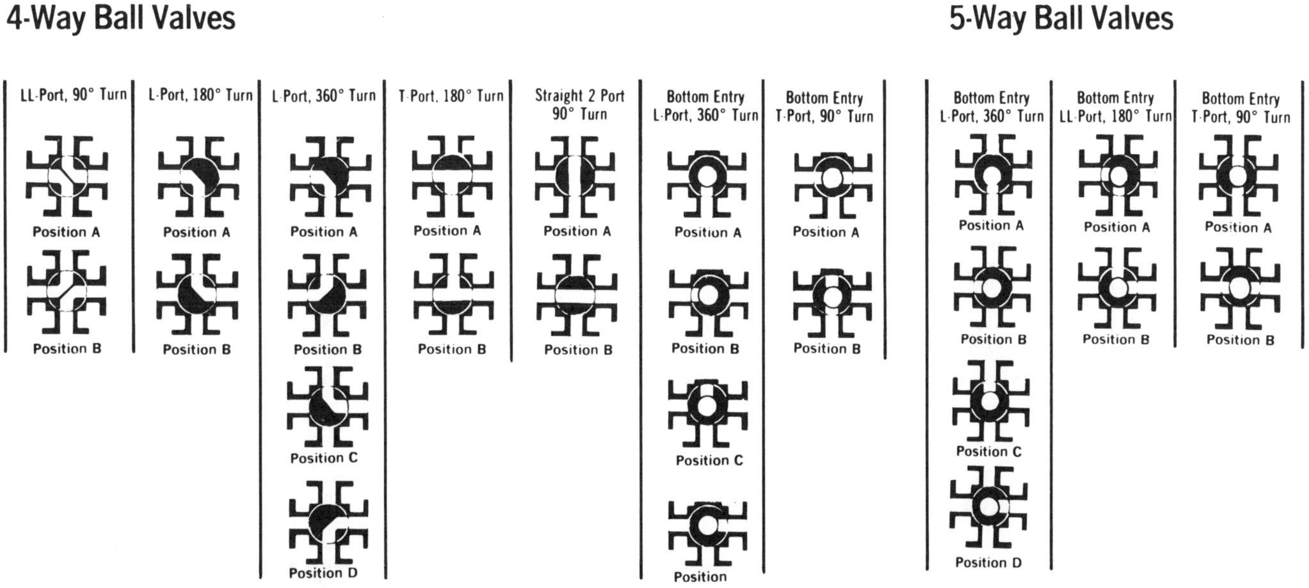

FIG. 3-26c. Four- and five-way ball valve configurations. *Courtesy Pittsburg Brass Mfg. Co.*

In many applications, a piping system is multipurpose, handling different fluids through the same piping network. Or various processes may require mixing or diverting to other areas. Rather than an array of valves, one or two multiport valves will do the same job.

Figure 3-26 shows multiport valves available from a single manufacturer, including two-, three-, four-, and five-way versions.

Three-way valves for heating/cooling terminals are quite common in buildings. Figure 3-27 shows a three-way valve used to proportionally control chilled water flow. Arrows on the body indicate proper flow for mixing service. Improper flow reduces performance, due to chatter, and creates forces on the disc that the actuator is not designed to handle.

Double-seated valves, Figure 3-28, are used by the air conditioning and heating industry for diverting or mixing cold or hot water. A three-way valve body is used. For diverting service, one port is the inlet while two ports are outlets. For mixing, two ports are inlets and one is the outlet. As one of the inlet ports is closed, the other opens proportionally.

Note that in the diverting arrrangement, the discs are mounted opposite to one another on the stem while in the mixing arrangement the discs are back-to-back. Because of flow parameters, a diverting valve cannot be used for mixing service and vice versa. As an external actuator moves the stem/disc axially, the dual discs move to open or close, depending upon their location in their respective ports.

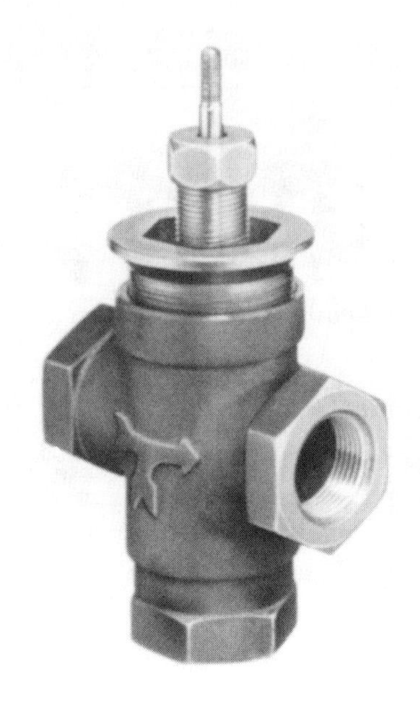

FIG. 3-27. Three-way mixing valve. *Courtesy Penn Controls, Div. of Johnson Service Co.*

In either service, the flow stream is designed to be under the disc, for better control. If chilled and hot water are mixed, to an intermediate temperature, the cold water may be at 40 °F while the hot water is at 180 °F — a 140 ° differential. The thermal stresses created by this differential may cause some problems in tight sealing.

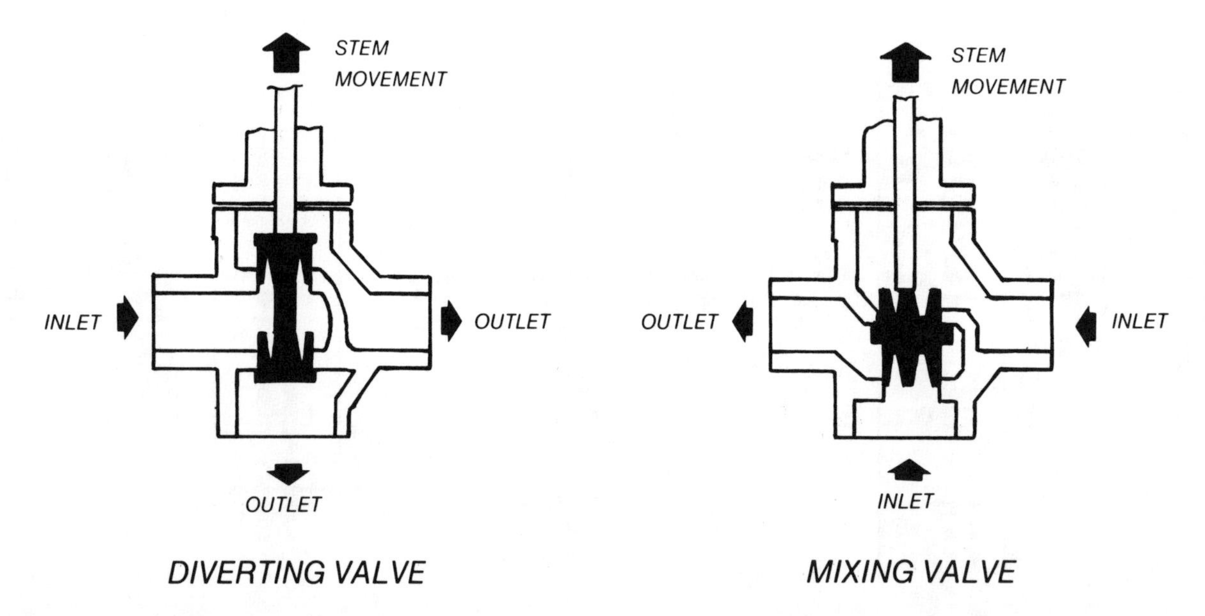

FIG. 3-28. Three-way valves may be used for either diverting or mixing service, depending upon the number of inlets.

Because an extended stem is required to handle both discs, and no bottom guide or cage is used, stem flexibility can cause disc misalignment. Therefore, the disc is equipped with an extended skirt that centers the disc in the port and limits problems arising from fluid forces and temperature differentials.

U-valve disc and seat The U-valve is unique in that its disc is a conical rubber plug that is raised or lowered like a gate valve disc. The plug seats against a matching cavity machined into the valve body and *flows* or *gives* to the shape of the seat to create a tight fluid seal.

Use of the U-valve is limited by the temperature, pressure, and chemical resistances of the plug. Having a nonmetallic disc, this valve often has the advantage of the gate valve in corrosive streams where a metallic disc may corrode and stick open or closed. The most popular use of this design is in hot and cold water systems.

Diaphragm disc and seat The flexible disc of a diaphragm valve is rubber, Teflon®, or Teflon-faced rubber. This flexibility is the key to its success and sealing properties, Figure 3-29.

The outer edge of the disc is bolted between the body and the bonnet and permanently secured. The valve stem is fastened to the top of the disc and is pushed down or pulled up for opening or closing the valve. Figure 3-30 shows both styles.

The stiffness of the diaphragm, which is substantial, limits its flexibility. Consequently, the seat or weir is raised, like a web, directly across the fluid path to minimize the diaphragm's travel when opening or closing.

Many sanitary applications utilize diaphragm valves because they have no cavities around the stem, do not expose fluids to a packing, and are easily cleaned by disassembling or by *clean in place* methods.

As with any elastomer — rubber, Teflon®, or whatever — temperature, pressure, and chemical resistances limit the range and application of this valve.

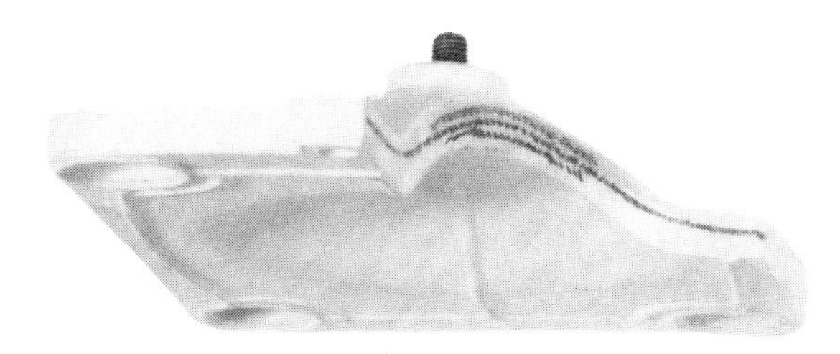

FIG. 3-29. Normally-open flexible diaphragm. Press down to close. *Courtesy Flow Control Div., Rockwell International.*

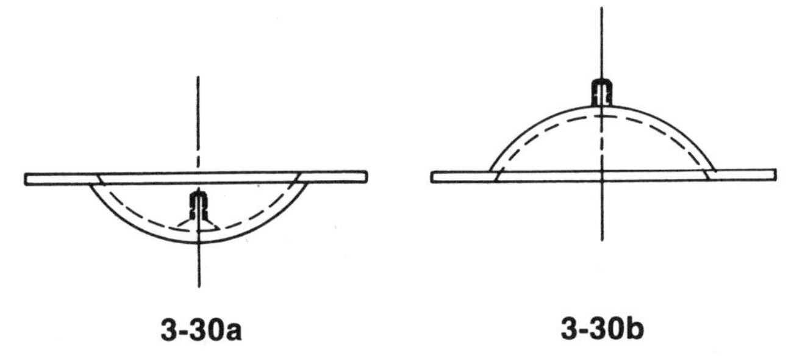

FIG. 3-30. Flexible diaphragms. **a.** Normally closed, pull up to open. **b.** Normally open, push down to close.

Solenoid disc and seats Three styles of solenoid discs and seats are available in refrigeration service, Figure 3-31.

- A tapered pin against a circular seat port, usually limited to 1/8″ diameter (direct acting).
- A diaphragm disc and a circular seat port that is over 1/8″ diameter and may be as large as several inches in diameter (pilot operated).
- Piston disc.

With the tapered pin design, the disc is machined on the tip of the steel stem. It is operated by an electromagnetic coil that causes the disc to snap upward

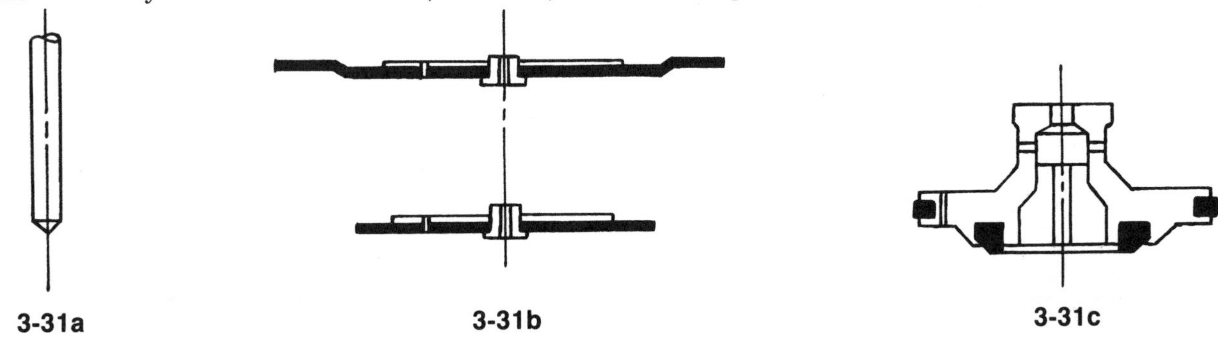

FIG. 3-31. Various solenoid discs. **a.** Direct-operated tapered pin. **b.** Nonmetallic flexible disc (top) and plain circular disc (bottom) are both pilot operated. **c.** Piston type is pilot operated.

into the coil due to magnetic attraction when the coil is energized. For normally closed (N.C.) valves, the magnetic field must be strong enough to overcome:

- The weight of the stem.
- The pressure differential across the valve disc.
- The force of the spring acting to close the port.

Since the magnetic field is not always this strong, the size of the port is usually limited to no more than 1/8″.

The pressure differential across the disc/seat is designated as MOPD (maximum operating pressure differential). The MOPD is the maximum pressure differential, $P_i - P_o$, against which the solenoid can safely operate and raise the disc. If the MOPD is too great, the magnetic force cannot overcome the pressure, the solenoid will not operate, and the disc will remain closed. The MOPD is usually listed in the manufacturer's literature.

In large size solenoids, the operating fluid assists in raising the disc. The solenoid raises the stem, opening a pilot orifice, that releases the pressure on top of the disc, Figure 3-32. This pressure release unbalances the forces acting above and below the disc, with the resultant force being upward to raise the disc.

When the coil is de-engergized, the pilot orifice is closed and full line pressure is trapped above the disc after flowing through the bleed orifice. This equalizes line pressure on both sides of the disc. Thereafter, the weight of the piston and the spring tension can close the disc, Figure 3-33.

Refrigerant solenoid valves employ a number of discs: metallic pistons, nonmetallic circular discs, and nonmetallic flexible diaphragms that are held between the bonnet and the valve body.

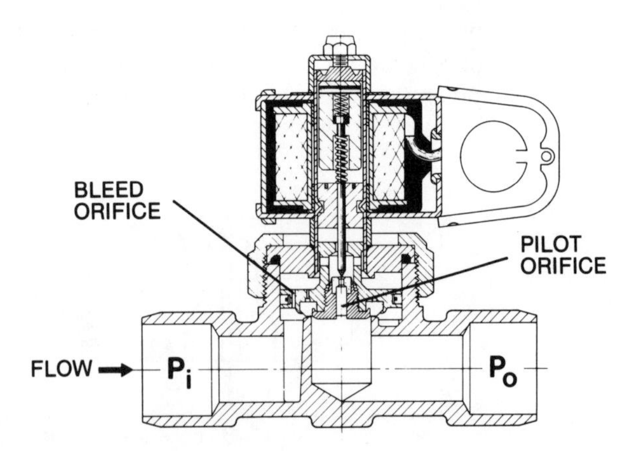

FIG. 3-32. Pilot-operated solenoid valve. *Courtesy Sporlan Valve Co.*

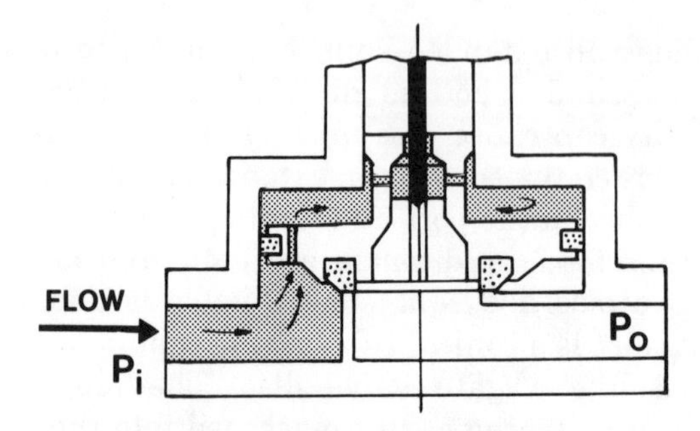

Normally-closed (NC) valve is unenergized. Pilot orifice is blocked by solenoid plunger.

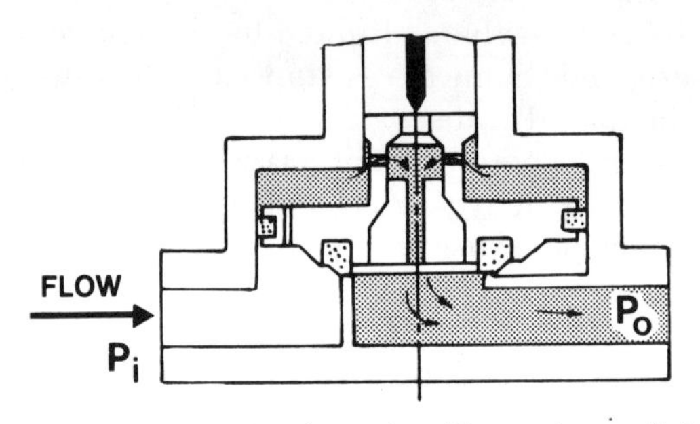

Valve is energized and opening. Plunger is up, pilot orifice is open. Fluid above the disc flows to outlet, reducing pressure above the disc. The smaller bleed orifice cannot replenish fluid as fast as it escapes through pilot orifice.

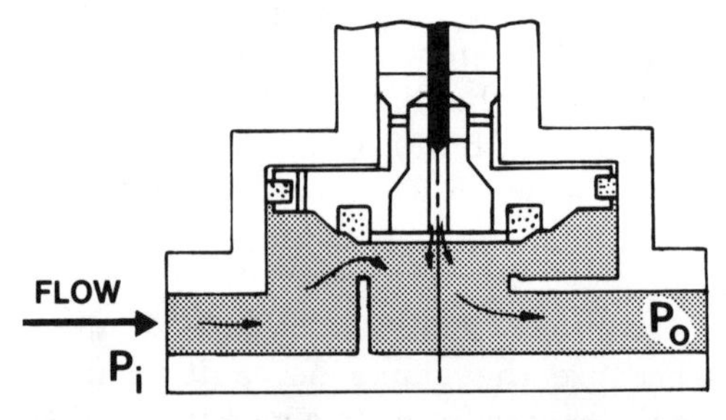

Valve is de-energized and plunger closes pilot orifice.

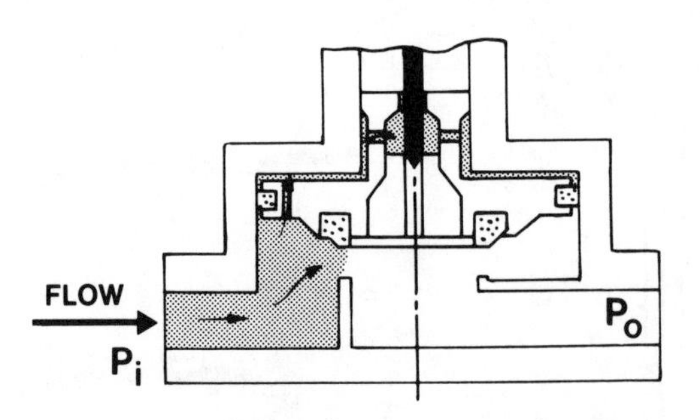

Bleed orifice allows high pressure fluid to accumulate above disc and close valve.

FIG. 3-33. Sequence of operation of a pilot-operated solenoid valve.

Stem

The stem is critical because it transfers the opening and closing forces from the valve operator to the disc. Selection is based on the forces created by the pressures within the piping system and the torque needed to turn the stem. There must be minimum stem deflection so that the disc will repeatedly reseat itself in a positive and reliable manner.

Materials are carefully selected and the stem is usually more noble than the valve body to minimize corrosion.

The stem material must be easily machined to a high polish and fine finish. In most valve designs, the stem must rotate, to open and close the disc, and the packing must withstand the rotation of the stem. The finer the finish, the longer the packing life.

In gate and globe valves, the stem and disc are fastened together by a screw thread, button on the end of the stem that allows the disc to *float*, or disc nut, all of which have special features for individual applications. Whichever is used, the stem and disc must be properly aligned to assure that the disc reseats itself again and again with perfect alignment.

Five stem designs are available for gate and globe valves and are selected to meet various corrosive, pressure, and temperature limitations, Figure 3-34.

- **Inside screw, rising stem** In this most common of the configurations, the hand wheel is fixed to the stem and turns with the stem, Figure 3-34a. The threads of the stem mate with the threads in the bonnet and the stem rises and lowers as it is turned. Since the threads are exposed to the flowing media inside the bonnet, corrosion often occurs which shortens valve life and hampers reliability. Extra head room is needed with this design to allow the stem to rise.
- **Outside screw, rising stem** In this design, the hand wheel is attached to the stem and turns with it, Figure 3-34b. However, the threads on the stem are outside the bonnet, not exposed to the flowing media, and less subject to corrosion. The exposed threads can be frequently lubricated to ease operation and prolong valve life. However, if the valve is in a corrosive atmosphere, it must be maintained to remain serviceable.
- **Inside screw, nonrising stem and hand wheel** In this design, the stem and hand wheel remain at

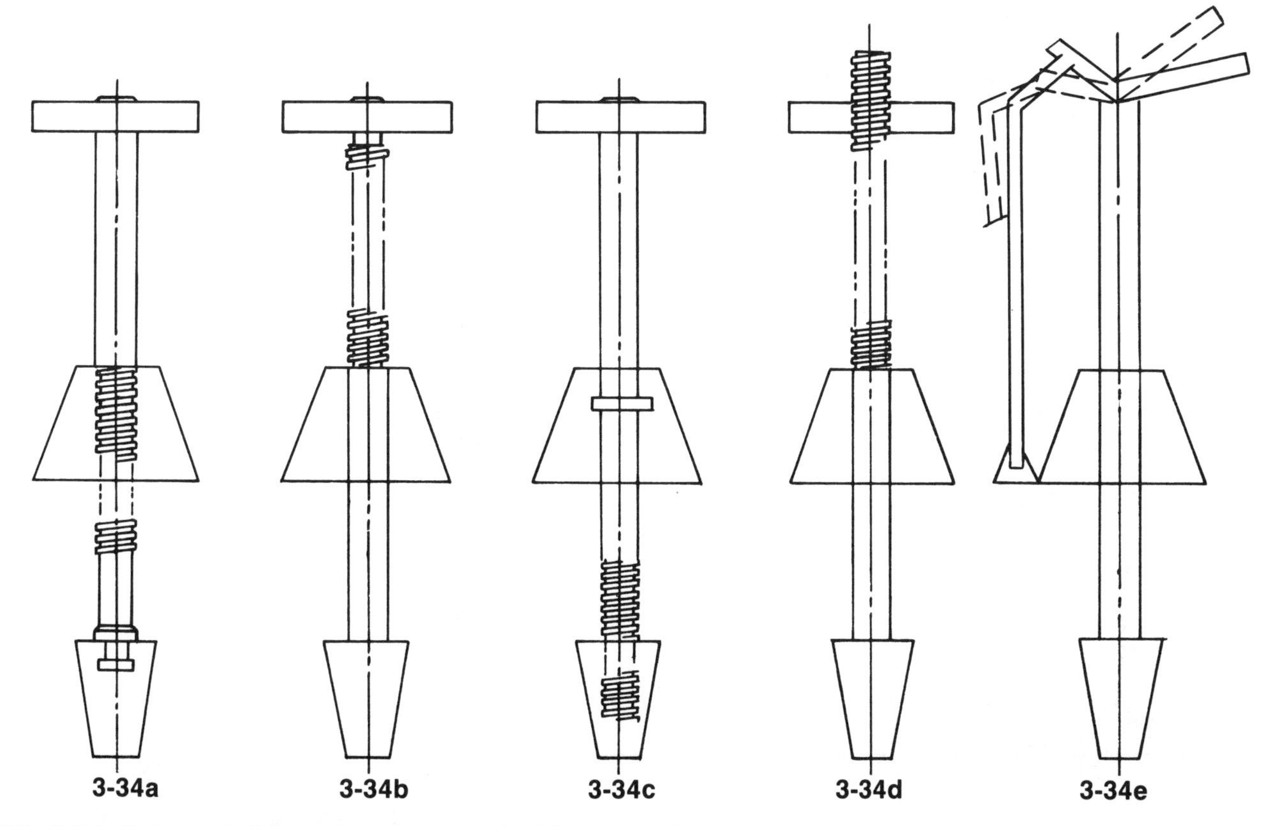

FIG. 3-34. Gate and globe valve stems. **a.** Inside screw, rising stem. **b.** Outside screw, rising stem. **c.** Inside screw, nonrising stem. **d.** Rising stem, nonrising hand wheel. **e.** Sliding stem.

a constant height as they are turned, Figure 3-34c. A stem collar, in the bonnet, prevents the stem from moving up or down and the disc is lowered and raised by riding on threads at the end of the stem. Packing wear is reduced because the stem rotates but does not travel. The working fluid, however, contacts the threads and corrosion may occur.

- Rising stem, nonrising hand wheel Here, the hand wheel operates a threaded yoke and the stem is raised and lowered without turning, Figure 3-34d. However, head room is still required for the rising stem. The threads in the stem are untouched by the flowing media and protected from that source of corrosion.
- Sliding stem This *quick opening* design has a lever-operated sliding stem in place of the hand wheel/screw stem combination, Figure 3-34e. Where quick opening and closing are needed, such as boiler *blow down*, this design is ideal. However, the quickness of fluid stoppage may subject the piping system to violent shock.

Stems for 90° actuated valves Plug and ball valves have short, stubby stems that transfer a 90° rotational motion to their discs. The valve stem is usually machined as part of the tapered plug and protrudes through the top of the body where the packing, packing gland, and operator are located. In lubricated plug valves, some manufacturers bore out the stem's center and use this path to feed lubricant to the plug/seat interface. On larger valves, the stem is keyed to the plug and, in high pressure applications, balance stems are attached to the bottom of the plug, Figure 3-35.

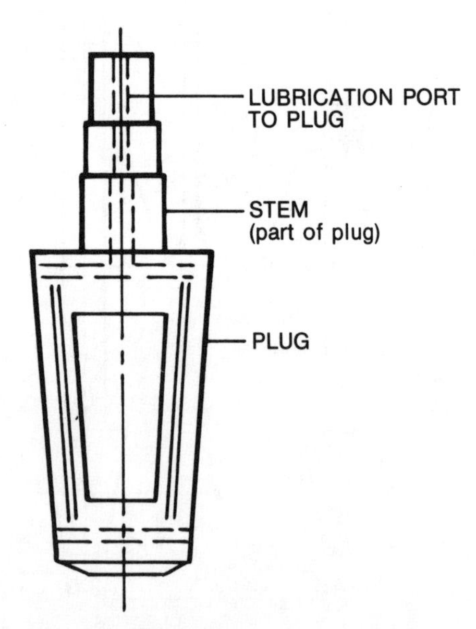

FIG. 3-35. Combination plug and stem for lubricated plug valve.

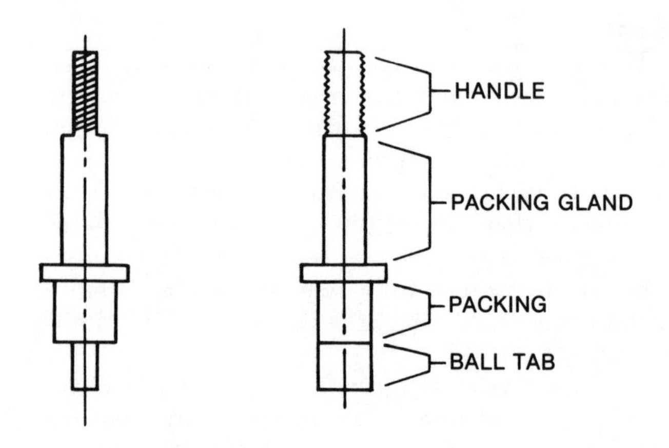

FIG. 3-36. Ball valve stem.

The ball valve has a groove machined into the top of the ball, into which fits a corresponding male tab on the bottom of the stem. This keying allows the stem to turn the ball. The packing rings are located directly above the tab, followed by the packing nut and handle, Figure 3-36.

In smaller size butterfly valves, the stem is centered in the disc and passes through holes in the top and bottom of the liner and valve body, Figure 3-37a. This extension of the stem eliminates canti-

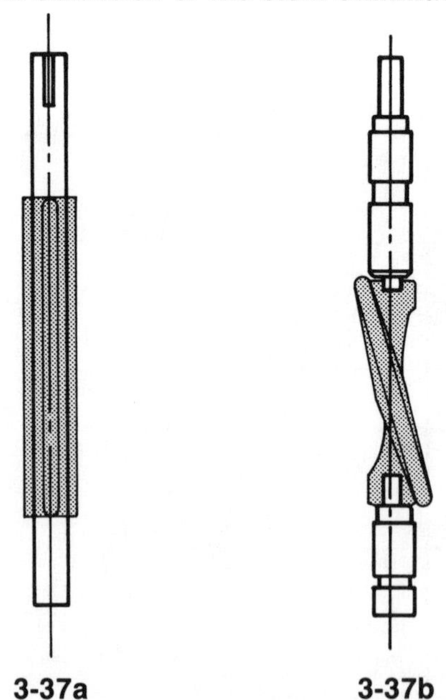

FIG. 3-37. Butterfly valve shafts. **a.** Straight shaft located directly through disc. **b.** Double shaft keyed to eccentric disc.

levering as the fluid forces try to unseat the disc and bend it toward the valve outlet. In some larger butterfly valves, the stem is eccentric to the disc centerline, Figure 3-37b. Bearings are mounted in the body above and below the disc to assure long operational life and good disc seating characteristics.

In all of the 90° rotational valves, the stem must have a generous diameter to minimize deflection by fluid forces. Good structural design assures that the discs will continue to maintain a good tight seal which is crucial to butterfly valves.

Diaphragm valve stems The stem of a diaphragm valve, Figure 3-38, is fastened to a flexible, nonmetallic disc. When the stem is turned, it pulls or pushes the diaphragm to open or close the valve.

There is a rounded metal disc, called a compressor, between the stem and the diaphragm. It serves to distribute the opening and closing forces over the diaphragm and sandwiches it between itself and the raised weir seat to assure positive closure.

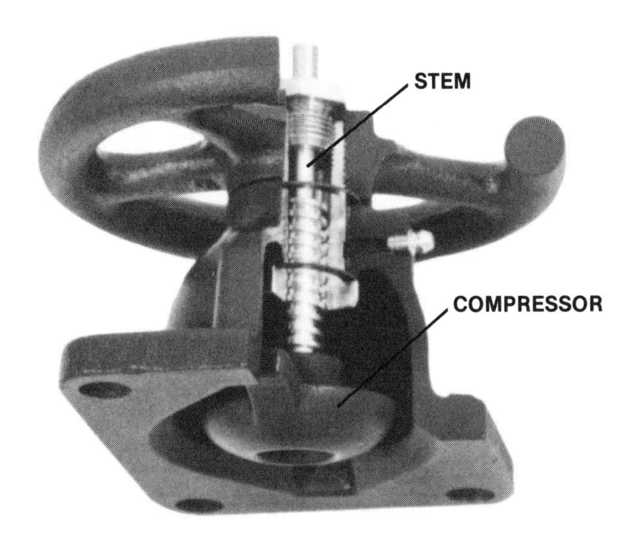

FIG. 3-38. Stem/compressor arrangement on a diaphragm valve. *Courtesy Flow Control Div., Rockwell International.*

Refrigerant valve stems Refrigerant valve stem configurations vary with the type of valve. There are three types of valves:

- Packed valves
- Diaphragm (packless valves)
- Regulators

Like gate and globe valves, packed valves use a screwed stem to open and close the disc. With smaller packed valves, for receivers and compressors, the disc and stem are machined as one piece, Figure 3-39. When front-seated, the valve is closed. When

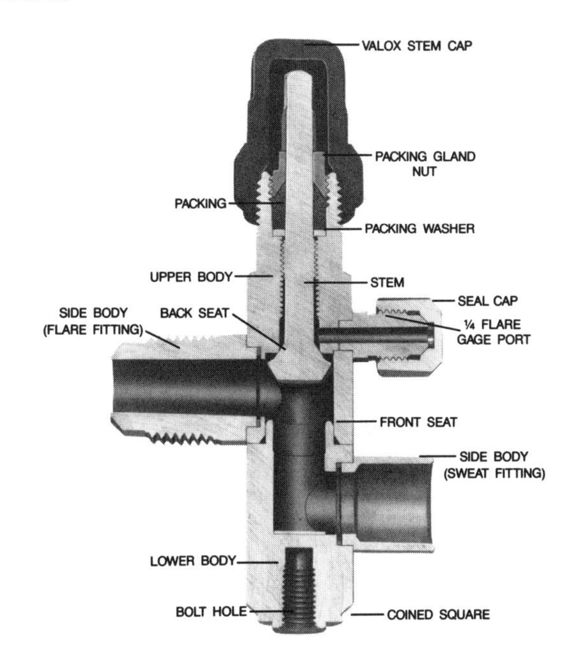

FIG. 3-39. Packed refrigerant valve is cut away to show internal arrangement. *Courtesy Primore Sales Inc.*

back-seated, it is fully open and the gage port is closed. With the valve back-seated, the packing may be replaced while the system is under pressure. If the disc is rotated a turn or two from the back seated position, the gage port can be uncapped to add refrigerant or measure system pressure.

In other packed refrigerant valves, usually 7/8" or larger, the stem and disc are separately machined. The disc is fastened to the stem with a snap ring and

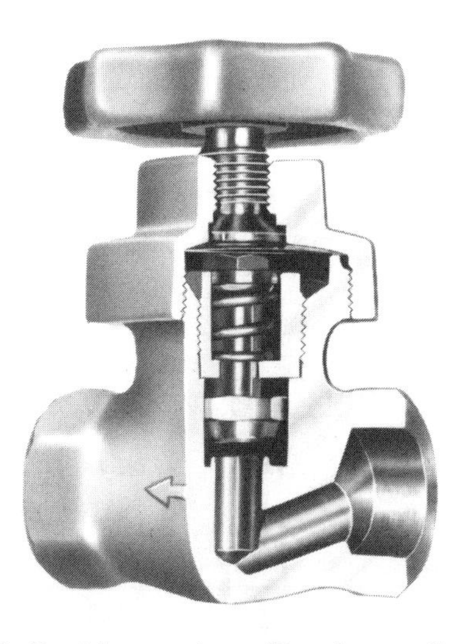

FIG. 3-40. Packless valve with a hermetic diaphragm sealing the fluid passage from the exterior environment. *Courtesy Kerotest Mfg. Corp.*

swivels to adapt to the seat for positive shutoff and long life.

The exterior portion of the packed valve's stem is covered by a seal cap that avoids leakage past the packing.

The packless diaphragm valves used with refrigerants should not be confused with the diaphragm valves described earlier. The stem in this valve, which is usually no larger than ¾″, is not connected to the diaphragm and is used to flex a metallic seal that closes a spring loaded disc, Figure 3-40. This hermetic seal is highly reliable in minimizing leakage of refrigerant to the atmosphere.

Regulator stems There are three types of refrigerant regulators:

- Spring loaded bellows
- Spring loaded diaphragm
- Pilot operated

Small, tubular refrigerant pressure regulators employ a bellows, rather than a diaphragm, to create a hermetic seal, Figure 3-41. The bellows is spring loaded on the atmospheric side and connected to the valve stem on the system side. The stem is fastened to the disc by a nut and guided by a web that is built into the body. The normally open pressure regulator is closed by excessive pressure on the outlet side of the disc.

Depending on the type of regulator, the system pressures will move the stem downward, opening the port, or upwards, to close the port. The pressure range can be changed by externally adjusting spring tension to alter the balance of forces.

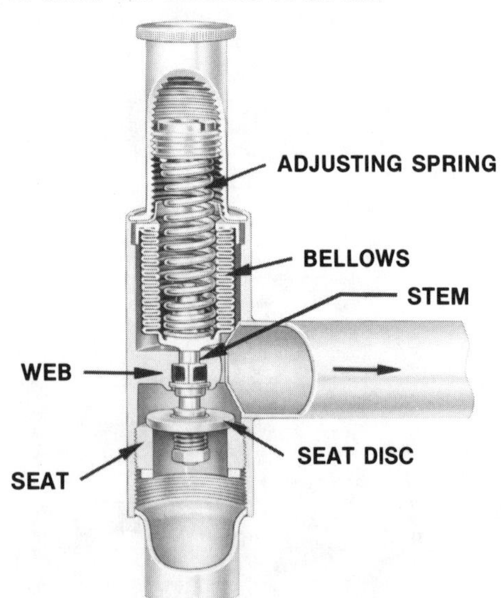

FIG. 3-41. Refrigerant regulator showing stem arrangement. *Courtesy Sporlan Valve Co.*

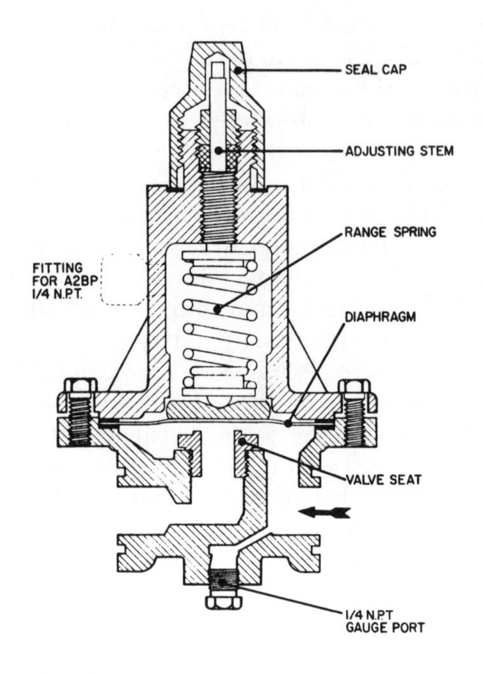

FIG. 3-42. Inlet pressure regulator. *Courtesy Refrigerating Specialties Div., Parker-Hannifin Co.*

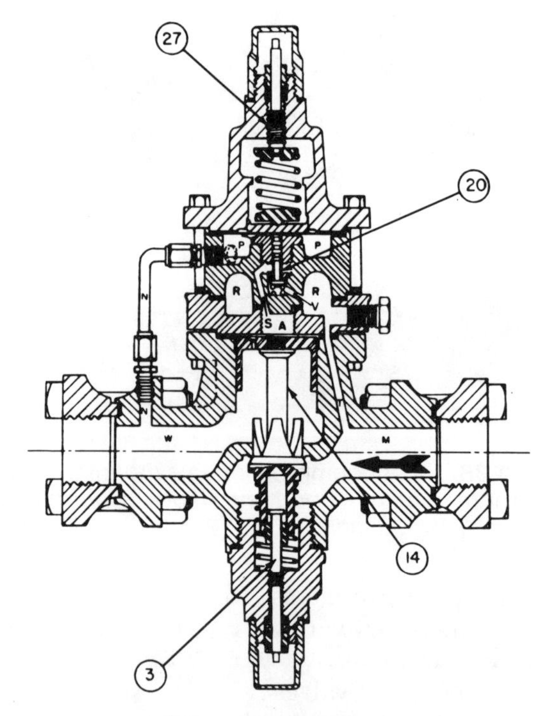

FIG. 3-43. Downstream sensing regulator. *Courtesy Refrigerating Specialties Div., Parker-Hannifin Co.*

Larger, cast body, refrigerant pressure regulators have no disc or stem, Figure 3-42. In their stead, there are a spring and diaphragm. Inlet pressure is applied under the diaphragm and is opposed by the downward spring force in this upstream or relief regulator. These regulators, although sufficient for their job, are not as accurate as the larger, pilot-operated valves.

Pilot operated regulators, Figure 3-43, may have as many as three or four stems, each with a separate function. They are interrelated through pressures opposing each other within the regulator. Since the regulator in Figure 3-43 is a downstream sensing regulator, pressure variations above and below the set point will close or open the regulator. The set point is established by the pressure adjusting stem (item 27) which increases or decreases the tension on the diaphragm.

The pilot or piggyback regulator has its own stem, disc, and seat. The stem of the pilot (item 20) is moved by the action of the diaphragm when it senses a difference between the spring pressure on top and the refrigerant pressure from below. As the pilot stem moves its disc, refrigerant flows into or out of the cavity above the power piston, raising or lowering the pressure. The power piston is attached to the stem of the main disc, which controls the flow of refrigerant in the system.

The disc lift stem (item 33) is used to manually lower the main disc to a wide open position. The disc is fastened to the main valve actuating stem (item 14) and it controls the fluid flow as it passes between the disc and seat.

Packing containment devices

The containment device restrains the packing and can be tightened, as the packing wears, to reduce leakage. Several designs are available for gate and globe valves, Figure 3-44:

- Simple packing nut The packing is placed around the stem and forced against the bonnet with a packing nut, Figure 3-44a. This design is most often found in small, inexpensive valves rated for light duty, low pressure service.
- Packing nut and gland A loosely fitting packing gland, Figure 3-44b, is placed between the packing and the packing nut. This allows the nut to turn against the gland for more and tighter adjustments with less maintenance.
- Bolted gland The bolted bonnet and gland, Figure 3-44c, is commonly used with higher temperatures and pressures and is paired with a deeper bonnet. A variation, the one-piece bolted gland, has swing type eye bolts for applying greater pressure on the packing. The upper portion of the gland is somewhat larger than the stem to prevent binding.
- Lantern ring design A lantern ring and bolted gland are used on larger valves and those in high pressure/temperature service, Figure

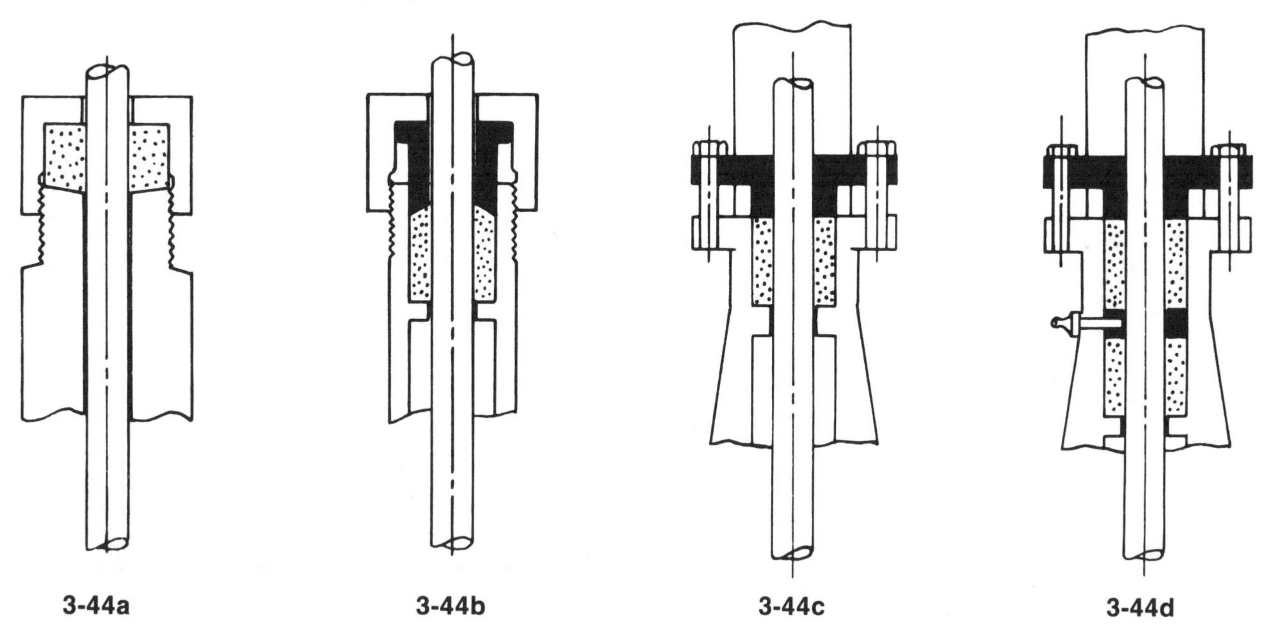

FIG. 3-44. Packing containment devices. **a.** Simple packing nut. **b.** Packing nut with gland. **c.** Bolted gland. **d.** Bolted gland and lantern ring.

3-44d. A double deep stuffing box is employed, with the first or lower portion of the packing taking most of the wear and pressure off the upper portion. The lantern ring, placed between the two sets of packing, creates space for the introduction of a lubricant for greater protection against leakage and wear.

Usually, the higher the temperature and pressure within the valve, the deeper the packing cavity.

Smaller refrigerant valves, such as service valves, have a packing gland that forces a nonmetallic, flexible packing against a washer, Figure 3-39. In addition, since refrigerants are so difficult to seal, a stem cap and copper gasket are used as a secondary seal. The stem cap is removed for service, then replaced.

No packing is used in small refrigerant regulators. Hermetic sealing is achieved by welding a flexible bellows to the top of the valve body and the stem Figure 3-41. As the disc moves up and down, the bellows accordions, often through millions of cycles.

Expansion valves, and other flow regulators used with refrigerants, have a flexible diaphragm that creates a hermetic seal.

Packless valves are becoming more common due to wider use of volatile, corrosive, and hard-to-confine fluids and gases. The packing is eliminated and replaced with a hermetically sealed diaphragm, Figure 3-40.

Packings

Packings are used to control leakage around the valve stem. The packing fits around the stem and is compressed between the packing gland and the bonnet.

Packing materials include braided asbestos, Teflon® yarn, or fibers and asbestos yarn impregnated with Teflon. They may be die-formed or extruded asbestos fibers, graphite, or wire-reinforced asbestos to handle extreme conditions, Figures 3-45a, b, and c.

Selection of correct packing depends upon the service conditions to which the valve will be subjected and the limitations of the packing material, including temperature and pressure limits and pH reactions.

For many years, the mainstay packing materials were asbestos based. However, government agencies are increasingly restricting its use. There are a number of substitute materials:

- Grafoil This is an all graphite packing with a temperature limit of 1000 °F in an oxidizing atmosphere and 6000 °F in a nonoxiding atmosphere. Its pH range is 0 through 14 except in

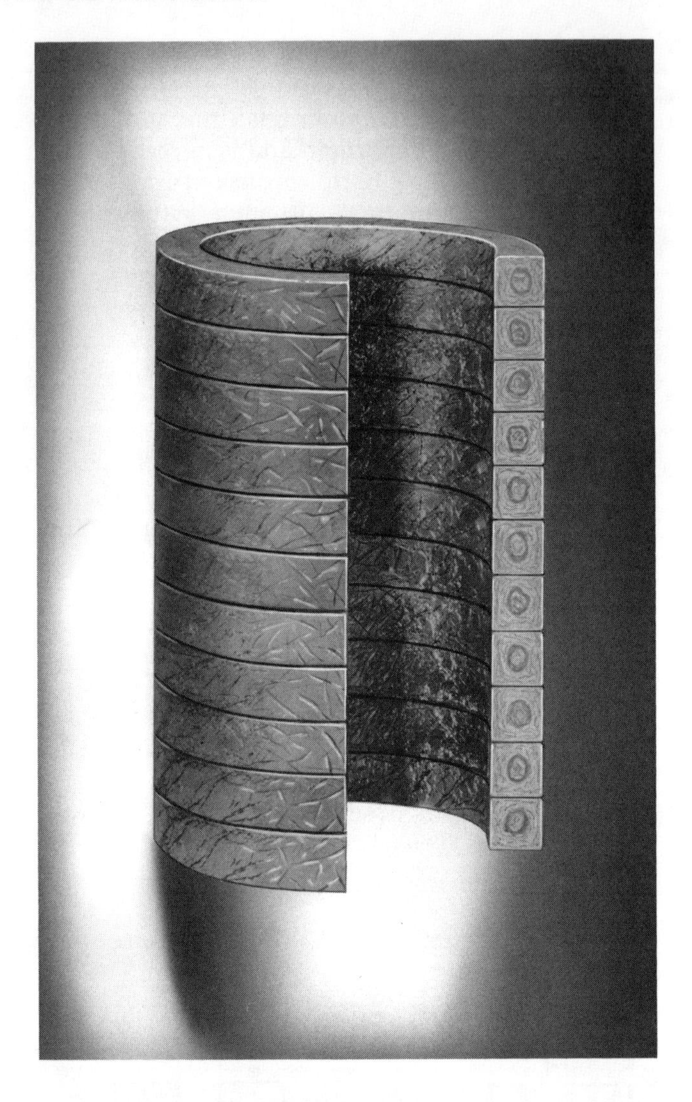

FIG. 3-45a. Braided and formed packing rings. *Courtesy Crane Packing Co.*

FIG. 3-45b. Pure PTFE fibers in a square plaited construction. Yarn is lubricated with PFTE before braiding. *Courtesy A.W. Chesterton Co.*

FIG. 3-45c. Die-formed, two-part graphite powder rings with minimal binder. *Courtesy A.W. Chesterton Co.*

1. Compressible extruded core material of 100% asbestos fibres and graphite is specially compounded to retard pitting of stainless steel equipment, yet it conforms easily under minimum gland pressures.

2. Finest quality braided Inconel wire reinforced asbestos jacket for increased temperature and pressure resistance, treated with tungsten disulphide which lubricates to 2400°F. (1320°C.).

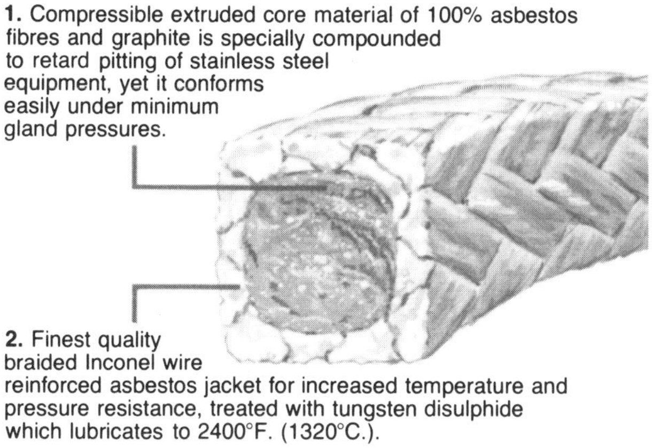

FIG. 3-45d. Packing for steam service with compressible core. Jacket is reinforced with braided wire impregnated with zinc particles to inhibit electrolytic corrosion of the valve stem. *Courtesy A.W. Chesterton Co.*

strong oxidizing acids. Grafoil usage can range from standard to nuclear service. It is available as tape, for making packings in the field, as die-cut rings, Figure 3-45c, or laminated sheet.

- Graphite filament yarn This is a braided material that can handle temperatures to 1000 °F in an oxidizing atmosphere and go somewhat higher in a nonoxidizing atmosphere. Its pH range is from 0 to 14 and it is basically a carbonized rayon.

  A second type, the carbon yarns, are made from either pitch or polyacrylicnitrile and have a temperature range of 650 °F in an oxidizing atmosphere. While the pH is from 0 to 14, certain acids, such as highly concentrated nitric acid, have a tendency to break down this material.
- Kevlar This material is an aramid fiber with a very high tensile strength and a pH range of 3 to 11. It is good up to 500 °F. Because of its very high tensile strength, it can withstand high pressures and can be used where stem support is required.
- TFE filaments TFE filaments, Figure 3-45b, can be furnished dry or impregnated with either TFE dispersion or a silicon lube. They have a temperature range to 500 °F and a pH range from 0 through 14. TFE packings are used in all types of chemical applications. Some TFE packings, if treated properly, can be used in oxygen service, but must be specified to the manufacturer.

  Another type of TFE filament is made with an ultra fine graphite filler that helps reduce stem friction and requires less gland pressure to effect a seal.

  TFE V-rings are often used in gas applications. They are classified as an automatic type packing because, as the fluid pressure contacts the lips, the V-rings automatically begin to seal.
- PTFE, solid polytitrafluoroethylene This material can either be machined or molded into V-rings, U-cups, or cup and cone sets for use in chemical or food handling valves with temperatures up to 500 °F.

These packings may be modified for special applications by adding fillers such as glass fiber, molybdenum disulphide or graphite.

Natural fiber packings, such as flax, have been used for years. Flax has a temperature range of under 150 °F and a pH range of only 6 to 8. It can be furnished with tallow lubrication, special mineral oils, or TFE dispersion. Flax packings can handle very high pressures, but are limited by a low temperature range and inability to tolerate various chemicals.

Cotton fiber has a temperature range of up to 250 °F and a pH range of 6 to 8. While very soft, it will give very good service in cold water applications such as sewage plants.

Metallic packings have been in use for many years. However, their limitations must be carefully considered.

- Babbitt This is a lead-based material that can withstand temperatures up to 500 °F and has a pH range of 3 to 10. Babbitt can be used as support rings in valves and pressure breakers and has been used up to 5000 psig in valves.
- Aluminum Aluminum has a temperature range up to 1000 °F and a pH range of 2 through 11. It is being used in valves handling heat transfer fluids as well as some selected acids. While aluminum is a hard packing, it will not score valve stems because the movement is slow.
- Copper Copper is another material that has a temperature range of up to 1000 °F and a pH range of 2 through 11. It is used primarily in hot water, steam, and hot oil valves. Copper provides support where the stem is not properly supported or where there are large clearances at the bottom of the stuffing box.

Asbestos packings are still available as braided asbestos or wire-inserted asbestos, with a variety of lubricants, and probably will continue to be used.

Wire-inserted asbestos packing, Figure 3-45d, can be braided entirely from wire-inserted asbestos or have a resilient core inside a jacket of wire-inserted asbestos for greater resiliency. The wire in the asbestos may be copper, brass, monel, or inconel, depending on the application. In high temperature/pressure applications, the most popular is inconel wire-inserted packing that has high tensile strength. It withstands temperatures up to 1200 °F, has a pH range of 3 to 10, and can be used with water, steam, petroleum distillates and many mild chemicals.

A similar material, monel wire-inserted asbestos has a temperature limitation of 850 °F. Brass and copper wire-inserted asbestos packings will have a number of lubricants and a pH range of possibly 4 to 8, depending on the lubricant. Their temperature range is limited to 600 °F.

Many of the higher temperature wire-inserted packings have a corrosion inhibitor to retard corrosion of stainless steel stems while in storage or when standing on line.

Plain asbestos packings have temperature ranges below 500 °F and a pH range of 4 to 8. They can be furnished as square braided, braid-over-braid, or as an interlace style packing. All are satisfactory in valve service due to slow movement of the stem. Plain asbestos is normally used with low pressure, low temperature valves where a more economical product is sought for general use.

Asbestos packing may be TFE coated to increase its pH and temperature ranges. TFE dispersion within asbestos yarns gives good chemical resistance and can be used with low, medium, and high pressures. It is a good, resilient, long lasting packing that can be used in many steam and water applications.

As the valve industry turns away from asbestos products, the initial cost of packings will rise considerably. However, the service life of the newer

FIG. 3-46. Lantern ring placed between two packing stages. Design excludes foreign particles from the stuffing box by maintaining a constant flow of clean flushing fluid past the bottom packing stage, reducing abrasion. *Courtesy A.W. Chesterton Co.*

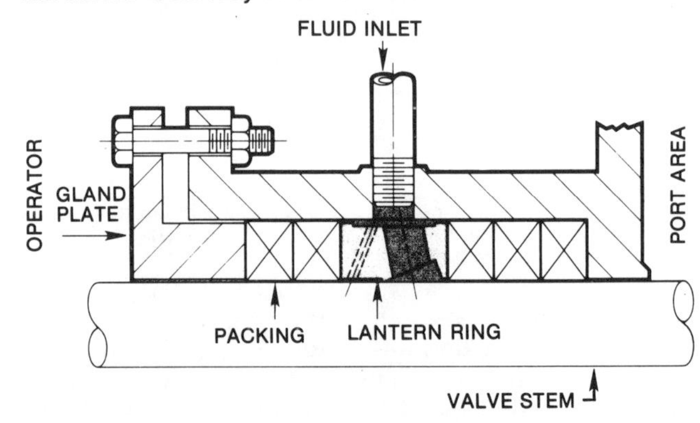

FIG. 3-47. Schematic showing lantern ring placement between two stages of packing. *Courtesy A.W. Chesterton Co.*

packings far exceeds that of the older materials. Asbestos packings tend to dry out and need to be replaced in 18 to 24 months. Grafoil or graphite/carbon packings have given service up to five years — saving maintenance and down time.

Many fluids carry abrasives, suspended solids, precipitating, evaporating, and congealing liquids. Because of this, much use is made of the *lantern ring*, Figures 3-46 and 3-47, in the chemical and industrial fields. The lantern ring separates two sets of packings, which may or may not be of the same type or style, and provides a means of introducing flushing material under pressure. This flushing material is usually a grease that keeps the packings lubricated and flushes out foreign material that could cause the valve stem to wear.

Operators

Operators open, close, divert, or throttle the flow of fluids through valves by repositioning the disc in relation to its seat. The operator is adfixed to the stem, above the bonnet, and can be classified as:

- Mechanical device
    - Hand operated
    - Automatic (regulators)
        - Temperature sensing
        - Pressure sensing
        - Expansion valves
- Externally powered
    - Electric
        - Solenoid coil
        - Motor
    - Pneumatic
    - Hydraulic

Why use actuators?

1. So that operators are not required to go to remote valve locations.
2. In environmentally hazardous areas where personnel could be subjected or exposed to dangerous gases or conditions.
3. In sanitary areas where personnel could contaminate sterile products.
4. Where computer-operated programs can actuate valves in a repetitive pattern of process controls.
5. For central control of plant operations.
6. Operate valves, in case of emergency, to shut down processes.
7. Valves over six inches in size are difficult to operate manually.
8. Preposition valves in partially open positions for throttling flow.
9. Fluid forces within a system may make manual operation difficult and, once the position of a butterfly valve is set, may move the disc away from the desired position.

It must be remembered that forces in a valve are created by a number of factors which must be overcome by operators:

- Forces created by the moving fluid.
- Differential pressure across the valve port.
- Friction of the stem, disc, and packing.
- Weight of the moving parts.

These factors determine torque requirements in sizing the operator. Torque requirements are in three categories:

- Breakaway torque
- Running torque
- Lockup torque

Different valve designs will have variations in each of the above torques, with some considerably greater than others.

Actuators are mounted on gate, ball, plug, and solenoid valves to fully open or close the valve to fluid flow. Globe and butterfly valves may be set in an infinite number of positions for extremely fine control.

Hand valves are the most common of the mechanically-operated devices.

Automatic, mechanical regulator operators allow valves to control the temperatures and pressures of fluids. For example, they control the temperature of water through air conditioning chillers, the head pressure in water cooled condensing units, and refrigerant flow to an evaporator. These regulators are not to be confused with those described earlier. Self-contained regulators sense fluid pressure *within* the regulator itself and control flow accordingly. Actuator-operated regulators sense conditions *external to* the regulator and transmit a signal back to the actuator to reposition the valve disc and adjust the flow.

Mechanical devices

Hand operators Hand operators are either handwheels or levers. The handwheel is used with gate or globe valves where circular motion of the stem is required to open or close the valve. Levers are used with ball, butterfly, plug or diverting valves where a 90° turn of the handle opens or closes the valve disc.

Manual valves are usually reserved for on-off applications. On larger size valves, or valves just out of the reach of personnel, geared operators, square shaft, crank or chain operated devices are used, Figure 3-48. Underground valve locations are handled by the extension of the stem for accessibility.

Globe, plug, and butterfly valves have high seat-

FIG. 3-48. Chain operator for valve locations that are out of the reach of personnel. *Courtesy DeZurik.*

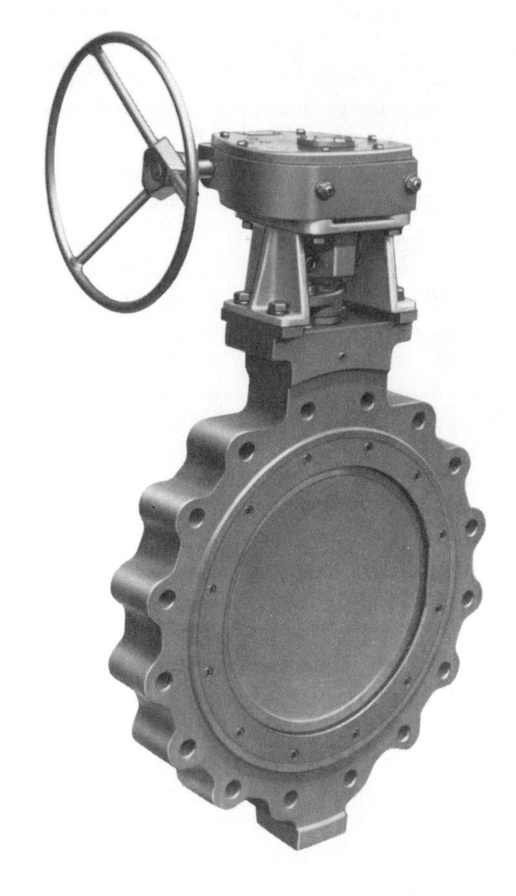

FIG. 3-49. Hand wheel operator for butterfly valve. Gearing inside the housing changes motion 90° and increases mechanical stability. *Courtesy Jamesbury Corp.*

ing torque requirements.

Some tapered disc (wedge style) gate valves also have high seating torques. The conical shape causes a wedging action between the disc and the seat, producing high metal-to-metal friction that must be overcome. The metal of both the disc and seat tend to *cold flow* into each other. Butterfly valves, on the other hand, have discs that *squeeze* into liners or against resilient seats during shutoff.

When a butterfly disc is opening, fluid flow can make the disc unstable and tends to slam it shut. Various butterfly disc shapes have different torque requirements at different positions of fully open. Butterfly operator torque requirements can vary considerably, depending on the style of seat, and care must be taken in operator selection, Figure 3-49.

Automatic regulators

Temperature sensing regulators Temperature regulators are controlled by a sealed bulb and capillary tube that contain a fluid or vapor known as the thermal system. The capillary tube is joined to the regulator bonnet, which is usually a hermetically sealed bellows that expands or contracts and transfers a force to the regulator that, in turn, opens or closes the disc. Figures 3-50 and 3-51 show examples of these temperature regulators.

The thermal system may contain a vapor, liquid, mercury, or any expandable, thermal-sensitive material. As the temperature increases above the set point, the pressure in the thermal system increases and causes the valve to close. This is typical of a direct-acting regulator in a heating application. A reverse-acting regulator, used in a cooling application, will open the valve when the temperature increases.

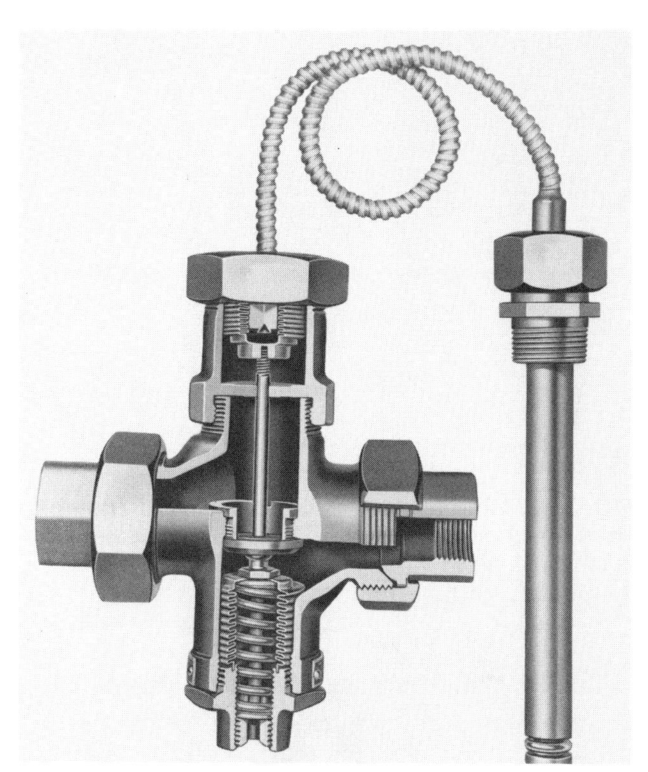

FIG. 3-50. Self-contained, temperature sensing control valve. *Courtesy Sterling, Inc.*

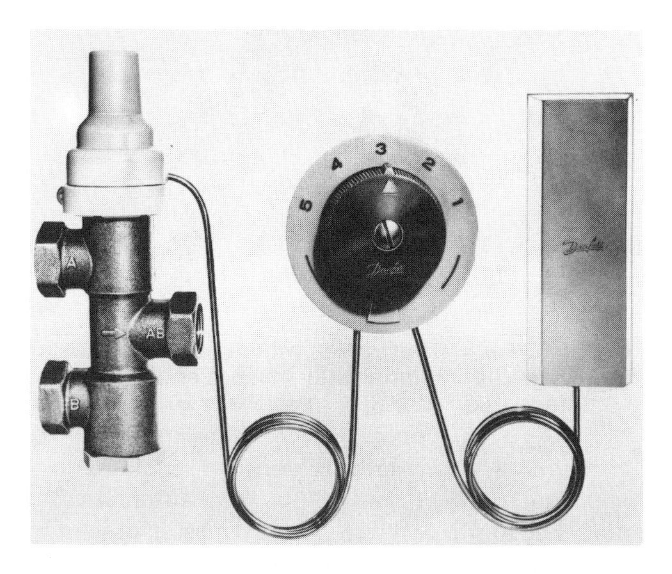

FIG. 3-51. Thermostatic elements for heating/cooling terminals consisting of sensor, control unit, and valve, connected by a liquid-charged capillary bulb system. *Courtesy Danfoss.*

The bulb of a vapor system, Figure 3-52, contains a volatile liquid that vaporizes upon a temperature increase. This vaporization is accompanied by increased pressure which is transmitted, through the capillary tube, to the bellows. When the bellows expands, due to increased pressure, it moves the stem down to close the valve and halt fluid flow.

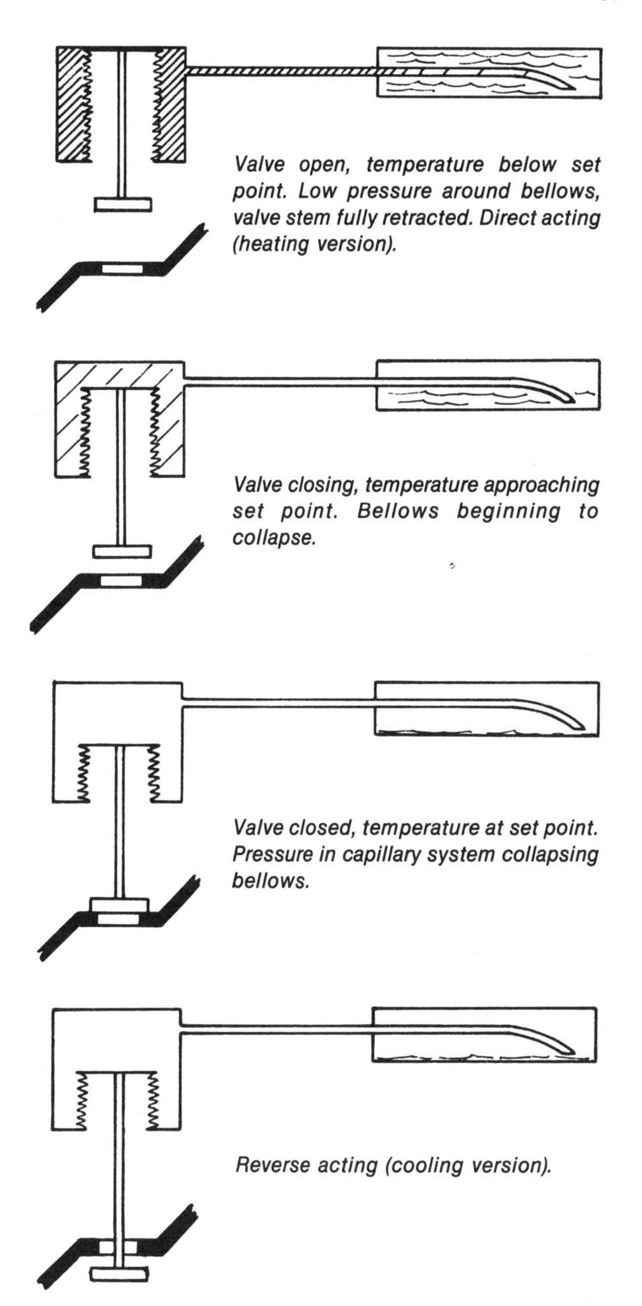

FIG. 3-52. Liquid/vapor system, showing an increase in remote bulb temperature and its effects on valve position for direct and reverse acting regulators.

The vapor system may contain alcohol, water, propane, ether, etc. The liquid selected depends on the temperature range which may vary from 700 °F to −40 °F.

In liquid and mercury bulb systems, temperature increases expand the liquid which exerts a force on the bellows. Mercury has a low coefficient of expansion, consequently it is used for pilot operation and very accurate control.

Gas filled systems employ a noncondensible gas. Again, expansion is less than with a vapor-liquid system and cannot be used to directly actuate the regulator. Pilot operation, to control the main regulator disc, is their most frequent use.

The compound actuator utilizes a heat sensitive, wax filled enclosure, Figure 3-53. As the temperature increases, the wax expands and exerts force on a pin that directly operates the regulator disc.

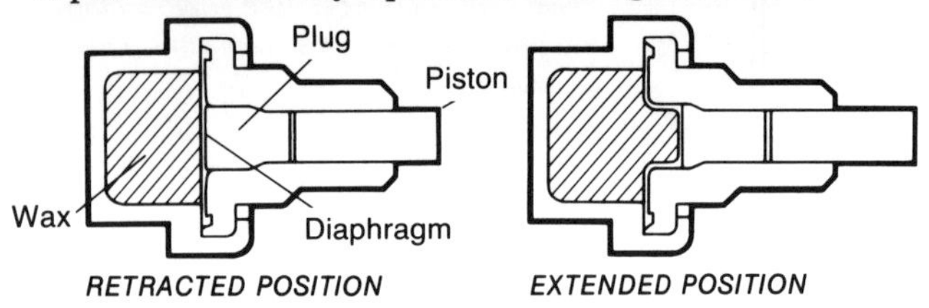

FIG. 3-53. Wax-filled temperature sensitive actuator. *Ogontz Controls Co.*

FIG. 3-54. Ambient temperature sensing control valve for automatic flow control in steam tracer lines. *Courtesy Ogontz Controls Co.*

The temperature-sensing control valve shown in Figure 3-54, uses a wax-filled sensor to keep fluid lines from freezing during cold weather.

Another application of the wax filled regulator is the relief valve found on all pressurized hot water heaters, Figure 3-55. This relief valve may be opened by either excessive internal pressure or temperature. The wax material in the tip of the sensing element expands at the design temperature, usually 210 °F, to raise the disc off its seat and allow pressure to escape. Upon cooling, the wax contracts and the disc reseats itself.

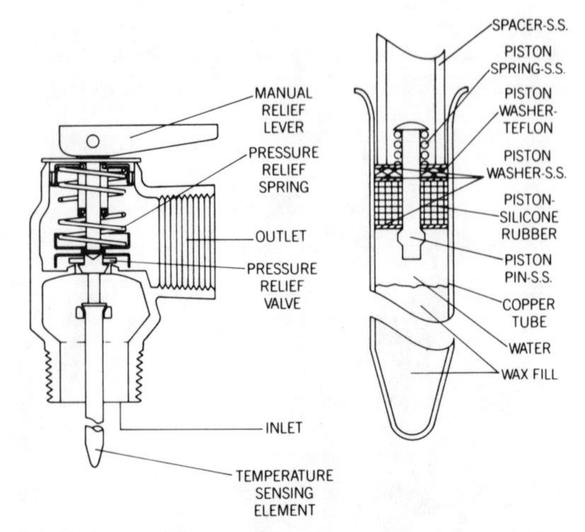

FIG. 3-55. Internal construction of a pressure/temperature actuated relief valve. *Courtesy Robertshaw Controls Co.*

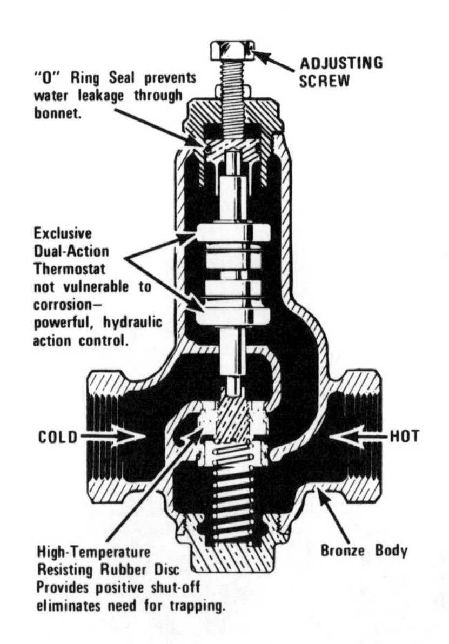

FIG. 3-56. Tempering valves with internal thermostat assemblies that automatically mix hot and cold water to a safe temperature. *Courtesy Watts Regulator Co.*

Still another temperature sensing regulator is a tempering valve, Figure 3-56, that automatically mixes hot and cold water. A thermostat assembly within the regulator senses fluid temperature directly and expands or contracts depending on the temperature setting. An adjustable knob will set a range of 40 ° or 50 °, that is, 120 ° to 160 °F or 130 ° to 180 °F. If the incoming water is hotter than the set temperature, the cold water port will open and mix cold water with the hot until the desired temperature is restored. Solar and heat recovery systems are good applications for this regulator.

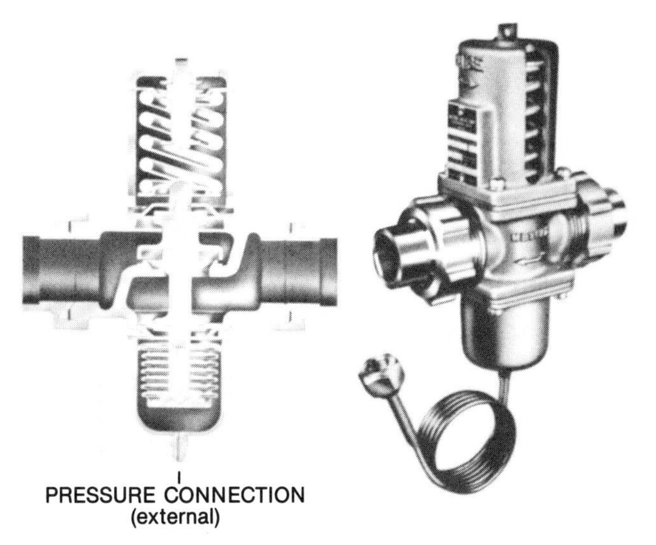

FIG. 3-57. Pressure actuated water regulating valve for water cooled condensers. *Courtesy Metrex Valve Corp.*

Pressure sensing regulators There are two basic types of pressure regulators: internal sensing and external sensing. The regulators already described were internal sensing devices that are controlled by the *same* fluid that the regulator is designed to control. They were upstream sensing, downstream sensing, and differential sensing, Figure 3-42. These regulators utilize the flowing fluid pressure, operating against a spring-loaded diaphragm, to balance forces for control.

The external sensing regulator uses an external source of pressure for control which may not be the pressure of the fluid being controlled.

Figure 3-57 shows an external sensing regulator. It controls water flow to a refrigerant condenser, in response to the pressure in the high side of the refrigeration system. Refrigerant pressure is applied to the bonnet of the regulator through a sensing tube that is open to the high side of the system. Increases in refrigerant pressure collapse the bellows, pushing the stem away from the port seat, to allow more cooling water to flow through the condenser. When the pressure falls, the bellows retracts the stem to close the valve.

Expansion valves Refrigerant expansion valves are regulators with special actuators that can sense both temperature and pressure. They can be:

- Constant pressure
- Thermostatic, internally equalized
- Thermostatic, externally equalized
- Electrically operated

The constant pressure expansion valve is a second generation device. It was developed after manually operated expansion valves of the late 1880's proved to be too cumbersome to operate and control. Thermostatic expansion valves are a third generation device designed to control variable loads.

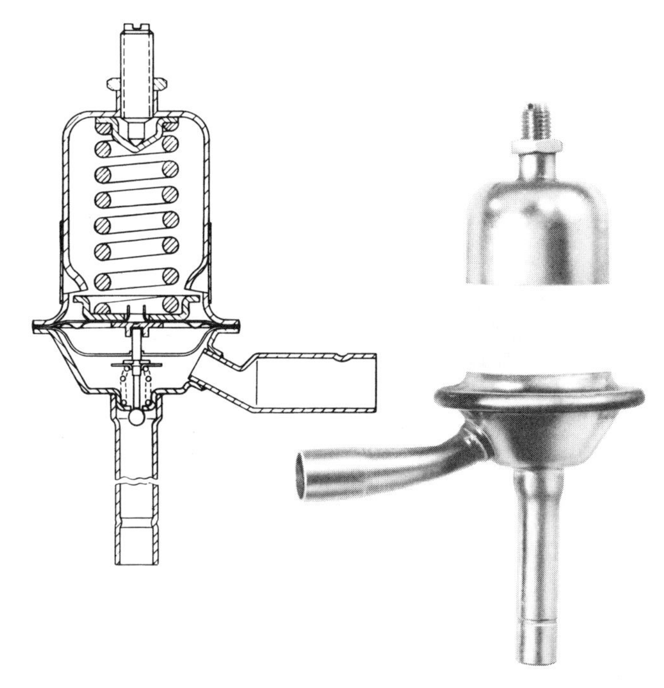

FIG. 3-58. Constant pressure expansion valve. *Courtesy Singer Controls Div.*

The constant pressure regulator, Figure 3-58, is best suited to systems that have nearly constant heat loads. Once set at the factory, to operate at a specific pressure, they cannot be adjusted.

The constant pressure device responds to the evaporator pressure downstream of the regulator and meters the refrigerant into the evaporator to maintain a constant pressure. As evaporator pressure falls, due to a light load, the regulator opens to add more refrigerant and thereby raise the refrigerant pressure and temperature. As the heat load increases, the evaporator pressure rises and the valve throttles the refrigerant flow to keep the pressure at the preset level. The constant pressure expansion valve keeps the evaporator from frosting during light loads and reduces power consumption during heavy loading.

The constant pressure expansion valve has a hermetically sealed diaphragm, a control spring atop the diaphragm that forces it downward, and a spring-loaded disc, Figure 3-59. The control spring tends to open the port while evaporator pressure beneath the diaphragm tends to close the port.

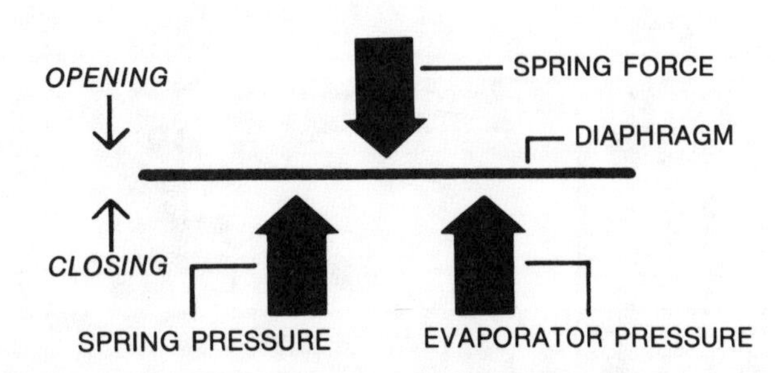

FIG. 3-59. Arrows show forces on the diaphragm of a constant pressure expansion valve.

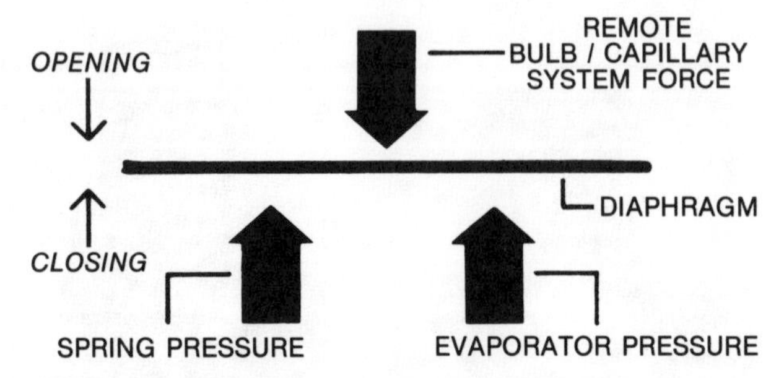

FIG. 3-61. Arrows show forces on the diaphragm of a thermostatic expansion valve.

Whenever evaporator pressure is above the factory set point, as it is during the off cycle or high load conditions, the port is closed. When evaporator pressure falls, due to light loading, the port is opened. The change in stem stroke is approximately 0.001 in. per psi pressure change.

Stabilization occurs when the refrigerant boils off at a constant rate which assumes a constant load.

The thermostatic expansion valve shown in Figure 3-60 is actuated by both temperature and pressure. The thermal bulb meters the flow of refrigerant to the evaporator while the external pressure-sensing line compensates for the pressure drop through the evaporator coil.

Three forces, Figure 3-61, actuate thermostatic expansion valves: In the internally equalized valve, the external force is applied by a remote thermal-sensing bulb that responds to refrigerant temperature at the outlet of the evaporator. This force is applied to the top of the diaphragm and tends to open the port. Two internal forces are applied to the underside of the diaphragm: spring pressure and the pressure in the evaporator. Together, they tend to close the port.

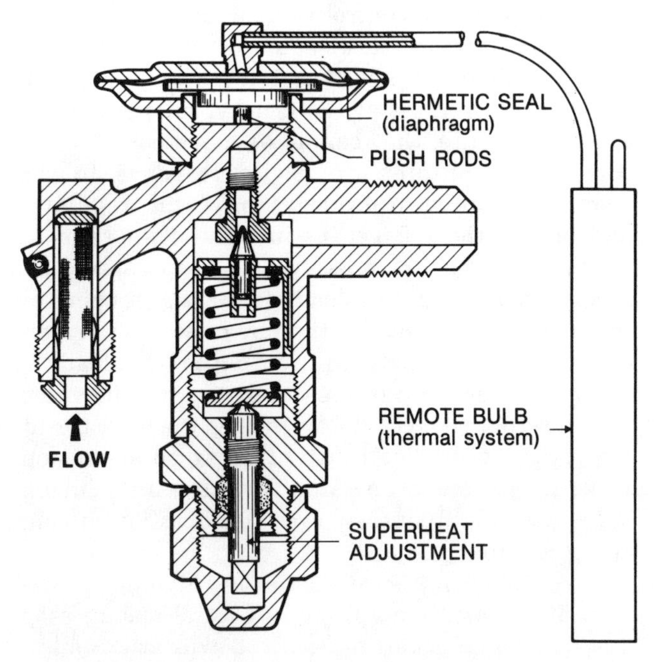

FIG. 3-60. Thermostatic expansion valve showing power head, external equalizer passage, and push rods. *Courtesy Sporlan Valve Co.*

In the externally equalized valve, two external forces act upon the diaphragm: the remote thermal bulb acting downward and the pressure at the evaporator outlet acting upward. The superheat spring still acts upward as in the internally equalized valve.

An externally equalized valve is employed with evaporators where the refrigerant pressure drop is large.

Since the valve operates by the balance of three forces, it feeds refrigerant based on the resultant force. Under light loading, the remote bulb temperature and pressure decrease and there is less pressure on the top of the diaphragm. The combined evaporator and spring pressures exceed the downward force and tend to close the port, to reduce the flow of refrigerant.

Conversely, when the load increases, the remote bulb pressure increases and its force exceeds the force of the spring and evaporator pressure. The resultant pressure on the diaphragm forces the port open to feed refrigerant to the evaporator. These reactions are more responsive to varying load conditions and are opposite to the response of the constant pressure expansion valve.

When the liquid in the remote sensing bulb is the same as the refrigerant in the system, the bulb and the evaporator will have identical pressure/temperature relationships. However, the bulb contents may be altered in several ways:

- *Gas charge* These are limited liquid charges using the same refrigerant as that in the system.

All of the liquid will vaporize at a predetermined temperature, thereby limiting the top force on the diaphragm, Figure 3-61.

- *Liquid charge* Also known as *cross ambient* charge. This charge is the same as that used in the system and there will always be some liquid under ambient conditions.
- *Liquid cross charge* The liquid charge is a different refrigerant than that used in the system. This practice produces a more consistent superheat at lower temperatures.

Thermal electric expansion valve The actuator of the thermal electric expansion valve is a thermistor, located in the suction line at the outlet of the evaporator, and a bimetal and heater hermetically sealed into the head of the valve, Figures 3-62 and 3-63. The bimetal is part of a heat motor operating against a spring-loaded needle valve that is infinitely positionable in response to 0 to 24 AC voltage input to the heater. The thermistor controls the voltage to the heat motor to maintain a saturated vapor condition in the suction line. At 0 voltage, the valve is closed. As voltage increases, the bimetal is warmed and deflects upwards, opening the needle valve. The greater the voltage, the greater the refrigerant flow.

A thermistor has a negative temperature coefficient. As it heats up, its resistance decreases and it allows more current to flow to the bimetal heater. With the system under load, the vapor in the suction line will be hotter than the vapor in the evaporator. Therefore, the thermistor is being warmed, reducing its resistance. More current will flow to the heat motor and the valve will open proportionately. The valve will meter refrigerant into the evaporator until the vapor in the suction line is at the temperature of the evaporator and the thermistor is cooled. With

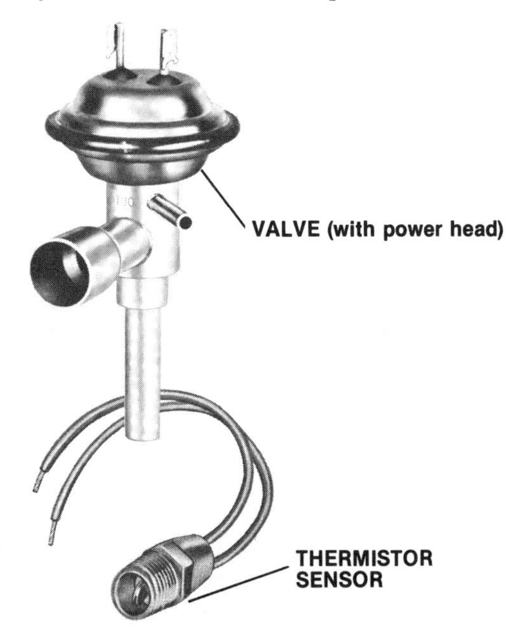

FIG. 3-62. Thermal electric expansion valve. *Courtesy Singer Controls Div.*

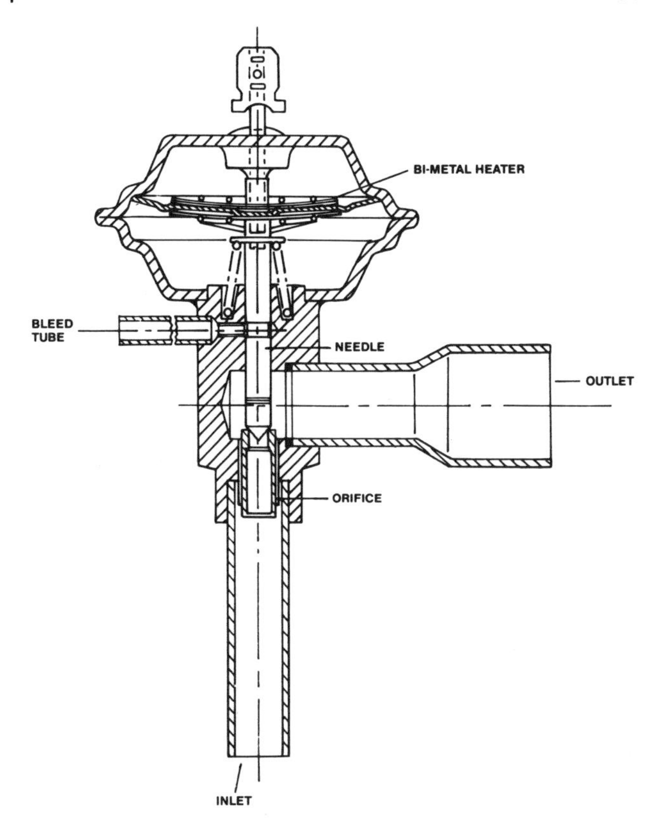

FIG. 3-63. Internal construction of thermal electric expansion valve. *Courtesy Singer Controls Div.*

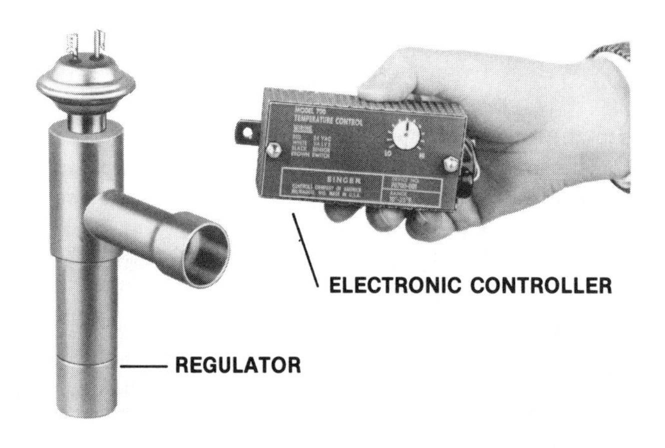

FIG. 3-64. Electronically controlled regulator for refrigeration systems. *Courtesy Singer Controls Div.*

no current to the heat motor, the bimetal will close the valve and throttle the flow of refrigerant.

This design may be used as an expansion valve in conventional cooling systems and heat pumps, in hot gas bypass valves and pressure limiting devices.

By replacing the thermistor with electronic controls, consisting of sensors and amplifiers, the electronic valve can control refrigerant flow by sensing air temperature, Figure 3-64.

Externally powered actuators

These actuators are controlled and powered by an external energy source. Depending on whether the stem movement is rotational or reciprocating, the actuator may be electrically, pneumatically or hydraulically driven.

Electrically powered actuators Solenoid valves are an on-off type of electric actuator. The core of the solenoid is a coil of insulated copper wire that produces a magnetic field when an AC or DC current passes through the coil, Figure 3-65. When the coil is energized, its magnetic field attracts and lifts an iron alloy plunger and opens or closes the valve depending on whether the valve is normally closed (N.C.) or normally open (N.O.).

The magnetic coil has limited power. Therefore, the size of the valve port and the operating pressure differential of direct-acting solenoid valves are also limited. The force required to lift the disc off its seat is directly proportional to the port area for any given pressure. Large electric currents are needed to generate the magnetic force required to move the disc in larger valves. However, winding larger coils is not economically practical.

The coil design depends on four factors:

- Ambient heat
- Heat from the fluid in the valve
- Heat from the electric current in the coil
- Duty cycle

Rating factors for solenoid coils depend upon the class of coil insulation which varies with the application. For dry, routine ambient conditions, paper section construction is adequate. With elevated ambients, moist environments, chemical or fungal exposure, and mechanical stress, epoxy-molded coils with higher grades of insulation, are employed.

Exposure to excessive heat causes coil insulation to deteriorate and requires the use of heat-resistant materials.

Figure 3-65 shows a molded solenoid coil consisting of insulated copper wire wrapped with glass tape and triple dipped in varnish and baked to resist moisture penetration.

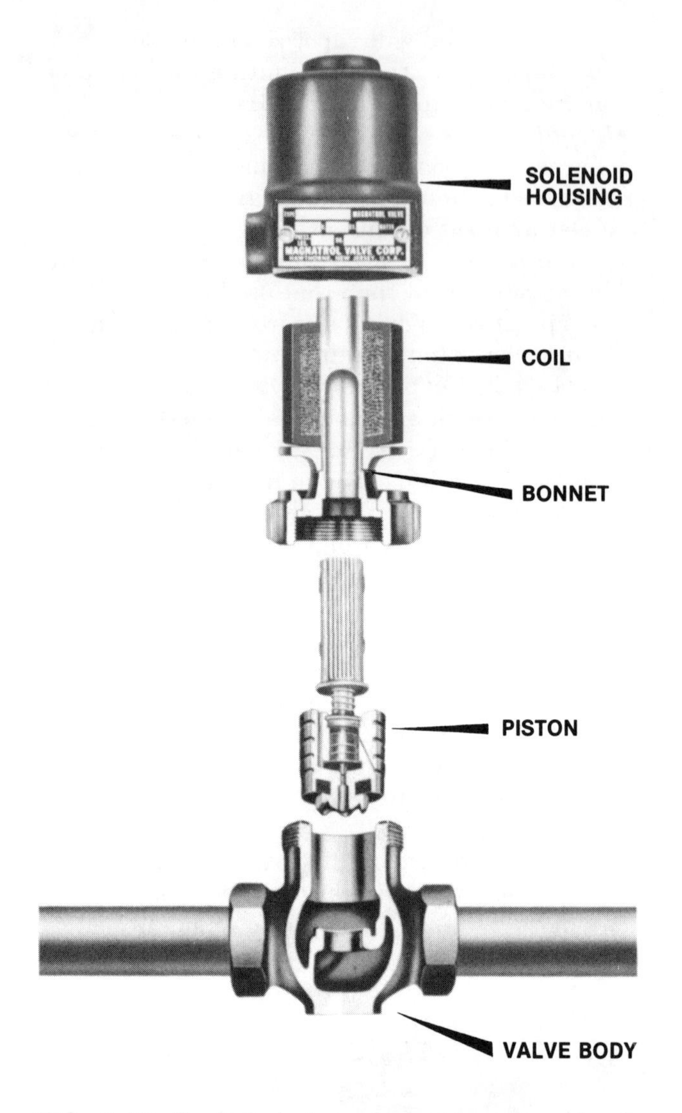

FIG. 3-65. Exploded view of electric solenoid valve with cross sectional view of coil actuator. *Courtesy Magnatrol Valve Corp.*

Solenoid enclosures vary with the environment. Dust laden or explosive vapor atmospheres will explode if an electrical short creates a spark. Therefore, the electrical enclosure is rated by NEMA and these ratings are described in the Appendix.

Special refrigeration designs Solenoid valves can be used with other valves and regulators to increase the scope of the original valves. The ability of the solenoid to alternate between open and closed is used to switch high side and low side pressures in pilot operations, to move main valve pistons, for regulating refrigerant flows. Understanding this particular application is important because it is so widely used

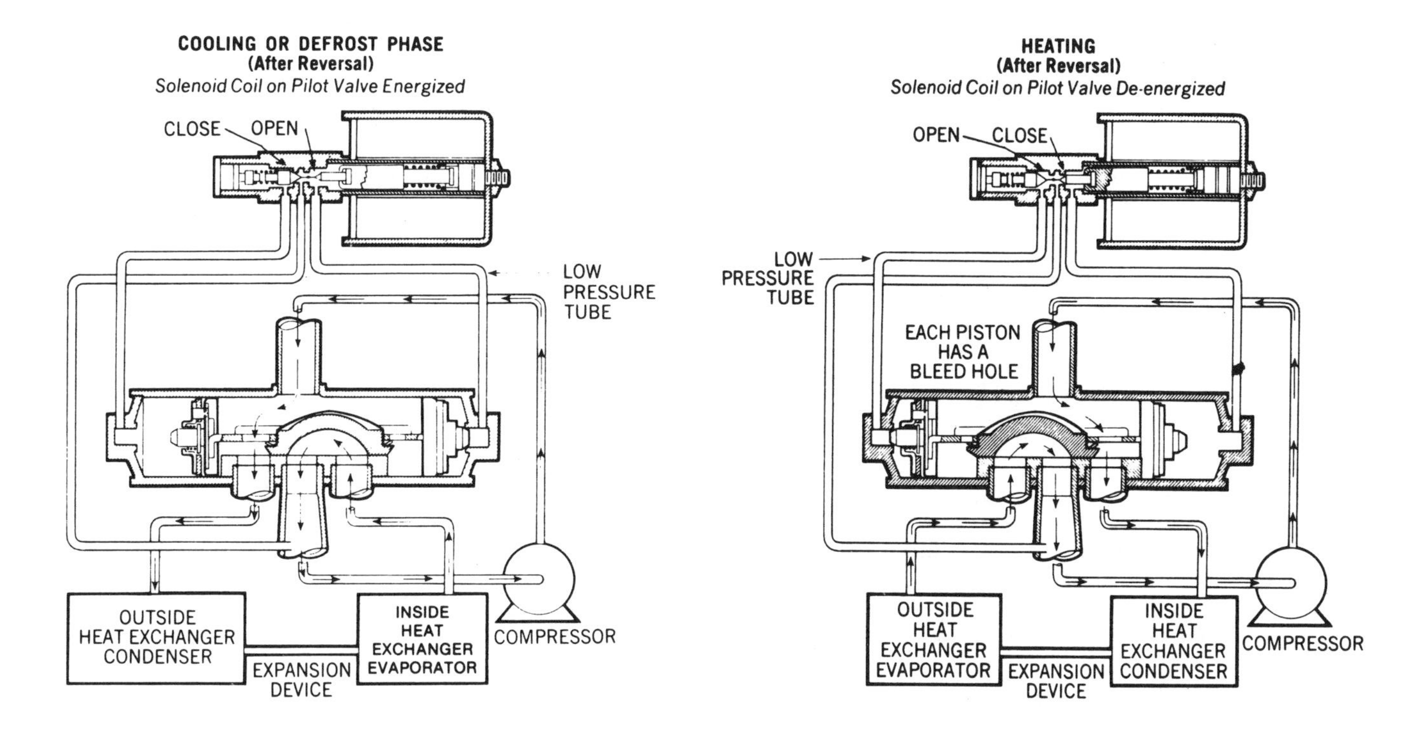

FIG. 3-66. Reversing valve for heat pumps. Solenoid valve, actuated by a pilot valve, drives the main piston which shuttles between ports. *Courtesy Ranco Controls Co.*

in the refrigeration industry. The versatility of the solenoid, even though it is only an on-off device, can be seen in the following examples.

Use of a solenoid valve as a pilot operator is best shown in multiport valves for switching flow from one piping system to another. In heat pump reversing valves, Figure 3-66, the solenoid coil is used to bleed refrigerant, through an orifice, to one side of a piston. The increased pressure drives the piston in the appropriate direction, either closing or opening a main port. This action, in turn, diverts refrigerant flow from one piping system to another to convert a cooling unit to a heat pump or vice versa.

Figures 3-67a, b, c, and d show four valves, from one manufacturer, that use a solenoid valve as a pilot. Figure 3-67a shows a control valve that is used in a heat reclaim system. The pilot solenoid opens or closes the cavity above the power piston to either allow suction pressure above the piston (normal condenser mode), which closes off the heat reclaim leg, or allows head pressure above the piston (heat reclaim mode) to move the disc downward closing off the normal condenser leg. With the suction pressure (low side) above the piston, the piston is in an up position. With discharge pressure above the piston (high side), the piston is all the way down. Chamber pressure above the piston is controlled by the solenoid valve operation.

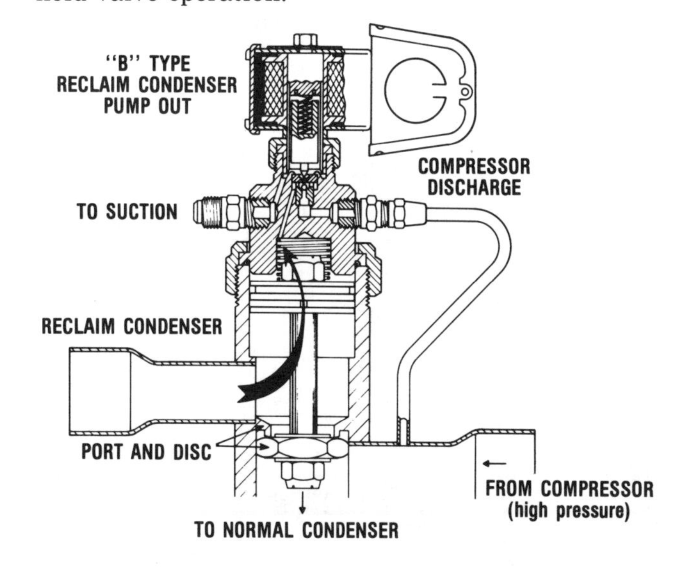

FIG. 3-67a. Three-way heat reclaim valve. *Courtesy Sporlan Valve Co.*

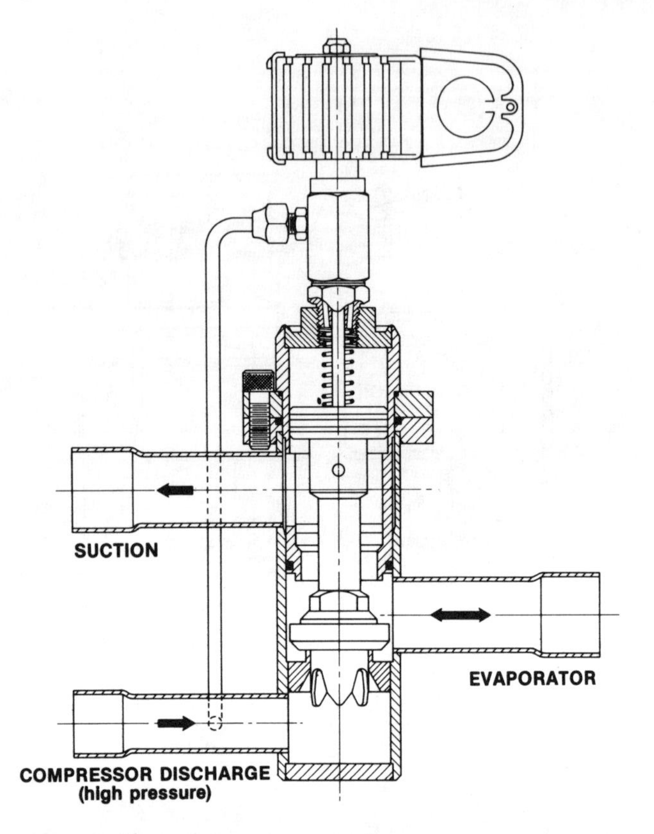

FIG. 3-67b. Three-way hot gas valve. *Courtesy Sporlan Valve Co.*

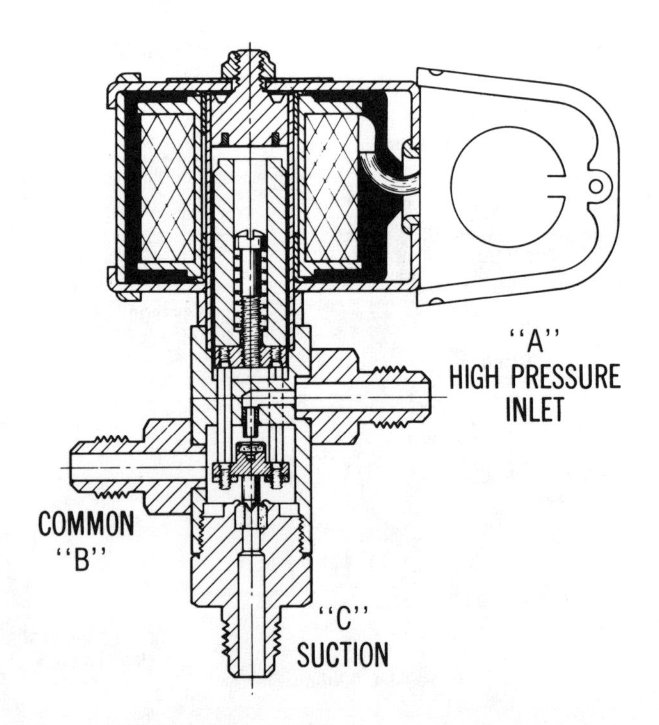

FIG. 3-67c. Pilot controlled selenoid valve for use with expansion valves. *Courtesy Sporlan Valve Co.*

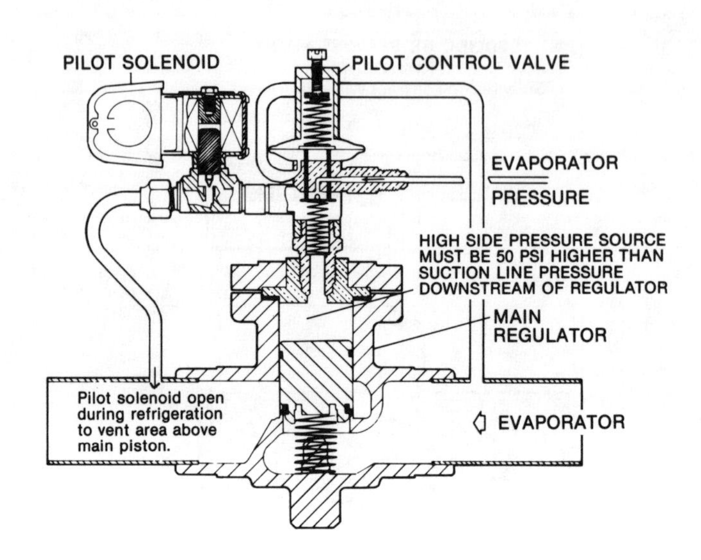

FIG. 3-67d. Evaporator pressure regulator with pilot control for defrost cycling. *Courtesy Sporlan Valve Co.*

The position of the disc in the three-way hot gas valve, Figure 3-67b, is controlled by the opening and closing of the pilot solenoid. The normal refrigeration mode uses high side pressure above the piston to close the lower hot gas port to the evaporator, keeping the suction line open. Upon defrost initiation, the solenoid energizes and bleeds the high pressure above the piston to the low side and the main piston moves upward, allowing hot gas to enter the evaporator for defrosting. When defrosting is complete, the valve moves back to its original refrigeration position as the pilot solenoid de-energizes.

Figure 3-67c shows another pilot operation where a solenoid valve controls high side pressure at the expansion valve. On large systems, where it would be costly to install a liquid line solenoid valve, this three-way valve is used as an alternative. While the refrigeration is operating, the solenoid is energized, closing the high pressure inlet port. The equalizer line, from the expansion valve to the suction line, is open through ports B to C. True suction pressure at the coil outlet is available at the expansion valve diaphragm. While de-energized, the low pressure port to the suction line is closed and full high side pressure is applied under the expansion valve diaphragm — overcoming the remote bulb pressure, closing the expansion valve, and preventing the refrigerant from migrating to the low side during the off cycle.

The pilot-operated evaporator pressure regulator

in Figure 3-67d is another example of a solenoid valve used to assist a main regulator's function. However, rather than two position control, the regulator throttles through an infinite number of positions. The regulator is operated by the pilot valve directly over the main piston. However, during a defrost mode, the regulator is shut down by the pilot solenoid. When the pilot solenoid is de-energized, full high side pressure is applied to the top of the main valve piston and immediately closes the regulator main piston. This mode is used only during a defrost cycle on evaporators requiring the defrost sequence. When defrost is complete, the pilot solenoid is energized, venting the high pressure from the top of the main piston and allowing the main port to open for continued refrigeration.

Motor Driven Valves Electric actuators have the following advantages:

- There is usually electricity somewhere nearby. Compressed air may not always be available.
- With computers, it is easier to set up electric systems than use pneumatic actuators.
- In larger sizes, electric actuators are more compact than their pneumatic counterparts.

Two types of electric motor actuators are most commonly used in the heating and cooling industry: unidirectional and reversing.

Electric motor actuators consist of the motor, gear train, limit switches and brake. They are designed to be used with both quarter-turn ball, plug or butterfly valves (rotary motion) or with globe style valves by raising or lowering a stem (linear motion). In either case, the motor's rotary motion travels through a gear train to an output shaft. For quarter turn valves, the rotary motion is used directly or through a linkage to the valve stem, Figure 3-68a. For linear application, the rotary motion is transferred to a cam that transforms the rotary output to a linear motion, Figure 3-68b.

Unidirectional motors rotate the stem through a full 360° in 180° increments for on-off control and are used on globe style valves. There is no stop to prevent 360° rotation. It is basically a two position, on-off control that runs in one direction in its output function. See Figure 3-69.

Reversing motor actuators use a split phase, capacitor start motor and have better energy efficiency and higher torque than unidirectional designs. Reversing motors do exactly as the name implies, they can be operated forward or backward from a neutral position. This gives a finer degree of infinite position control — from fully open to fully closed. In some of the reversing designs, brakes are utilized to minimize coasting, hunting or flutter, depending upon the valve design.

Butterfly valves frequently utilize brakes because they tend to be most unstable. The brake allows the disc to hold its position in the moving fluid against forces that can cause flutter, setting up pulsations in the piping system in addition to varying fluid flow. A reversing motor system is shown in Figure 3-70.

Since the motor drive actuator is an electrical

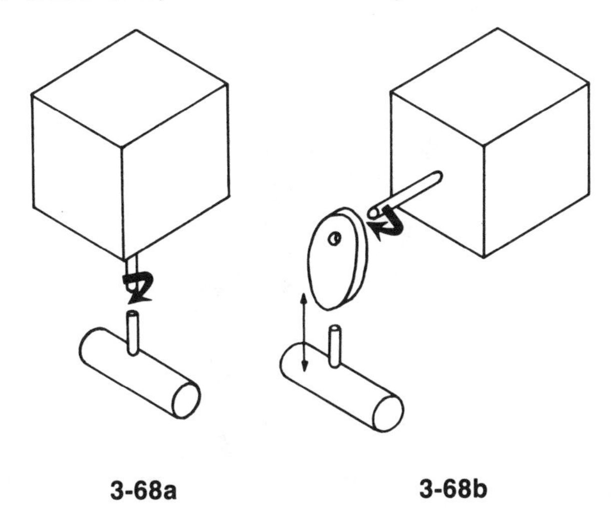

FIG. 3-68. Electric actuators. **a.** Rotary output for quarter turn valves. **b.** Rotary-to linear output for rising stem valves.

FIG. 3-69. Unidirectional electric actuator. *Courtesy Barber-Colman Co.*

FIG. 3-70. Reversing electric actuator. *Courtesy Barber-Colman Co.*

device, an appropriate enclosure must be selected. NEMA 4 watertight enclosures are usually specified for wet environments. If the motor is located in a hazardous location, the proper enclosure must be selected to prevent an explosion if electrical failure occurs.

Optional features for motor actuators include limit switches that remotely signal valve position to an operator, internal space heaters to keep actuators warm, mechanical brakes, and overriding manual operators.

The electric actuator, Figure 3-69, is a unidirectional actuator used in heating/cooling terminals in buildings. They have a two position, three wire, motor driven gear train that will either open or close a valve by operating through a 180° clockwise rotation. It may be operated on 24 or 120 volts AC and have operating torques to 220 in-lbs. The actuator may be used on 2-way or 3-way style valves. The force of the actuator is modified to turn rotary motion to linear motion. This is accomplished by a cam operated plunger connected to the valve stem and, depending on the motor operation, raises or lowers the stem to open or close the valve port.

Figure 3-70 shows a reversible electric motor actuator for proportional flow control in heating/cooling water systems. Extreme care must be taken to see that the fluid flow is properly orientated, that is, the pressure is under the seat of globe valves, for good control, or in the direction of the arrow cast into the body of 3-way mixing or diverting valves.

Pneumatic Actuators Pneumatic actuators are recommended in plants with a highly moist atmosphere where electric actuators may prove too unreliable. Explosive and hazardous environments, such as dust laden areas, are also good places for pneumatic systems.

The pneumatic systems are cylinder, vane, piston, spring-diaphragm, scotch yoke, and rack and pinion. The major task of any actuator is to overcome the internal fluid forces and frictional forces and to convert an external linear force into rotary motion needed to turn the valve stem through a 0° to 90° arc for quarter turn valves. The normal linear motion of

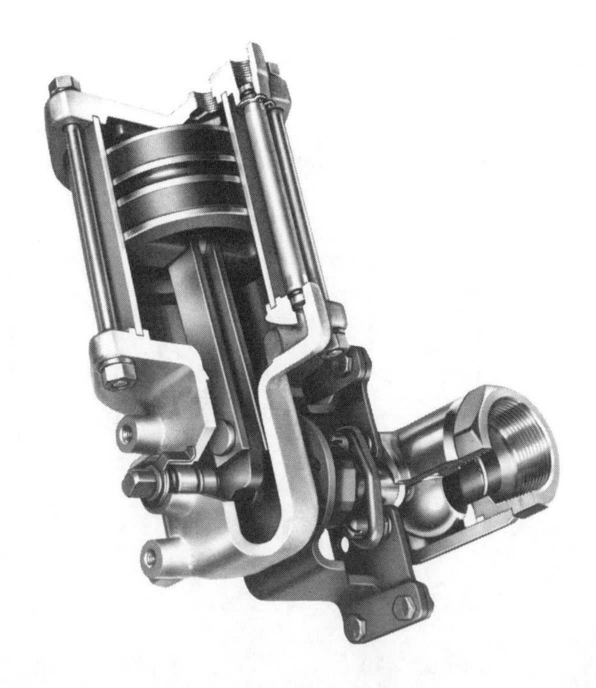

FIG. 3-71. Piston driven pneumatic actuator with enclosed and housed linkages. *Courtesy Jamesbury Corp.*

some pneumatic actuators is used directly with globe style valves.

Adequate air pressure is needed to actuate the valve and is usually under 100 psig. If only a low pressure source is available, oversized actuators are used to produce the forces needed for rotation. Pneumatic operators usually do not have the high torque ratings of their electric or hydraulic counterparts, consequently they are limited to lighter loads and smaller valves.

Cylinder actuators consist of a separate pneumatic cylinder, on a mounting plate, that is connected to the valve stem via clevis and lever. Air pressure is applied on either side of the piston to open or close the valve. The cylinder can be readily applied to a valve on the job site, however, its shortcomings include exposed linkages and a good possibility that it will not meet the valve's torque requirements. Buying a complete package eliminates the possibility of a mismatch.

A vane style actuator consists of a sector, usually 90°, that houses an oscillating vane. Rotation is caused by air pressure on one side of the vane, turning it through 90°. Return is either by a spring and the removal of inlet air pressure or by air pressure on the other side of the vane.

The piston style actuator is similar to the separate pneumatic cylinder except that the piston and linkages are all enclosed in a housing that is designed

FIG. 3-72. Flexible diaphragm, pneumatic operator style actuator converts linear to rotary motion. *Courtesy Jamesbury Corp.*

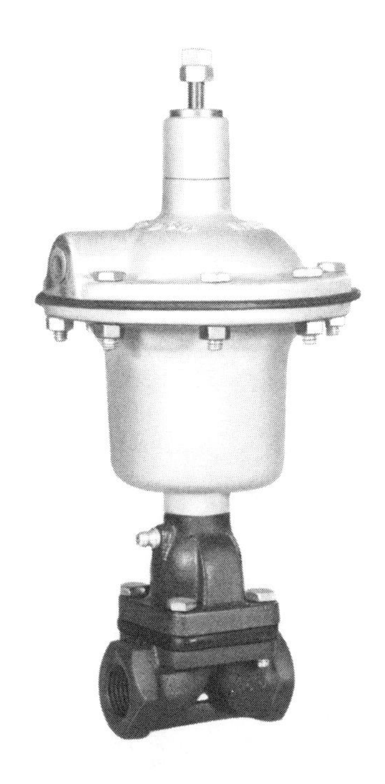

FIG. 3-73. Pneumatic flexible diaphragm operating strictly on linear motion as mounted on a diaphram valve. Air to close. *Courtesy Flow Control Div., Rockwell International.*

FIG. 3-74. Air-to-open pneumatic system. *Courtesy Flow Control Div., Rockwell International.*

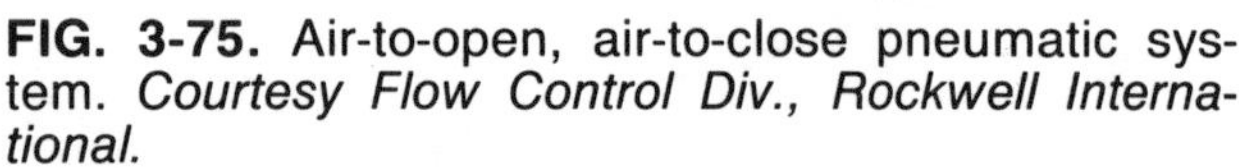

FIG. 3-75. Air-to-open, air-to-close pneumatic system. *Courtesy Flow Control Div., Rockwell International.*

FIG. 3-76. Family of pneumatically operated valves for sanitary applications. *Courtesy Ladish Co., Tri-Clover Div.*

specifically for the valve on which it is mounted, Figure 3-71.

The spring diaphragm system utilizes an oversized flexible rubber membrane, Figure 3-72, working against a reset spring. This system comes in three versions:

- Air pressure-to-close, spring return, (N.O.) normally open. This is also known as a *fail open* valve if pneumatic pressure is lost, Figure 3-73.
- Air pressure-to-open, spring return, (N.C.) normally closed. This design is known as a *fail closed* valve when pneumatic pressure is lost, Figure 3-74.
- Air pressure-to-open, air pressure-to-close, Figure 3-75.

The scotch yoke design utilitzes a piston in a cylinder, operated by air pressure, similar to those described above. However, the linkage consists of a pin moving in slots machined in the arms of a yoke. This design minimizes side loading on the stem. Its greatest advantage is that its torque characteristics more closely match the ball and butterfly valve torque requirements.

The double rack and pinion actuator has an advantage over all other actuators in that its force, as applied to the stem, is symmetrical. No side loading and subsequent bending forces are applied to the stem.

Figure 3-76 shows a complete line of air operated valves for automatic control of food products.

Hydraulic Actuators Hydraulic actuators can be systems that include pumps, cylinders, reservoirs, filters and hoses. They are usually restricted to larger valves and can handle torques upwards of 1,000,000 in-lbs and pressures up to 5000 psi. Actuators for hydraulic systems have designs similar to their pneumatic counterparts, however, many of the parts are built stronger to accomodate the higher pressures, torques, and usually larger sizes of valves.

4
Selection Parameters

Valve performance and reliability are dependent upon the selection of appropriate component materials that resist the physical and chemical effects of the various gases and fluids and the exterior environments. Top performance also hinges upon the proper selection of a valve's style, rating and function.

Selection parameters must be identified and the valve choice made for proper application and subsequent procurement. The parameters include:

- Fluids
  - ☐ Fluids to be handled
  - ☐ Corrosive effects (internal)
  - ☐ Corrosive effects (external)
  - ☐ Fluid properties
- Pressure/temperature (P-T) ratings
- Valve markings
- Material selection
- Fluid flow
- Pressure drop
- Flow coefficient, C_V
- Service (duty cycle)
- Normal position
- Leakage
- Internal forces
- Size
- Maintenance
- Cost
- Safety
- Actuator
  - ☐ Fluid forces
  - ☐ Pressure drop
  - ☐ Frictional factors
  - ☐ Torque
  - ☐ Opening and closing speed (fail safe)
  - ☐ Manual
  - ☐ Powered
    - Pneumatic
    - Electric
  - ☐ Accessories

Fluids to be handled

The fluids that are controlled by valves are almost infinite in number. Fluids can affect basic body integrity as well as the disc, seat, stem and packings. Incorrect valve materials can allow valves to be destroyed in short order when used with the wrong fluid. Even among compatible fluids, if the pH level, temperature, pressure, or presence of entrained air or moisture is changed, the once compatible fluid may soon become destructive and a valve could be put out of service.

Hot water that is close to boiling, at atmospheric pressure or a slight vacuum, or cold water with entrained air or a high mineral content can cause corrosion problems. The problems can be mineral deposits, electrolysis, impingement attack or metal removal (cavitation) by collapsing air bubbles.

In the refrigeration and air conditioning industry, fluids include treated brines in closed cooling systems, glycols for heating and cooling systems, city or well water for condenser cooling, ground water, including brackish or salt water, for use with thermally assisted heat pumps, solar energy fluids, at elevated temperatures, for solar heating and cooling applications, water from open type cooling towers, and refrigerants in evacuated, moisture-free systems. Fluid temperatures may range anywhere from −100 °F to 600 °F.

Corrosive effects (internal)

Water systems Water problems in the heating and cooling industry are as varied as day and night because water is a universal solvent and each area of the country has water of different quality.

Water problems can be divided into three categories, based on the impurities in the water, and each can affect valves, associated piping, and heat transfer surfaces to a degree that can make them inoperative shortly after installation. They are: fouling, scaling, and corrosion.

Fouling is literally a dirty business! It is caused by leaves, sludge, algae or slime that clog the flow passages of piping. Open systems, such as cooling towers and evaporative condensers, are the most susceptible. Clogged piping not only stops fluid flow, but decaying organic matter can cause further problems with direct and indirect corrosion.

Scaling is caused by minerals from the water depositing onto the internal surfaces of equipment. This usually happens, at elevated temperatures, in water cooled condensers or refrigeration and air conditioning machinery. Minerals include calcium and magnesium carbonates along with iron and silica. All these are found in the average water supply and contribute to the hardness of water.

Limestone ($CaCO_3$, calcium carbonate) is the most abundant mineral in water and is responsible for about 99% of the scaling problems.

Corrosion, on the other hand, is a highly complex and not so easily identified problem. Many variables are intertwined. The most common corrosion factors, however, are fluid acidity, dissolved gases, such as carbon dioxide and oxygen, galvanic potentials, biological organisms, fluid velocity and temperature. Valve deterioration is frequently due to corrosion.

Fluid acidity Fluid acidity or basic conditions are compared on the pH scale. A neutral fluid has a pH value of 7, Figure 4-1. Anything less than pH 7 is acidic and anything over pH 7 is basic. Strong acidic solutions will have pH values of 2 to 4 while strong caustic (basic) solutions will have pH values of 9 and above.

Water can be mildly acidic, as it is received from city pumping stations, due mainly to chemicals used in its treatment. If ground water is being utilized directly, its acidity can be caused by atmospheric chemicals that are washed into the soil by rain or from discarded chemicals from dump sites. Ground water can also contain dissolved minerals and grit held in suspension.

Brackish and seawater sources have dissolved salts that increase corrosion rates by enhancing the conductivity of the fluid. Higher conductivity makes electron flow easier and encourages electrolysis between valve components or piping of different materials.

In closed loop heating/cooling systems, makeup water is not used. Therefore, once the system water is chemically treated, its pH value should not change rapidly and the water should not create problems, if monitored on a regular basis. However, closed loop, secondary brine systems, for refrigera-

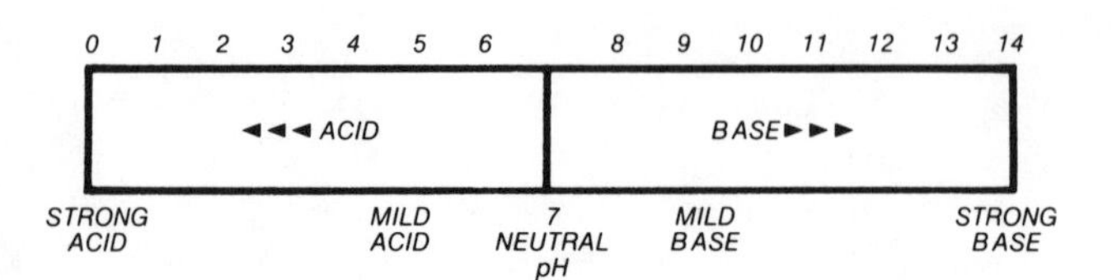

FIG. 4-1. Scale of pH values.

tion systems used in ice rinks and the like, have their own unique problems. Brines are good electrical conductors, similar to seawater, and encourage electrolytic corrosion by galvanic action.

Swimming pools and cooling tower systems, on the other hand, are open systems. Swimming pool water can be acidic, due to chlorine treatment for health purposes, and pH adjustment, to levels above pH 7, is necessary. This adjustment is a must with metallic solar heating devices, to minimize corrosion. Many solar heating components are fabricated of plastic materials when used in swimming pool applications.

Cooling tower water is subjected to all types of fouling, scaling and corrosion. Atmospheric dust, chemicals, debris and even acidic rain, organic and biological organisms can be found in cooling tower water systems. pH and biological control are needed to extend component life.

Solar system corrosion results from the use of dissimilar metals, corrosive liquids and the presence of air in the system.

Heat transfer fluids used in solar systems can lead to special corrosion problems. These fluids contain corrosion inhibitors that can degrade at high temperatures and render the fluid more corrosive.

Some waters are highly aggressive and promote pitting. Aggressive water usually has a high mineral content, high levels of total dissolved solids (TDS) including sulfates and chlorides, high carbon dioxide content and dissolved oxygen.

Many of these fluids are not corrosive to copper-base alloys or to copper itself. However, if less noble metals are used in the same system, for example zinc or aluminum with copper or brass, the less noble materials will corrode and eventually be destroyed. Do not mix metals if at all possible.

Ground water heat pumps can be affected by high mineral content, carbon dioxide and oxygen in the water. Many heat pump users along sea coasts or

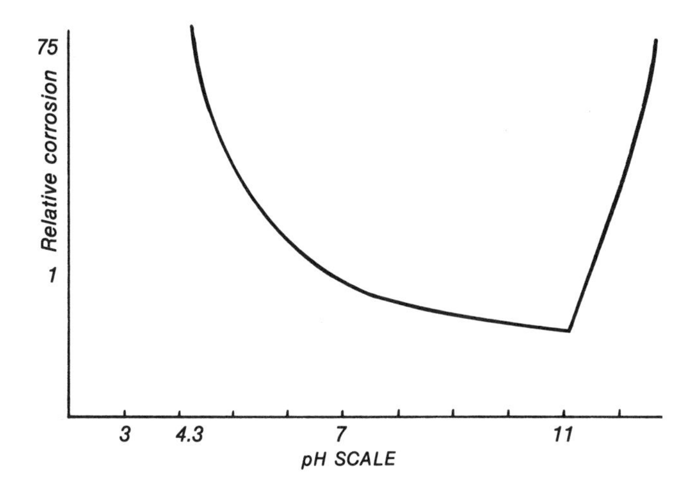

FIG. 4-2. Curve showing the effect of pH on corrosion of mild steel. *William J. Scott: Handling Cooling Water Systems during a Low pH Excursion (Chemical Engineering, Feb. 22, 1982)*

near rivers and lakes are now taking advantage of this low grade heat source. However, many of these water sources are chemically polluted, making them brackish and very corrosive.

Figure 4-2 shows the effect of pH on the corrosion of steel. The relative corrosion rate rapidly increases as the pH is allowed to fall below neutral pH 7. When the pH level is raised above 7 (basic), corrosion is reduced until a pH of 11 is created at which point corrosion rapidly increases again. Most systems operate well and have a longer life if the pH is maintained at a value of 8 or 9.

All these acidic problems can be corrected by proper pH control. Water treatment chemicals and automatic injection systems are readily available. Care in selecting and applying them can keep heating/cooling system piping, valves and heat exchange devices clean and operating at full efficiency.

Dissolved gases Dissolved oxygen and carbon dioxide, in excess of 50 parts per million (ppm), accelerate corrosive conditions. Both gases rapidly corrode iron, zinc and brass.

Carbon dioxide (CO_2) in ground water averages about 50 ppm and can go as high as 300 ppm. CO_2 is held in equilibrium but will not remain in equilibrum if the temperature or pressure changes. The higher the temperature, the lower its solubility, Figure 4-3. The higher the pressure, the higher the solubility, Figure 4-4.

If ground water contains both dissolved oxygen and carbon dioxide, aeration will remove the carbon dioxide and increase the oxygen content. Oxygen can be removed by heating the water, to drive the oxygen out of solution, or by applying a vacuum or adding chemicals.

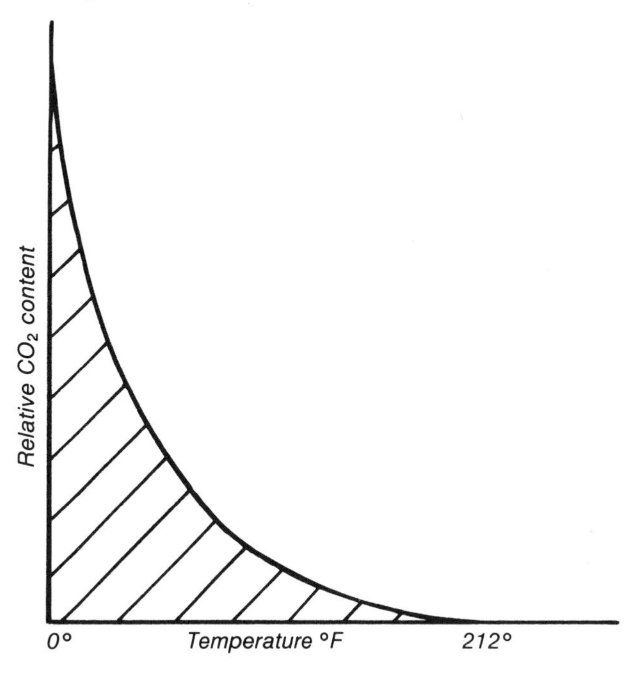

FIG. 4-3. CO_2 solubility vs. temperature. *(Ground Water Heat Pump Journal, 1980, p. 13)*

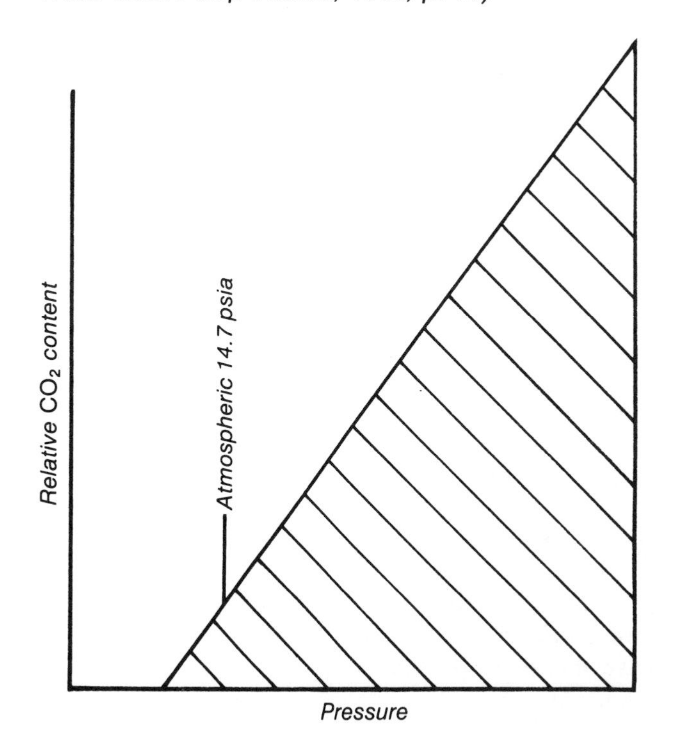

FIG. 4-4. CO_2 solubility vs. pressure. *(Ground Water Heat Pump Journal, 1980, p. 13)*

Galvanic corrosion Galvanic corrosion occurs when dissimilar metals are used in a piping system. Metals high on the *electromotive series*, Table 4-1, are anodic and will corrode, by sacrificial action, to the more noble cathodic metals that are lower in the series. When two or more different metals are used in the system, the one higher in the electromotive series is corroded. The greater the separation between the metals in the series, the more rapid the rate of corrosion. Galvanized, zinc-coated pipe and bronze valves are widely separated in the series and rapid corrosion of the zinc can be expected.

The electromotive series must be interpreted with care. Slight variations in service conditions, due to aeration, temperature, and fluid chemistry, may reverse some of the positions in this series! Knowledge of the type of fluid and its use can be exceedingly helpful.

Dielectric isolators, used between widely separated metals in the electromotive series, will help to reduce or slow down the corrosion rate. With strong brackish or seawater systems, it is necessary to keep metal combinations as close together as possible, such as cupro-nickel alloys, bronzes and monels.

For valves constructed of nonferrous metals, corrosion effects are based upon the galvanic series, Table 4-1. Many of the nonferrous alloys are combinations of elements such as brass (60% copper, 38% zinc and 2% lead) or bronze (85% copper, 5% tin, 5% lead and 5% zinc). Stem materials for inexpensive valves may utilize 60% copper and 40% zinc. Special alloys of 80% copper, 16% zinc and 4% silicon have better corrosion resistance and toughness.

A corrosion problem commonly found in valves used in water systems is dezincification. The elemental zinc (high in the electromotive series) is leached out of solution and a porous, spongy, structurally weak, and leaking valve remains. Critical components, such as the disc, seat and valve stem must be fabricated from the more noble metals (lower in the series) for long life expectancy.

Many of the nonferrous metals will form self-protective films of oxide layers of only several mils thickness or less. These films, such as aluminum oxide on aluminum valves, are very hard and insoluble in water. They are excellent corrosion protection as long as the film is not marred or scratched and the base metal exposed. Stainless steel will passivate itself under high fluid velocities but will readily pit under stagnant conditions. Nitric acid will chemically passivate stainless steels with a protective film.

For valve bodies and components fabricated of ferrous metals, corrosion depends on the chemical composition of the steel, the fluid pH, fluid hardness and chemical nature, dissolved gas level, fluid velocity, temperature and biological organisms.

Table 4-1. Electromotive series

Corroded End (Anode)	Magnesium Magnesium Alloys
↑	Zinc
	Aluminum 25
	Cadmium
	Aluminum 17ST
	Steel or Iron Cast Iron
Decreasing Corrosion Resistance	Chromium-Iron (Active)
	Ni-Resist
	18-8 Chromium-Nickel-Iron (Active) 18-8-3 Chromium-Nickel-Molybdenum-Iron (Active)
	Led-Tin Solders Lead Tin
	Nickel (Active) Inconel (Active)
	Brasses Copper Bronzes Copper-Nickel Alloys Monel
	Silver Solder
	Nickel (Passive) Inconel (Passive)
	Chromium Iron (Passive) 18-8 Chromium-Nickel-Iron (Passive) 18-8-3 Chromium-Nickel-Molybdenum-Iron (Passive)
↓	Silver
Protected End (Cathode)	Gold Platinum

Groundwater Heat Pump Journal, 1980, p. 14

Microorganisms Microorganisms in cooling systems can be yeasts, molds (plant matter), algae (sunlight is necessary for its growth) and bacteria that cause slime, indirect corrosion, gas production and wood decay. Slime can clog valves and shield metal surfaces from corrosion inhibitors, that were added to the fluid, allowing corrosion to take place underneath. Hydrogen sulfide gas, produced by certain bacteria, is acidic and can attack steel, stainless steel and copper alloys. Microorganisms, if not checked by proper chemical and biological treatment, can attack steel at the rate of 100 mils per year and pit stainless steel in 60 to 90 days.

Incrustation is due to the activity of iron-sulfur bacteria. These bacteria colonize wells and their byproducts form a tough scale and a gelatinous mass that clog valves and filters.

Fluid impingement The rate of fluid flow in a piping system will determine the rate of erosive and galvanic effects. Different metals and alloys are affected differently by changes in velocity. When velocity is extremely high, mechanical effects may add to the corrosion rate. Globe valves are especially susceptible to corrosive attack under these conditions. The *Z* style of flow pattern creates considerable turbulence and pressure drop. The flow cavities in the body should be as smooth as practical to minimize the turbulent effects.

Most metals have fluid velocity limits. Below these limits, corrosive rates due to velocity are considered normal and within reason. Above the velocity limits, accelerated corrosion takes place and metal destruction is rapid. Copper, brasses, and steel are usually kept at three feet per second maximum, cupro-nickels at about 10 to 12 feet per second, while stainless steels and titanium have no upper limit if the system is free of abrasive particles.

Stainless steels are one alloy that reverses its corrosion rate at low velocity. At high velocity, stainless surfaces are passivated and a protective film is created. At low velocity or stagnant conditions, however, the passive film is easily breached and rapid pitting follows.

It must be remembered that, at low velocity, water-borne inhibitors cannot be distributed or carried to the various valves and component surfaces. The protective film produced by the inhibitors is a function of the velocity of the fluid.

High velocity, combined with temperature, affects metal at different rates. At cold water temperatures, the concentration of dissolved oxygen is higher than in hot water systems.

Temperature Temperature affects the solubility of dissolved gases. At high temperatures, about 150 °F to 200 °F, dissolved gases lose solubility, water viscosity decreases and conductivity increases. The corrosion rate can double for every 40 °F temperature rise.

Refrigeration systems In refrigeration systems, corrosion can be attributed to oils, moisture, air, solder fluxes, excessive discharge temperatures and electric motor insulation in hermetic and semi-hermetic systems.

Oil is constantly being circulated in a closed refrigerant system and refrigerant and oil react with one another at high temperatures. Some oils are more stable with refrigerants while others are more reactive. The reaction of refrigerant and oil at high temperature produces hydrochloric acid, HCl. Once the acid exceeds a critical concentration, copper is no longer stable and its ions can be deposited on the iron surfaces of bearings and pistons. There have been cases where enough copper was plated out to the compressor's cylinder walls that, when scraped from the walls by the pistons, it actually bound the system's valves.

FIG. 4-5. Varying degrees of corrosion of expansion valve parts due to oil sludge from moisture and high temperature. Valve failure resulted from the corrosion. *Courtesy Sporlan Valve Co.*

Moisture and air cannot be tolerated. These contaminants can produce direct corrosion, oil sludging, excessive pressures and ice formation, all of which attack regulators and control valves. Moisture in refrigerants will produce corrosive acids and, if air (oxygen) is present, corrosion will be accelerated. The greatest single cause of chemical action within a refrigerant system is moisture. Figure 4-5 shows valve components corroded by excessive moisture and temperature.

Temperature accelerates the chemical reactions, solvent effects and solubilities of various acids. Combined with oxygen from the air, higher temperatures bring about chemical decomposition of refrigerants and oils. Ammonia systems, for example, could theoretically use copper pipe if air and moisture never entered the system. In practice, this is not the case. Steel piping is used with ammonia systems to eliminate the corrosion problem.

High discharge temperatures can also cause the oil to decompose, depositing gum and carbon. Heat accelerates elastomeric breakdown of motor winding insulations and varnishes. Discharge temperatures should be limited to about 200 °F, measured 6 inches downstream from the discharge service valve.

Solder fluxes remove oxide films, preparatory to soldering copper, brass, bronze and steel tubing and valves. However, many technicians believe a lot is better than a little. Consequently, some of the fluxes enter the refrigerant system and, in excess, can create an aggressive atmosphere.

Corrosion inside refrigerant systems can be due to:

- Moisture
- Oxygen
- High temperatures
- Poor or sloppy installation

Corrosive effects (external)

External corrosion is mostly due to atmospheric conditions. Polluted industrial areas, with high humidity levels, and seacoast environments can cause rapid metal deterioration. In arid, pollution-free areas, corrosion is almost nil.

Valves with exposed stems, that is, gate or globe valves with rising stems or actuators with exposed linkages, can be made inoperative by severe corrosion. Selection of the proper component materials and designs, to withstand these external environments, is required.

The addition of minute amounts of copper, phosphorus or nickel improves the corrosion resistance of basic steel. However, steel valves that are subjected to constant humidity, where the passive corroded layer cannot dry out, will continue to show rapid corrosion. Rust must be allowed to dry periodically if it is to serve as a protective film.

Coatings, from sacrificial metals to paints, are used to protect against exterior corrosion. Organic and inorganic films are used in conjunction with sacrificial zinc (high in the galvanic series). Epoxy topcoats, vinyls, acrylic and silicons are useful, but each must be compatible with the base primer and the corrosive atmosphere.

Fluid properties

The more important physical properties of fluids are:

- Viscosity
- Density and specific gravity
- Specific volume

There are two factors that resist the movement of flowing liquids and gases: frictional resistance by the pipe walls, fittings and valve internal passages; and the viscosity of the fluid which is the internal fluid friction tending to oppose movement.

The viscosity of a fluid defines the fluid's ability or inability to flow when pumped and varies with temperature. The viscosity of liquids decreases with an increase in temperature; the viscosity of gases increases with an increase in temperature.

Molasses, for example, is a high viscosity fluid while water is a considerably less viscous fluid. Gases have very low viscosity values.

Unfortunately, viscosity values can be expressed in a variety of ways, most of which are confusing. Depending on the industry, that is, the oil industry, chemical industry, scientific laboratories, etc., the units of viscosity will vary.

Viscosity tables are available for most fluids and care must be taken to see that the *proper units* are used in making calculations. Kinematic viscosity and absolute viscosity are two terms that form the base for the tables. Kinematic viscosity is the ratio of absolute viscosity to the mass density. Viscosity terms can be listed as:

Kinematic	Absolute
Stoke	Poise
Centistoke	Centipose
Saybolt universal (SSU)	

For most materials used in the air conditioning/refrigeration industry, the absolute viscosity is tabulated in *centipoise* units. All the viscosity values in Table 4-2 are given in centipoise (cp) units.

The frictional pressure drop of a flowing fluid is directly proportional to the viscosity. The higher the viscosity of the fluid, the greater the pressure drop in the system.

The density of a fluid is its weight per unit volume and in most cases is given as pounds per cubic foot (lbs/ft^3). The density of liquids is fairly constant over the range of pressures and temperatures normally encountered in the heating and refrigeration industry.

Gases are, however, another proposition. Their density will vary greatly, depending upon pressure changes. Table 4-3 shows variations of density for both water and air over a range of pressures and temperatures.

In turbulent flow conditions, the pressure drop is directly proportional to the fluid density. The greater the density, the greater the pressure drop. Fluids such as glycols and concentrated brines have

Table 4-2. Viscosity of common fluids.

Refrigerant 12 (vapor)	.0125 cp at 100°F
Refrigerant 12 (liquid)	.250 cp at 100°F
Dowtherm A	2.5 cp at 100°F
Water	.68 cp at 100°F
SAE 30 lubricating oil	100 cp at 100°F
Air	.018 cp at 100°F
Air	.028 cp at 500°F

Table 4-3. Density of water and air.

Water (lb/ft³)	Air (lb/ft³)
@ 32°F density is 62.41	@ 30°F and 0 psig .081
@ 200°F density is 60.11	@ 30°F and 100 psig .633
	@ 200°F and 0 psig .060
	@ 200°F and 100 psig .470

high viscosity and density, consequently have higher pressure drops than water at the same flow conditions.

The specific gravity of a liquid is the ratio of the density of a substance to that of water at 60 °F. Brines are heavier than water and their specific gravity is over 1. A 10% calcium chloride brine has a density of 68.05 pounds per cubic foot while water has a density of 62.4 lb/ft³. Therefore, 68.05/62.4 yields a specific gravity of 1.09. SAE 10 oil, at 60 °F, has a density of 54.64 lb/ft³, therefore, its specific gravity is 54.64/62.4 or .876.

The specific volume of a fluid is the reciprocal of the density and is measured in cubic feet per pound (ft³/lb). Specific volume is used most frequently with gases since gas density is such a small number. Air, for example, has a density of .0752 lb/ft³ at 68 °F. The specific volume is the reciprocal, 1/.0752 or 13.29 ft³/lb.

Pressure/temperature (P-T) ratings

The pressure-temperature (P-T) ratings of valve materials, whether metal or plastic, are directly dependent upon the fluid temperature. At cryogenic temperatures, materials become brittle and excessive shock may rupture a line or valve. At high steam temperatures, materials may fail due to the erosion of valve internal components or by losing tensile and yield strengths.

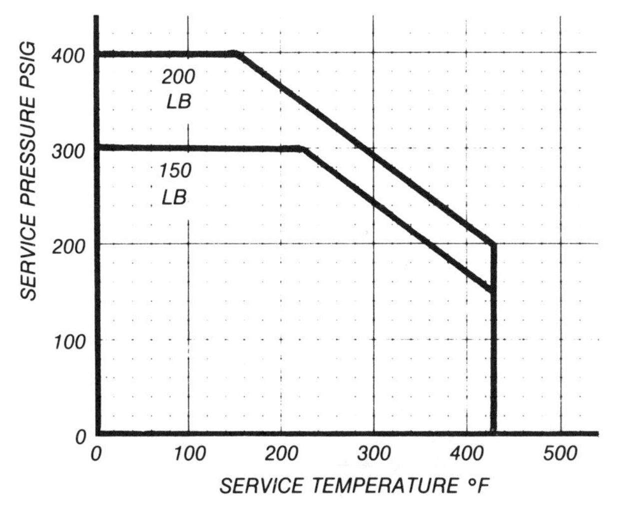

FIG. 4-6. Pressure-temperature ratings for bronze solder end and union end valves. *(Heating/Piping/Air Conditioning, May, 1980)*

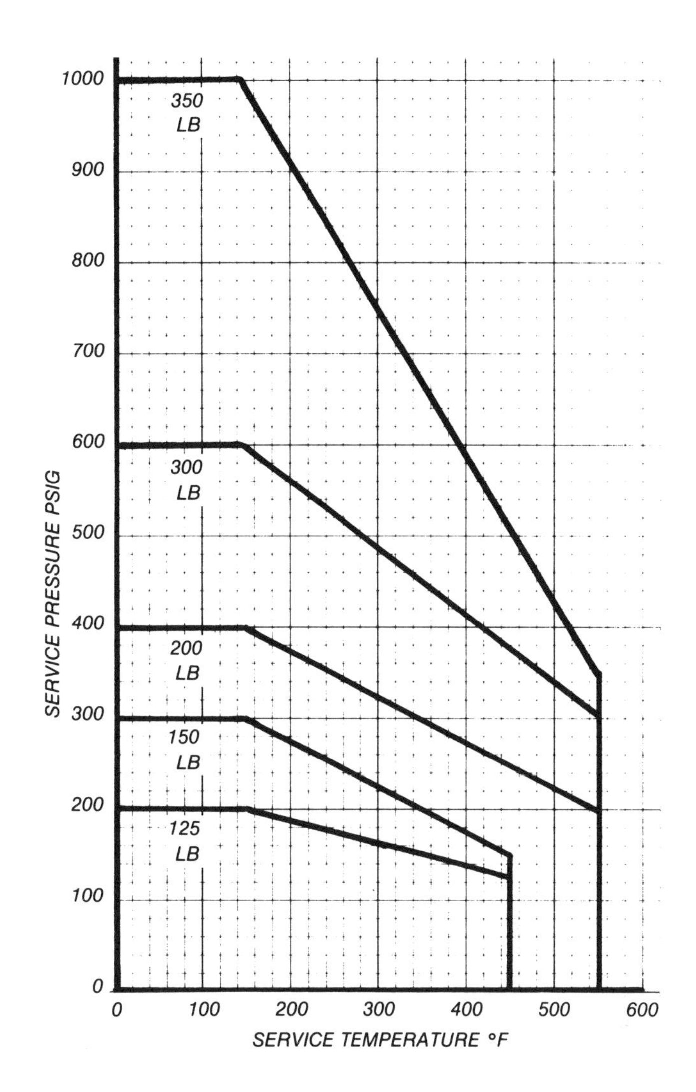

FIG. 4-7. Pressure-temperature ratings for bronze screw end valves. *(Heating/Piping/Air Conditioning, May, 1980)*

Both pressure and temperature determine final valve selection and rating. Usually, the higher the temperature, the lower the allowable operating pressures. It is assumed that ambient temperatures are in the range of −20 °F to 100 °F.

There is a set of P-T ratings for each material: bronzes, steel, iron, plastic, etc. The P-T ratings also vary with the end connections, that is, soldered, screwed, flanged, etc. P-T curves, showing operating pressures vs. temperature, are given for three types

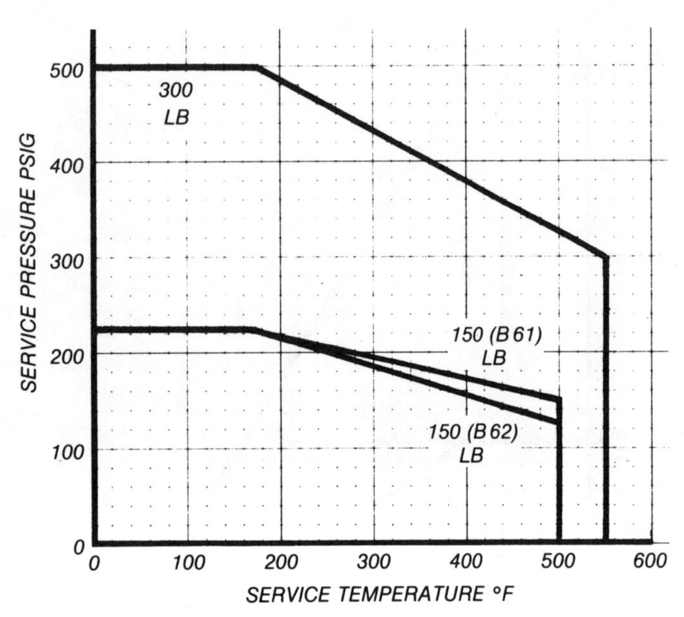

FIG. 4-8. Pressure-temperature ratings for bronze flanged end valves. *(Heating/Piping/Air Conditioning, May, 1980)*

of end connections and various valve pressure ratings for bronze valves in Figures 4-6, 4-7 and 4-8. Care must be taken, however, in interpreting the P-T curves. For example, the allowable body-bonnet ratings (upper limits) for a 300 pound rated bronze valve with screwed ends is 350 psig at 500 °F, Figure 4-7. However, if the disc, seat or packing cannot withstand the 500 °F limit, then the valve must be limited to that of the weakest material!

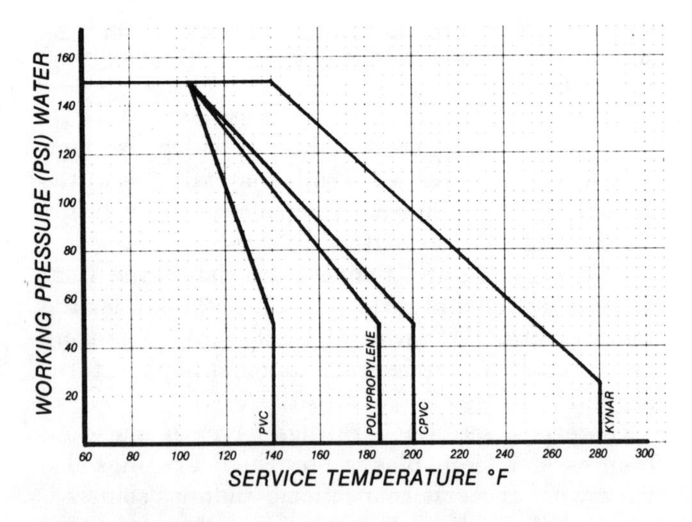

FIG. 4-9. Pressure-temperature ratings for plastic valve materials.

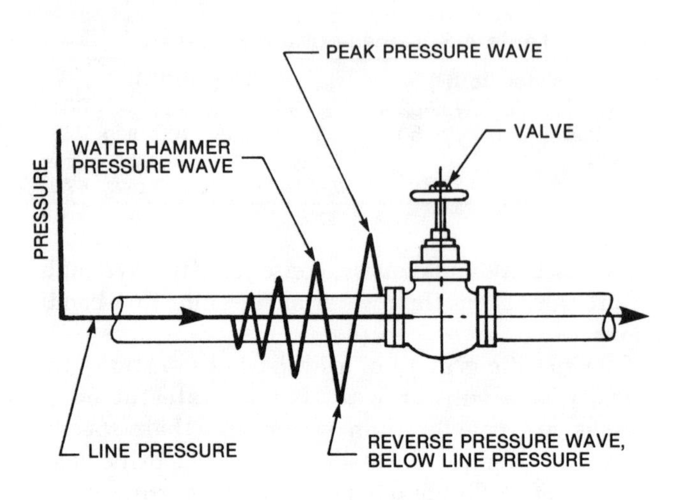

FIG. 4-10. Water hammer pressure waves from a fast-closing valve. *Courtesy Fibercast Co.*

The P-T curves also hold true for plastic valves, with considerably lower ratings. Figure 4-9 outlines the upper limits of a number of plastic materials.

The P-T ratings of all materials are based on *non-shock* conditions. Figure 4-10 visualizes the shock waves within a fluid piping system caused by water hammer. Water hammer is a function of the fluid's physical properties and is produced by abrupt changes in fluid velocity. A shock force may be many magnitudes beyond the valve rating. Shock waves are usually created by rapid valve opening or closing and can cause a system rupture at a considerable distance from the source.

Valves should not be overrated to compensate for the pressure surges caused by water hammer. Valve operators should open and close valves slowly. Accumulators and feedback loops can substantially reduce shock forces.

Table 4-4. Pressure ratings for low temperature solder joints, bronze valves.

Material used in joints	Working tempera-tures, F	Maximum working pressure, psi		
		1/4 to 1 inclusive	1-1/4 to 2 inclusive	2-1/2 to 3 inclusive
50-50 tin-lead	100	200	175	150
	150	150	125	100
	200	100	90	75
95-5 tin-antimony	100	500	400	300
	150	400	350	275
	200	300	250	200
	250	200	175	150

Heating/Piping/Air Conditioning, May, 1980

Selection of a properly rated valve may eliminate valve failure, however, the end connections and the piping itself must be adequately designed to withstand the P-T ratings. Soldered end connections on bronze valves are a good example. If the improper solder is used, the piping system could be weak. Table 4-4 shows the P-T ratings for soldered joints on bronze valves and must be considered along with the valve body, disc, seat and packing.

The P-T ratings assume that the fluid is compatible with the valve materials. Fluids that corrode the valve materials or attack them chemically will substantially derate the valve's P-T levels.

In review, the P-T rating curves assume the following:

- Ambient temperature is −20 °F to 100 °F
- Nonshock operation
- Use of compatible fluids

Valve markings

Valves are covered by a number of codes and standards. Organizations and standards governing the valve industry are:

MSS Manufacturers' Standardization Society of the Valve and Fitting Industry
ASME American Society of Mechanical Engineers
ASTM American Society of Testing Materials
ANSI American National Standards Institute
NACE National Association of Corrosion Engineers

Many of the valve ratings, testing procedures, material designations and types of end fittings are covered by standards that can be obtained by contacting the above organizations.

The valve body must contain the fluid pressure and temperature, withstand the flowing fluid effects, and have a built-in safety margin. Valves are marked to indicate their ability to withstand certain pressures, temperatures and fluids. Figures 4-11 and 4-12 show two valves that are marked to show their operational service limits. Rating designation markings consist of letters and numbers. Under normal operating conditions, using compatible fluids in nonshock service, and −20 °F to 100 °F ambients, the valve may be labeled for CWP or S service.

Cold Working Pressure, CWP, is designated by the following letters:

WO Water, oil pressure
OWG Oil, water, gas pressure
WOG Water, oil, gas pressure
GLP Gas, liquid pressure
WWP Working water pressure
W Water pressure

Steam application symbols or steam working pressure SWP is shown by:

SP Steam pressure
WSP Working steam pressure
S Steam

Figure 4-11 shows a globe valve rated for 150 pounds steam, S, and 300 pounds water, oil or gas, WOG. The arrow indicates the proper direction of flow. The valve in Figure 4-12 is rated for 300 pounds steam, S, and 1000 pounds water, oil or gas WOG.

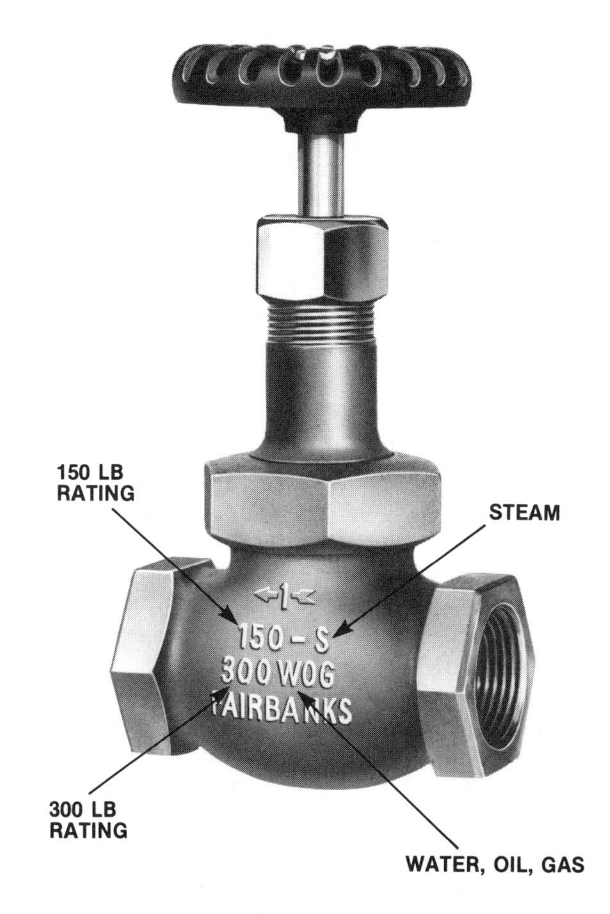

FIG. 4-11. Globe valve with body markings showing service limits of 150 psig with steam and 300 psig with water, oil or gas. *Courtesy The Fairbanks Co.*

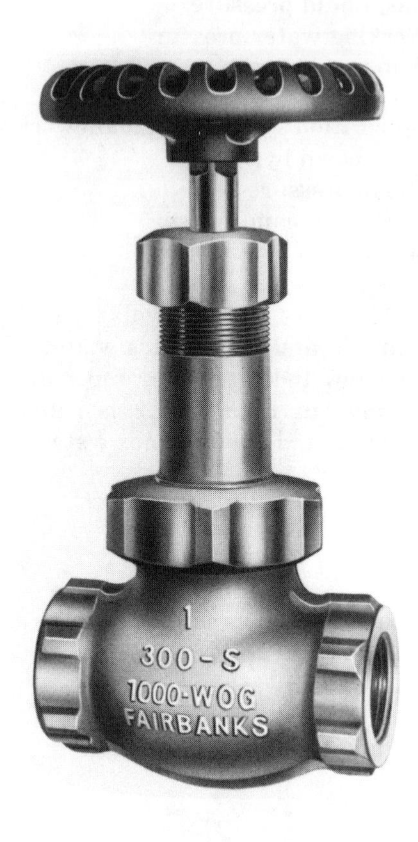

FIG. 4-12. Gate valve with body markings showing service limits of 300 psig with steam and 1000 psig with water, oil or gas. *Courtesy The Fairbanks Co.*

Relief Valves Relief valves have special markings that should be understood because of their safety applications.

The capacity of relief valves used in refrigerant service is rated in pounds of air per minute to eliminate marking individual valves for different refrigerants. The refrigeration code lists the factors that are used to convert the various refrigerant flow rates into equivalent air flow rates.

The UV symbol, within the *clover leaf*, Figure 4-13, indicates that the valve was made in accordance with the provisions of the ASME Boiler Pressure Vessel Code. The letters NB indicate that the capacities stamped on the name plate have been certified by the National Board of Boiler and Pressure Vessel Inspectors. The letter C, with the letters SA inside, is the insignia of the Canadian Standards Association which is equivalent to UL in the United States.

ASME (American Society of Mechanical Engineers) has specified the relief valve markings in

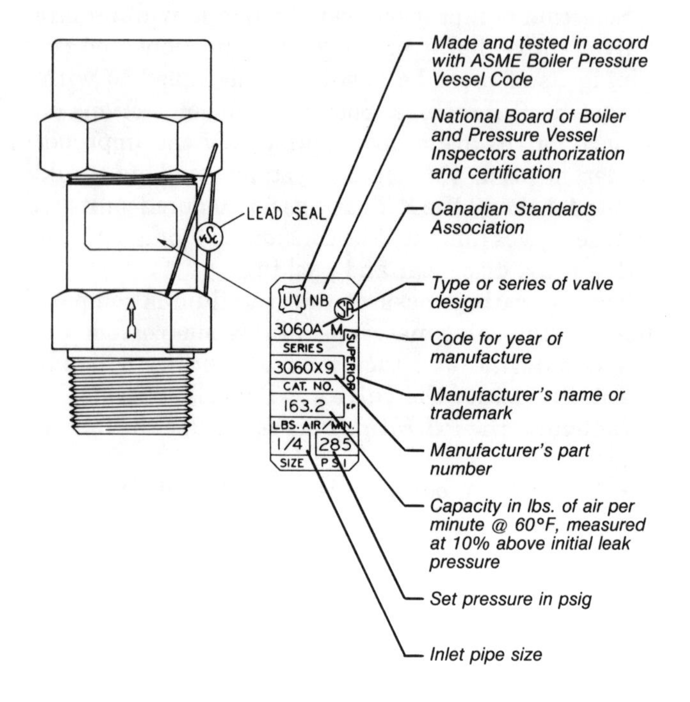

FIG. 4-13. Nameplate designates capacity of refrigerant safety relief valve. *Courtesy Superior Valve Co.*

their ASME Code, Section VIII, Division I. The entire marking per the ASME code shall consist of:

1. Manufacturer's name or trademark
2. Manufacturer's part number
3. Pipe size of the valve inlet in inches
4. Set pressure in psi
5. Capacity in pounds of air per minute at 60° F and 14.7 psia.
6. Year of manufacture of the valve
7. ASME clover leaf symbol

Figure 4-13 shows a nameplate and its designations.

ASME establishes performance characteristics and construction materials for safety relief valves. The National Board (NB) tests valves and certifies the capacities. The ASME codes are Section I for fired steam power boilers, unfired waste heat boiler drums, steam superheaters, economizers and reheaters. Section VIII is for unfired pressure vessels larger than six inches inside diameter with pressures above 15 psig.

Safety valves for steam, under Section I, must be sized for 90% of rated capacity and at 3% overpressure. These ratings are for saturated steam and most manufacturers of safety valves will give tables in pounds per hour at set pressures plus 3% overpressure.

Section VIII safety valves should be selected with a greater differential between the normal operating system pressure and the safety valve set pressure. There are three reasons for this:

- Greater operating pressure fluctuations.
- Larger spectrum of fluids being handled.
- Foreign contamination that could allow the valve to stick partially open and leak.

Refrigeration compressors supply pulsating discharge gases to the condenser and any safety valves at the receiver could experience pressure fluctuations. Therefore, it is recommended that the relief valve settings be 25% above the system operating pressure.

The type of relief device depends on the size of the pressure vessel. For vessels of 10 ft^3 or more, the ANSI/ASHRAE 15 code requires a three-way valve and two parallel relief valves, with each valve capable of complete system protection.

The discharge capacity formula for safety relief valves, based on ANSI/ASHRAE 15, is:

$$C = fDL$$

where C = *the minimum discharge capacity of the relief valve in pounds of air per minute*
D = *Vessel outside diameter in feet*
L = *Vessel length in feet*
f = *Refrigerant factor, as given below:*

refrigerant 717 (ammonia)	*0.5*
refrigerants 12, 22, 500	*1.6*
refrigerants 502, 13 and 14	*2.5*
all other refrigerants	*1.0*

The sizing calculations do not assume any downstream piping from the relief device. Back pressure will develop while the relief valve is discharging and will change the differential setting of the valve. The ANSI/ASHRAE 15 code permits a maximum back pressure, through any downstream piping, of 25% of the inlet pressure. Table 4-5 shows the maximum length of downstream discharge piping allowed for relief valves in refrigerant service.

Relief devices should be set to provide an adequate differential between the relief device and system operating pressures. R-22, for example, operates at pressures substantially higher than R-12 or R-502 and receiver pressure limits will also be higher.

A *pop* safety valve with top outlet, Figure 4-14, is designed for gas service. The disc has two areas for the gas or steam to work against. If the valve is set to open at 100 lb, the lower disc, which is beneath the seat, will start to rise at 100 lb. When the lower disc clears the seat, the internal pressure will act against a larger, second disc, above the seat, that increases the force on the spring and pops the valve wide open. If the pressure falls below 100 lb, the

Table 4-5. Maximum feet of discharge piping for 300 psig pressure setting.

RELIEF DEVICE CAPACITY LBS. AIR/MIN.	SOFT COPPER TUBE O.D.			SCHEDULE 40 PIPE					
	3/8"	1/2"	5/8"	1/2"	3/4"	1"	1-1/4"	1-1/2"	2"
2	49	262							
4	12	65	221						
6	5½	29	98	173					
8	3	16	55	97					
10	2	11	35	62	254				
12	1½	7	25	43	176				
14	1	5½	18	32	130				
16	1	4	14	24	99				
18		3	11	19	78				
20		2½	9	16	63	212			
25		1½	5½	10	41	136			
30		1	4	7	28	94			
35		1	3	5	21	69			
40			2	4	16	53	209		
45			1½	3	13	42	165		
50			1½	2½	10	34	133		
60			1	1½	7	24	93	200	
70				1½	5	17	68	147	
80				1	4	13	52	113	
90				1	3	10	41	89	
100					2½	8	33	72	252
125					1½	5½	21	46	161
150					1	4	15	32	112

Courtesy Henry Valve Co.

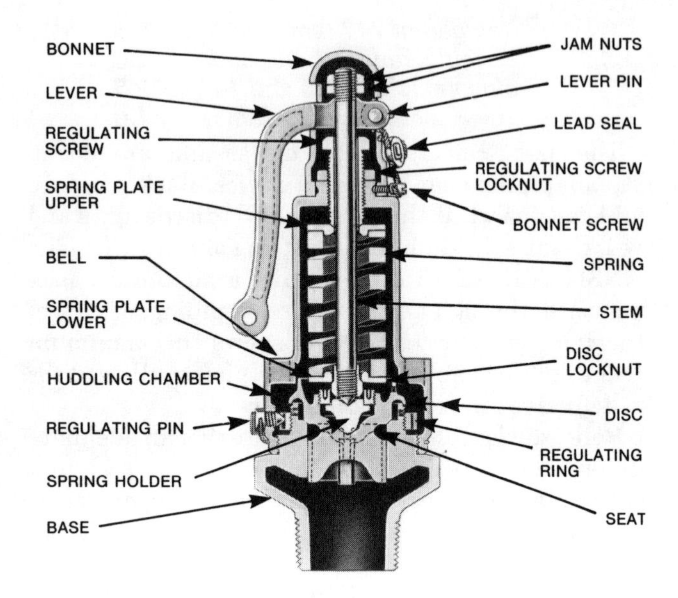

FIG. 4-14. A pop safety valve used with gases, steam and air. Not recommended for use with liquids. *Courtesy The Lunkenheimer Co.*

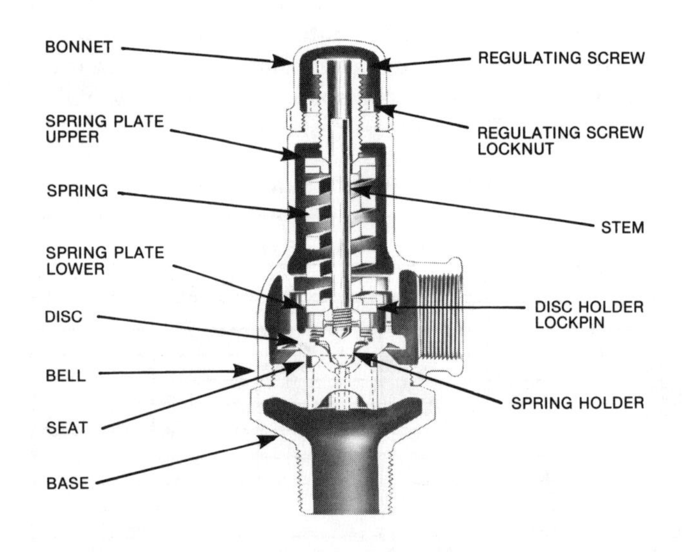

FIG. 4-15. A relief valve designed for use with liquids. *Courtesy The Lunkenheimer Co.*

valve will reseat.

Figure 4-15 shows a relief valve designed for liquid service. It differs from the pop valve in that it has only one disc. The valve will partially open at the preset pressure and will have maximum discharge when the pressure reaches 120% of the valve's rating. When the pressure is reduced, the disc will ease back into its seat. There is never any pop action.

Material selection

Refrigeration systems The commonest valve materials found in the air conditioning, heating and refrigeration industry are brass, bronze, copper, ductile iron and some plastic.

Forged brass is usually reserved for sizes under one inch. Manufactured in high volume, forged shutoff and charging valves have a heat-stabilized nylon disc for positive shutoff and are usually a packless design. Corrosion resistance is high and, if the system is kept free of moisture, these valves will last the life of the system. Solder and flare ends are the most common connections. The working range is from −40 °F to 200 °F, with 275 °F as the upper limit for short periods of time. 500 psi is the maximum working pressure in the direction of the flow (arrow) and 350 psi with reverse flow.

Cast bronze is usually reserved for valves over one inch and up to 4 inches in size, with soldered ends. Heat stabilized nylon discs are used for temperatures up to 275 °F, although lead alloy or Teflon® discs, for hot gas applications up to 400 °F, are available on special order. Maximum working pressure is 425 psig in the direction of flow (arrow) and 300 psi with reverse flow.

Ductile iron valves, for ammonia service, have thinner walls, a greater shock resistance and higher working pressures than cast iron valves. Ductile iron valves are smaller and lighter than their cast iron predecessors. 400 psi is the upper working limit of the ductile iron valve, with a safety factor of 10 to 1. Discs are of a high-melting-point lead alloy. Valve sizes range from ¼″ to 6″ and have either female pipe threads, in the smaller sizes, or flanged, bolted unions, with nuts and gaskets, in the larger sizes.

Regulators used with either ammonia or halogen refrigerants have cast iron or ductile iron bodies and all internal parts are either steel or stainless steel. A V-port disc permits operation at reduced capacity without chatter.

Usually, regulators rated from 5 to 3000 tons have ductile iron valve bodies and may be used with any of the halogen refrigerants or ammonia. Most of these valves have flanged ends and are rated to 400 psi. Useful temperature ranges are from −40 °F to 60 °F for low side service and up to 250 °F for high side application.

When used with halogen refrigerants, such as R-12,- 22 and -502, large cast iron valves have copper ends sweated into the body for convenient connection to the copper piping system.

Regulators for applications under 5 tons usually

Table 4-6. Nickel bronze trim material for bronze valves used in moderately corrosive service.

CHEMICAL COMPOSITION (%)			PHYSICAL PROPERTIES	
	Minimum	Maximum		Minimum
Copper	63.00	67.00		
Tin	3.50	4.50	Tensile strength	40,000 PSI
Lead	3.00	5.00		
Manganese		1.00	Yield point	17,000 PSI
Nickel	19.00	21.50		
Iron		1.50	Elongation in 2 in.	10%
Zinc		Remainder		

Courtesy Stockham Valve & Fitting

Table 4-7. Stainless steel trim material for bronze valves used in corrosive service.

CHEMICAL COMPOSITION (%)			AVERAGE PHYSICAL PROPERTIES AT 500 BRINELL	
	Minimum	Maximum		
Carbon	0.15		Tensile strength	230,000 PSI
Manganese		1.00		
Silicon		1.00	Yield point	195,000 PSI
Phosphorus		0.04	Elongation in 2 in.	8%
Sulfur		0.03		
Chromium	12.00	14.00	Reduction of area	25%

Courtesy Stockham Valve & Fitting

have copper and bronze or copper and brass bodies. Stainless steel stems and brass discs are housed in a bronze bellows that is hermetically sealed to the adjusting stem housing.

Hot and chilled water systems Valves are used to control hydronic heating-cooling system flows and steam, in conjunction with boilers, centrifugal or absorption chillers, solar systems, brine circulating refrigeration systems and for water-cooled condensers. Valves used in these applications can be subjected to the severe corrosive effects discussed earlier.

Bronze, brass and iron are the common materials. Bronze and brass can be subject to severe corrosion, mostly due to dezincification. On the other hand, iron, with a tensile strength of 31,000 psi, and ductile iron, with a tensile strength of 60,000 psi, are excellent body materials and exhibit corrosion resistance as well as good physical strength, with ductile iron being superior.

Trim materials Valve trim, consisting of disc, seat and stem, can be the same material as the body, but in most cases are constructed of more noble metals. Bronze, iron and ductile iron valve trim is usually constructed of machinable materials that yield a super-fine finish, especially on stems. Cupronickel alloys, containing some zinc, are used where mild corrosion is expected, but in highly corrosive conditions, stainless steels are used. See Tables 4-6 and 4-7 for bronze and stainless steel specifications.

Many valves are manufactured with replaceable seats. The seat is in intimate contact with the body, but being of more noble material, will have a good useful life with little corrosion. Seats and discs of stainless steel are made from a highly hardenable alloy that is resistant to corrosion, erosion and wire drawing. Some stainless alloys may have a Brinell hardness as high as 500. For valve stems, the material must not only be capable of a high finish, but must exhibit good resistance to shear and torsional forces.

Two data sheets from the Stockham design catalogue, Figures 4-16 and 4-17, show the *bill of materials* for a globe and a gate valve. Note that the bronze valve in Figure 4-16 has stainless steel trim with a silicon-bronze stem. The pressure-temperature ratings are shown in Table 4-8. One should become skilled in reading and evaluating manufacturer's specification sheets since almost all the construction data will be shown in charts, tables and outline drawings.

Bronze valves are usually limited in size to ¼″ to 3″ and to pressures of 350 pounds steam. Iron valves are generally larger than 2″, with pressures of 125 and 250 pounds steam. Forged steel, available in sizes from ¼″ through 2″, is extremely strong and is reserved for high pressure and temperature applications. Cast steel, like forged steel, is rated for high pressure and temperature, but is generally found in

Table 4-8. Pressure-temperature ratings of bronze valves shown in Figures 4-16 and 4-17.

	PRESSURE – PSI						
Press. Class	125	150		200	300		2000 HYDL
End Conn.	THD	THD	FLG	THD	THD	FLG	THD
Temp.[1]	MATERIAL						
(Deg. F)	ASTM B-62			ASTM B-61			
−20 To 150	200	300	225	400	1000	500	2000
200	185	270	210	375	920	475	–
250	170	240	195	350	830	450	–
300	155	210	180	325	740	425	–
350	140	180	165	300	650	400	–
400	–	–	–	275	560	375	–
406	125	150	150	–	–	–	–
450	120	145	–	250	480	350	–
500	–	–	–	225	390	325	–
550	–	–	–	200	300	300	–

Stockham Valve Co.

NO.	DESCRIPTION	MATERIAL	ASTM SPEC.
1	BODY	BRONZE	B-62
2	BONNET	BRONZE	B-62
3	BONNET RING	BRONZE	B-62
4	DISC	STAINLESS STEEL 500 BHN	A-276 Type 420
5	DISC NUT	BRASS (¼-½)	B-16
		STAINLESS STEEL (¾-2)	A-276 Type 416
6	HANDWHEEL	MALL. IRON	A-197
7	HANDWHEEL NUT	STEEL-ZINC PLATED	
9	PACKING	TEFLON® IMPREGNATED ASBESTOS	
10	PKG. GLAND	BRASS	
11	PKG. NUT	BRONZE	B-62
12	SEAT RING	STAINLESS STEEL 500 BHN	A-276 Type 420
13	STEM	COPPER-SILICON ALLOY	B-371 Alloy 694

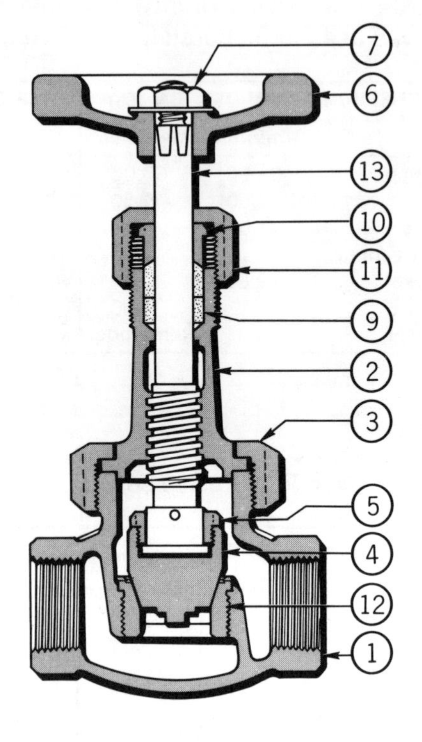

FIG. 4-16. Bill of materials for a globe valve. *Courtesy Stockham Valve Co.*

NO.	DESCRIPTION	MATERIAL	ASTM SPEC.
1	BODY	BRONZE	B-62
2	BONNET	BRONZE	B-62
3	BONNET RING B-105	BRONZE	B-62
4	DISC	BRONZE	B-62
5	HANDWHEEL	MALL. IRON	A-197
6	HANDWHEEL NUT	STEEL-ZINC PLATED	
8	PACKING	TEFLON® IMPREGNATED ASBESTOS	
9	PKG. GLAND	BRASS	
10	PKG. NUT	BRONZE	B-62
11	STEM	COPPER-SILICON ALLOY	B-371 Alloy 694

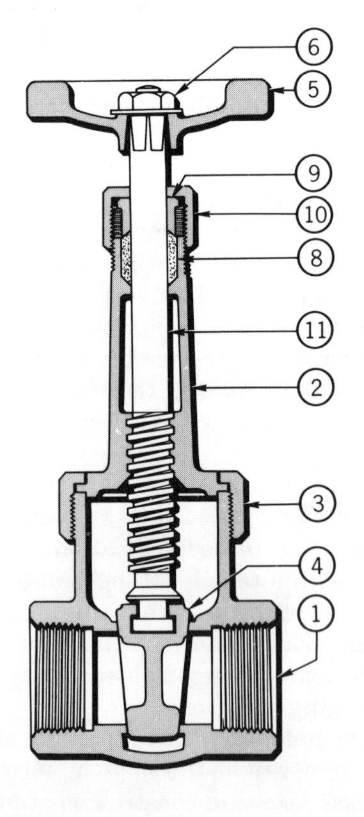

FIG. 4-17. Bill of materials for a gate valve. *Courtesy Stockham Valve Co.*

Table 4-9. Properties of elastomers.

COMMON NAME	ASTM D 1418 DESIGNATION	DESCRIPTION, USAGE
Natural rubber	NR	Excellent physical properties. Resists cutting and abrasion. Not recommended for petroleum use.
Neoprene	CR	Excellent ozone, heat and weathering resistence. Recommended for use with refrigerants, petroleum oils, mild acids. Temperature range is −65°F to 330°F.
Buna N (Nitrile)	NBR	Excellent oil resistance. Temperature range is −65°F to 275°F. Good resistance to compression set, tear and abrasion. Poor in sunlight, weather and ozone.
Butyl	IIR	Excellent resistance to gas permeation, good for vacuum service. Temperature range is −65°F to 225°F. Poor with petroleum oils.
Silicone	Si	Excellent resistance to temperature, range is −175°F to 450°F. Poor tear and abrasion resistance. Good in dry heat, poor in water and steam service.
Ethylene propylene	EPDM	Excellent in steam, water and silicone oil service, poor with petroleum oils. Temperature range is −65°F to 300°F.
Flourocarbon rubber (Viton)	FKM	Wide spectrum chemical compatibility. Low compression set. Temperature range is −15°F to 400°F.

2″ sizes and larger. Special alloys, such as stainless steel, can be obtained in just about all sizes and are good in corrosive applications at high pressures and temperatures. Nonmetallic materials, such as PVC and Teflon®, are available in all sizes and are good in corrosive service, but are limited in their pressure and temperature applications.

Elastomers Butterfly valves utilize elastomers for tight closure and the type of elastomer will define the temperature range of the valve. Subfreezing temperatures may cause the elastomer to harden, excessive temperatures may cause deterioration.

The usage range of elastomeric compounds is being extended by a number of methods, including fillers, reinforcing materials, aging inhibitors and curing agents. The most common materials are bunas, butyl, viton, neoprene, hypalon, silicone, natural rubber and ethylene propylene. Table 4-9 shows the general properties of some of these elastomers.

Elastomers are *alive* and have a limited usage period. Certain conditions can shorten their life span considerably. This aging process can be attributed to a number of conditions, namely, temperature and fluids that cause chemical reactions in the elastomers. The major effects of aging are brittleness, hardening, and compression set. Fluids may cause some elastomers to swell, others to shrink.

Plastic materials Of the nonmetallic materials, polyvinyl chloride (PVC), Teflon® (TFE) and nylon are the most common. Corrosive fluids, that normally would attack metallic valves, have little or no effect on their plastic counterparts as long as the pressure-temperature limits are observed. Entire valves may be made of these materials, while some metallic valves are lined with them.

One of the most commonly used plastics is a fluoroplastic, Teflon® or TFE. It is used in valve seals, liners and gaskets. The outstanding properties of TFE are its chemical inertness, high and low temperature stability, excellent electrical properties and low coefficient of friction. The major disadvantage, however, is poor wear and creep resistance. Fillers and reinforcing fibers will increase the wear resistance substantially. In one case, a butterfly valve, reinforced TFE extended the temperature limits of the valve as much as 50 °F, over the entire range of pressure ratings.

Properties of Elastomers Elastomers meet the ASTM D 1418 designation. Therefore, any specifications for valves should include the ASTM material designation. The valve manufacturers will have this information available.

Figure 4-18 shows one manufacturer's approach to solving corrosive problems, using a graphite base, non-metallic ball in lieu of exotic metals. Even though the corrosive fluids shown may not be readily seen in the air conditioning and refrigeration industry, Figure 4-18 shows that solutions to many

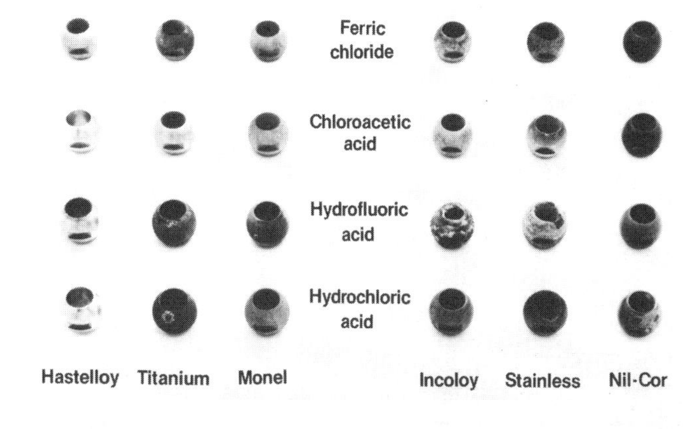

FIG. 4-18. Comparison of the corrosion resistance of nonmetallic Nil-Cor balls and metallic balls. *Courtesy Dresser Industries Inc., Advanced Valve & Composite Operation.*

stubborn corrosion problems can be overcome with modern technology.

Because the cost of corrosion resistant and exotic materials continues to increase, many valve manufacturers are turning to plastic-lined, metal valves that combine less expensive base metals, for physical strength, with the corrosion resistence of nonmetallic liners. Plastic liners and all-plastic valves must be compatible with the corrosive fluids and within the temperature ranges of the materials.

Packings Most packings are precut circular rings compressed into a stuffing box by packing glands. The traditional material has been asbestos but, because environmental regulations are becoming more strict, its use is declining. Table 4-10 shows a list of packing materials, their temperature and pH ranges.

Tables 4-11 and 4-12, from the Crane Packing Company, classify packings as a function of pressure, temperature, and pH.

Asbestos is still the best answer to a wide spectrum of packing problems, however. Some substitutes have certain good characteristics, but are poor in others. Whatever the substitute for asbestos, its cost is greater; from 2 to 20 times greater. The most usable at this time, for heating/cooling applications, are:

TFE Excellent chemical pH range and can have a braided square or rectangular cross section.

Graphite As excellent a chemical pH range as Teflon®, but superior for extremely high temperatures.

Asbestos/TFE A combination of asbestos and Teflon® is now employed in many valves. Temperature limits, due to the Teflon, are lower than those of pure asbestos.

Table 4-10. Temperature limits and pH ranges of packing materials.

Fiber	Maximum temperature, °F	Fluid resistance, pH
Asbestos a	1,200	2—12
Teflon	500	0—14
Graphite b	1,200	0—14
Carbon b	1,200	0—14
Aramid a	500	3—11
Glass a	500	2—12
Teflon-graphite b	500	0—14
Glass-polyamide a	500	2—12

a With TFE impregnation
b Except strong oxidizers in the 0 to 2 pH range

Machine Design, March 11, 1982

Table 4-11. Packing selection based on temperature and pressure. *Courtesy Crane Packing Co.*

PSIG \ °F	100 TO 300	300 TO 450	450 TO 600	600 TO 750	750 TO 850	850 TO 950	950 TO 1200
UP TO 300	A THRU M	A THRU M	A THRU M EXCEPT G	C, D, F, H, I, J, K, L	C, D, F, H, I, K	C, F, H, I, K	C, F, I, K
300 TO 600	A THRU M	A THRU M	A THRU M EXCEPT G	C, D, F, H, I, J, K, L	C, D, F, H, I, K	C, F, H, I, K	C, F, I, K
600 TO 900	A THRU L	A THRU L	A THRU L EXCEPT G	C, D, F, H, I, J, K, L	C, F, H, I, K	C, F, H, I, K	C, F, I, K.
900 TO 1500	A THRU L	A THRU L	A THRU L EXCEPT G	H, I, J	C, F, H, I, K	C, F, H, I, K	C, F, I, K
1500 TO 2000		A, B, F, G, H, I, L	A, B, F, G, H, I, L	F, H, I, L	F, H, I, L	F, H, I, L	F, H, L

GENERAL RECOMMENDATIONS ONLY FOR SPECIFIC APPLICATIONS CONSULT MANUFACTURER

NON ASBESTOS
A. KEVLAR *
B. TFE. FILAMENT
C. GRAPHITE YARN
D. CARBON YARN
E. SOLID TFE.
F. GRAFOIL **

METALLICS
G. BABBITT
H. COPPER
I. ALUMINUM

ASBESTOS
J. WHITE ASBESTOS (VARIOUS COATING)
K. WIRE INSERTED ASBESTOS (LOOSE CORE / NO CORE)
L. WIRE INSERTED ASBESTOS (BINDER CORE)
M. PLASTIC (ASBESTOS, BINDER, GRAPHITE)

* E.I. DU PONT (T.M.)
** UNION CARBIDE (T.M.)

Table 4-12. Packing selection based on pH factors. *Courtesy Crane Packing Co.*

pH range	Packings
pH 6 to 8 – NEUTRAL	Grafoil, Graphite filament, Carbon filament, TFE filament, Solid TFE, Kevlar, White ASB TFE treated, Blue ASB, White ASB graphite treated, Copper, Aluminum, Babbitt, Flax, Cotton, Jute, Plastics.
pH 2 to 6	Grafoil, Graphite filament, Carbon filament, TFE filament, Solid TFE, Kevlar, White ASB, TFE treated, Blue asbestos
pH 10 to 12	Grafoil, Graphite filament, Carbon filament, TFE filament, Solid TFE, Kevlar, White ASB, TFE treated, Blue asbestos
pH 0 – SEVERE ACID pH 14 – SEVERE CAUSTIC	Grafoil, Graphite filament, TFE filament, Solid TFE

To assist in selecting the proper valve materials, the Chemical Resistance Tables are provided in the Appendix. These tables yield a range of material adaptability to various fluids used in industry. The tables are only a guide and extreme caution should be taken in their interpretation.

The materials are listed in four categories:

Excellent-Unrestricted This category covers a range of pressures, temperatures and concentrations where valve materials are not affected and will stand up under extended use.

Excellent Excellent is a rating given to materials, with some restrictions on temperature, pressure, fluid concentrations, fluid velocities and the passivity of the fluid film. As long as a material is used within the specific range limit, it will give a long service life.

Good Reduced valve life can be expected and

periodic replacement may be necessary .

Limited to Unsatisfactory These ratings indicate the unsuitability of certain materials or require approval by the engineering department of the valve manufacturer prior to selection. Expected valve life will be poor with these materials.

The tables are divided into three sections:

- Metals
- Plastics
- Nonmetallic elastomeric materials

Fluid flow

Fluids flow in two different manners: with laminar flow, the fluid flows in a smooth streamline fashion; with turbulent flow, the fluid flows in a random, chaotic fashion within the pipe.

Laminar flow is seldom encountered in the heating-cooling industry except when the flow in solar panels is very low. In laminar flow, the fluid may be thought of as flowing in smooth concentric parallel layers with no intermixing of the individual layers. High viscosity fluids and fluids at very low velocity tend to exhibit the laminar flow condition, Figure 4-19.

Laminar flows are usually found in long, straight, smooth piping systems at low flows where few obstructions exist to upset the streamline conditions. Wide open gate, ball, plug and even butterfly valves offer little resistance, because of their straight-through flow pattern, and laminar flow can be achieved.

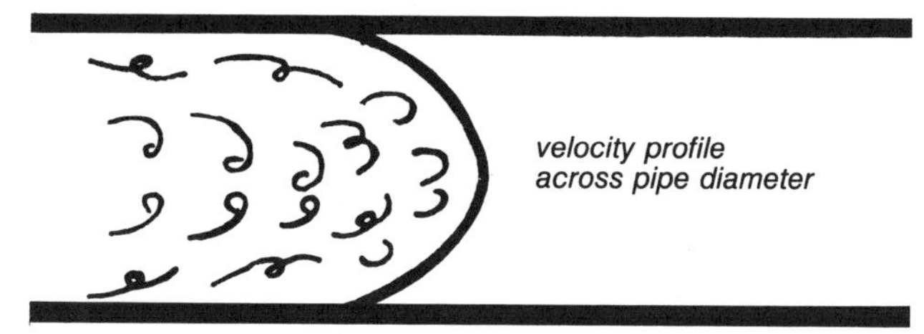

Turbulent flow mixes fluid across the pipe diameter. Fluid layer at the inside wall is broken up.

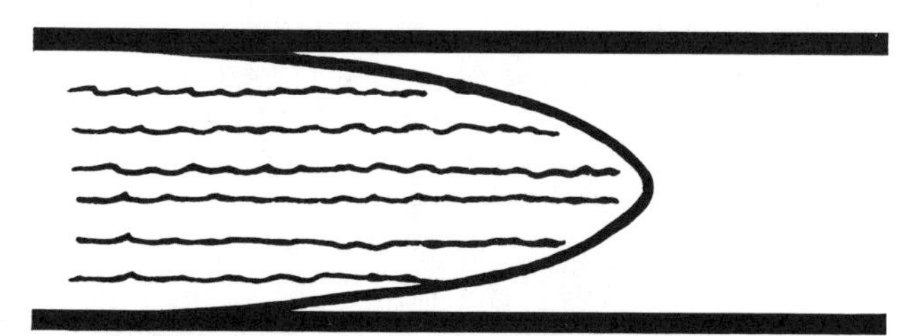

Laminar flow does not mix, but flows in individual parallel layers.

Fig. 4-19. Visualization of turbulent and laminar flow within a pipe.

With laminar flow, if the flow rate is increased a little at a time and the fluid is allowed to stabilize after each increase, at some point there will be a critical flow condition where the fluid will begin to fluctuate. The smooth flowing, concentric layers will begin to break down and become random, cross currents and eddies will appear, and turbulent flow will result.

Turbulent flow is the most common condition in the heating-cooling industry and is a function of four basic parameters: It is directly proportional to density, fluid velocity and pipe diameter. The larger the value of each of these parameters, the more chance that turbulent flow will result. The fourth parameter is the fluid's viscosity and the lower the viscosity, the greater the chance of turbulent flow.

Most heating/cooling applications will be in the turbulent flow area, for these reasons:

- Economics. If the pressure drop can be tolerated, smaller pipes can be used. The smaller the pipe, the lower the initial cost.
- Maximum heat transfer. Turbulent flow is essential for maximum heat transfer which reduces the size and intial cost of heat transfer equipment. Some solar applications may run contrary to this philosophy.
- Most halogen refrigerants have low viscosities. Consequently, turbulent flow is a normal condition.
- Refrigerants will condense and evaporate during their process. This condensation or boiling action causes turbulence.
- Many control or shutoff valves are globe valves that induce turbulence due to the abrupt directional flow changes and obstruction by the disc and stem.

Pressure drop

Pressure drop is crucial in hot and chill water systems if pumping horsepower is to be kept to a minimum. However, control valves require a certain percentage of the total system pressure drop for proper control. Insufficient pressure drop across a control valve will result in erratic operation, too much drop will result in excessive pump horsepower. Field experience has shown that control valves are usually oversized, rarely undersized.

The capacity of refrigerant systems is directly affected by excessive pressure drops, especially in suction lines. The lower the suction temperature, the more critical the pressure drop. Wherever possible, control devices should be kept on the high side of the refrigeration system.

Pressure drop through valves is directly related to the laminar-turbulent flow condition as well as the density, viscosity and velocity of the fluid. For practical purposes, assume that all flow is turbulent. The pressure drop though a valve, due to turbulent flow, is, in most cases, directly proportional to the square of the velocity. If the flow is doubled, the pressure drop will increase by four, if all other parameters remain constant. Laminar flow may be encountered in solar systems and care must be taken in calculating the pressure drops in these systems.

Knowing the pressure reduction and fluid velocity through a valve or regulator is important. Each valve style will have a different flow profile and those with the greatest resistance (globe) will have the greatest pressure drop.

Figure 4-20 shows what happens to a liquid stream as it passes through a valve. As the liquid flows from point 1 to point 4, both the velocity and pressure change values. Fluid velocity and pressure are stable at point 1, as the fluid in the pipe approaches the valve. When the restriction in the valve port area, point 2, is encountered, the pressure drops off substantially and the velocity increases rapidly. At point 3, slightly downstream of the port, the velocity reaches a maximum value and the pressure is at a miniumum. The unique point is called the *vena contracta*. Beyond the vena contracta, point 4, the fluid velocity returns to that of point 1, but the pressure does not. The differential pressure ΔP is the pressure drop or loss across the valve, $P_1 - P_4 = \Delta P$ or inlet pressure minus outlet pressure.

What must be remembered is that, at the vena contracta, the pressure of the fluid could be below its vapor pressure and the liquid could *boil* and form vapor cavities or bubbles even at low temperatures. Water, for example, will boil at 100 °F if the pressure is reduced to .95 psia. Cavitation is the result. This phenomenon is not too critical at point 3, but can become disastrous at point 4, as the vapor bubbles are carried downstream into the higher pressure zone. At point 4, fluid velocity is decreasing, pressure is rising, and the vapor bubbles begin to collapse or implode. The collapsing bubbles can cause serious cavitation damage, to the point where valve and piping surfaces will be violently eroded away. Local stress, due to cavitation, can be in the neighborhood of thousands of psi. Noise and vibration can also accompany cavitation.

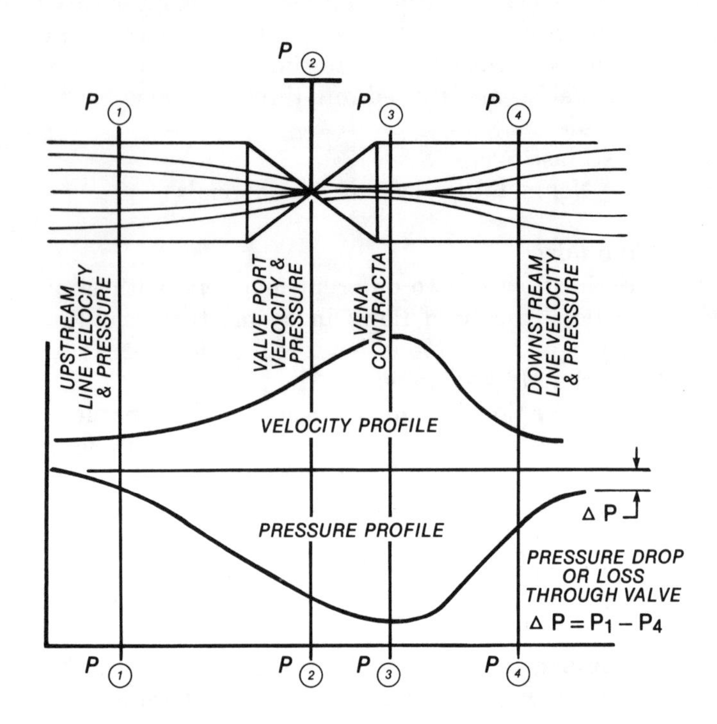

FIG. 4-20. Velocity and pressure profile through a valve. *Courtesy Jordan Valve.*

Cavitation can be eliminated by opening the valve wider, to reduce pressure drop, or by reducing the size of the control valve so that the disc will open farther and not operate so near its seat.

If cavitation cannot be avoided, hardened disc and seat materials will resist the damaging effects.

Pressure drops in water systems Pressure drops in chilled and hot water systems can be easily determined using charts and graphs tabulated by a number of manufacturers. The most widely used is Technical Paper No. 410, published by the Crane Co. The basic approach is to convert pressure drops of the various valves into the *equivalent length* of pipe.

The pressure drop in hot/chilled water systems is determined by using the tables shown in Figures 4-21, 4-22 and 4-23. The resistance of each valve and fitting in the system is given as *Equivalent Length in Pipe Diameters* in Figure 4-21 and converted to *Equivalent Length in Feet of Pipe* using Figure 4-22. The total eqivalent feet of pipe, for all the valves and fittings, are then added to the actual feet of straight

<table>
<tr><th colspan="4">Description of Product</th><th>Equivalent Length In Pipe Diameters (L/D)</th></tr>
<tr><td rowspan="2">Globe Valves</td><td>Conventional</td><td>With no obstruction in flat, bevel, or plug type seat
With wing or pin guided disc</td><td>Fully open
Fully open</td><td>340
450</td></tr>
<tr><td>Y-Pattern</td><td>(No obstruction in flat, bevel, or plug type seat)
– With stem 60 degrees from run of pipe line
– With stem 45 degrees from run of pipe line</td><td>
Fully open
Fully open</td><td>
175
145</td></tr>
<tr><td>Angle Valves</td><td>Conventional</td><td>With no obstruction in flat, bevel, or plug type seat
With wing or pin guided disc</td><td>Fully open
Fully open</td><td>145
200</td></tr>
<tr><td rowspan="3">Gate Valves</td><td>Conventional Wedge Disc, Double Disc, or Plug Disc</td><td></td><td>Fully open
Three-quarters open
One-half open
One-quarter open</td><td>13
35
160
900</td></tr>
<tr><td>Pulp Stock</td><td></td><td>Fully open
Three-quarters open
One-half open
One-quarter open</td><td>17
50
260
1200</td></tr>
<tr><td colspan="2">Conduit Pipe Line</td><td>Fully open</td><td>3**</td></tr>
<tr><td>Check Valves</td><td colspan="2">Conventional Swing
Clearway Swing
Globe Lift or Stop
Angle Lift or Stop
In-Line Ball</td><td>0.5†. . . Fully open
0.5†. . . Fully open
2.0†. . . Fully open
2.0†. . . Fully open
2.5 vertical and 0.25 horizontal†. . . Fully open</td><td>135
50
Same as Globe
Same as Angle
150</td></tr>
<tr><td colspan="2">Foot Valves with Strainer</td><td>With poppet lift-type disc
With leather-hinged disc</td><td>0.3†. . . Fully open
0.4†. . . Fully open</td><td>420
75</td></tr>
<tr><td colspan="3">Butterfly Valves (6-inch and larger)</td><td>Fully open</td><td>20</td></tr>
<tr><td rowspan="2">Cocks</td><td>Straight-Through</td><td>Rectangular plug port area equal to 100% of pipe area</td><td>Fully open</td><td>18</td></tr>
<tr><td>Three-Way</td><td>Rectangular plug port area equal to 80% of pipe area (fully open)</td><td>Flow straight through
Flow through branch</td><td>44
140</td></tr>
<tr><td rowspan="4">Fittings</td><td colspan="3">90 Degree Standard Elbow
45 Degree Standard Elbow
90 Degree Long Radius Elbow</td><td>30
16
20</td></tr>
<tr><td colspan="3">90 Degree Street Elbow
45 Degree Street Elbow
Square Corner Elbow</td><td>50
26
57</td></tr>
<tr><td>Standard Tee</td><td colspan="2">With flow through run
With flow through branch</td><td>20
60</td></tr>
<tr><td colspan="3">Close Pattern Return Bend</td><td>50</td></tr>
</table>

**Exact equivalent length is equal to the length between flange faces or welding ends.

†Minimum calculated pressure drop (psi) across valve to provide sufficient flow to lift disc fully.

FIG. 4-21. Equivalent length, in pipe diameters, of various valves. *Courtesy The Crane Co., Technical Paper 410, 1980*

pipe in the system and the sum is the total adjusted pipe length.

Using Figure 4-22 and the design flow rate in gallons per minute (gpm), the pressure drop is found per 100 feet of pipe.

Figure 2-1 compared the flow resistances (pressure drop) of gate, butterfly and globe valves. The pressure drop was lowest for the straight, flow-through types and largest for the globe styles.

For example, if a wide open, 2″ globe valve is used in a system pumping 70 gallons of water per minute, the pressure drop is found as follows: From Figure 4-21, the equivalent length in pipe diameters, L/D, for a fully open globe valve is 340. The 340 L/D value is found, in Figure 4-22, at point 1. Since the valve size is given as 2″, its diameter is found at point 2. A line run between points 1 and 2 crosses the L line at 60, point 3, which is the *Equivalent Length in Feet of Pipe.*

The pressure drop through the valve, not includ-

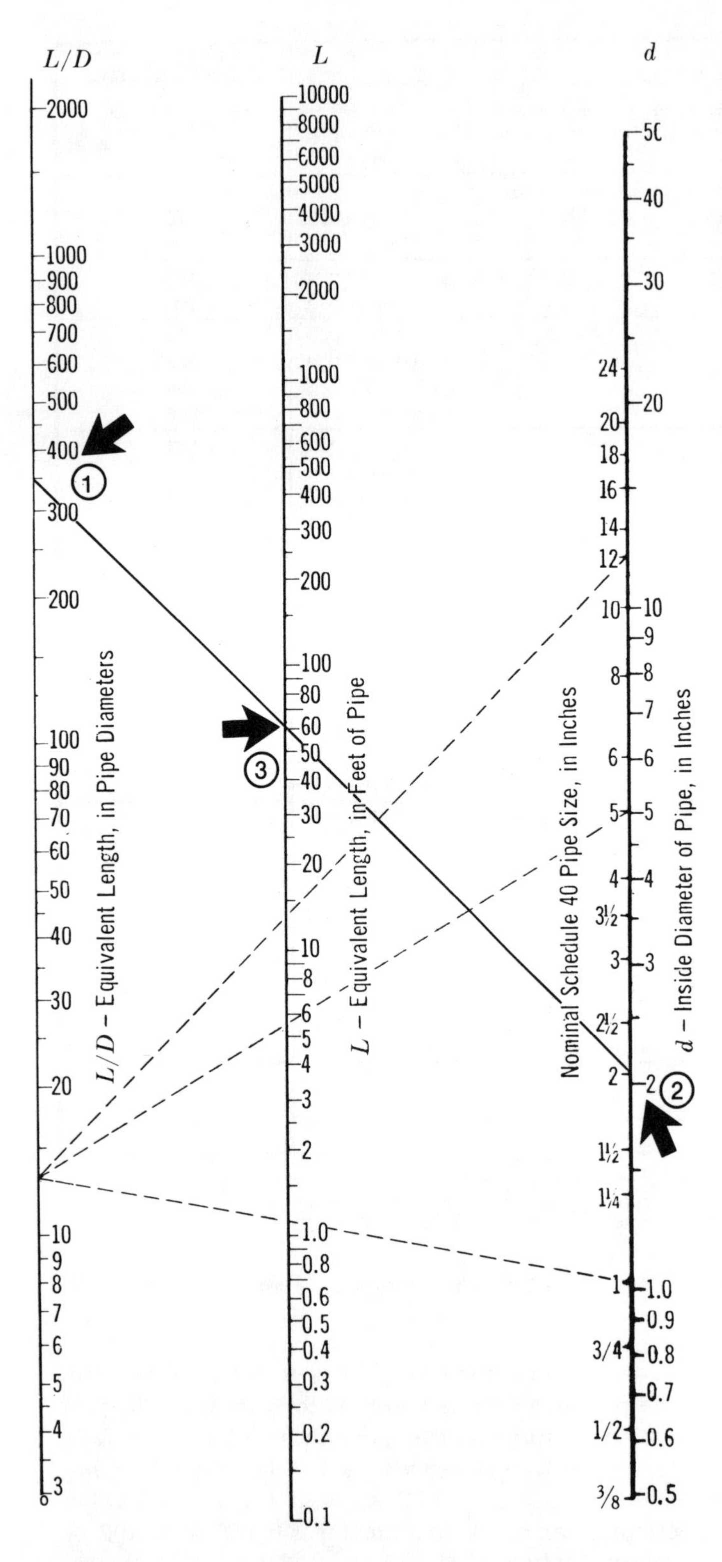

FIG. 4-22. Nomograph used to determine equivalent lengths of various valves and pipe fittings. *Courtesy The Crane Co., Technical Paper 410, 1980*

ing adjacent piping or fittings, is found in Figure 4-23. The flow rate, 70 gpm, is found at point 1. Reading horizontally to point 2, for a 2″ size, the velocity is found to be 6.7 feet per second and the pressure drop is 3.84 psi per 100 feet of pipe. With 60 feet of equivalent length, the pressure drop is 60/100 × 3.84 or 2.3 psi. Since there are 2.31 feet of *head* per psi, the pressure drop can also be given as 5.3 feet of head. Many pump curves are given in feet of water head rather than in psi.

In an overall system, the other components, including tees and elbows, are converted to equivalent feet and added together. This sum is then added to the length of pipe used in the system and a final result is obtained in equivalent length of pipe. Using Figure 4-23, the flow rate in gpm, and the pipe diameter, the pressure drop is found for each 100 feet of pipe. The final pressure drop, in psi, is obtained by dividing the total equivalent length by 100 feet and multiplying by the pressure drop.

These tables are for water only and cannot be used with any accuracy if the fluid in the system varies substantially from the density and viscosity of water. Tables for the specific fluid would have to be calculated or a correction factor applied.

In most cases, the pressure drop in a hot/chilled water system is minimized to reduce the horsepower needed to pump the required flow. This does not always hold true in systems that require control valves for fluid throttling to, or around, heating-cooling terminals. Here, at least 50% of the total system pressure drop must be through the valve for proper, smooth control.

Too low a pressure drop results in erratic operation and poor temperature control. If the pressure drop is minimized, the terminal and associated piping will take the majority of the system pressure drop and the control valve will not regulate, in the necessary linear fashion, and fine control will be lost. Also, the control valve, in trying to do its job, may operate at or near its close-off point, with the disc near its seat. As a result, the seat will *wire-draw* due to the extremely high fluid velocity near the closeoff point.

The worst control problem is with hot water terminals. The objective is to obtain as nearly a linear output as is possible, Figure 4-24.

With low pressure steam, below 10 psig, pressure drop across the control valve should be 80% or higher to achieve linear control.

Figure 4-25 shows the output of a hot water coil at a 20 °F, 40 °F and 60 °F temperature drop (TD)

Discharge		Pressure Drop per 100 feet and Velocity in Schedule 40 Pipe for Water at 60 F.															
Gallons per Minute	Cubic Ft. per Second	Velocity Feet per Second	Press. Drop Lbs. per Sq. In.	Velocity Feet per Second	Press. Drop Lbs. per Sq. In.	Velocity Feet per Second	Press. Drop Lbs. per Sq. In.	Velocity Feet per Second	Press. Drop Lbs. per Sq. In.	Velocity Feet per Second	Press. Drop Lbs. per Sq. In.	Velocity Feet per Second	Press. Drop Lbs. per Sq. In.	Velocity Feet per Second	Press. Drop Lbs. per Sq. In.	Velocity Feet per Second	Press. Drop Lbs. per Sq. In.
		⅛″		¼″		⅜″		½″									
.2	0.000446	1.13	1.86	0.616	0.359												
.3	0.000668	1.69	4.22	0.924	0.903	0.504	0.159	0.317	0.061	¾″							
.4	0.000891	2.26	6.98	1.23	1.61	0.672	0.345	0.422	0.086								
.5	0.00111	2.82	10.5	1.54	2.39	0.840	0.539	0.528	0.167	0.301	0.033						
.6	0.00134	3.39	14.7	1.85	3.29	1.01	0.751	0.633	0.240	0.361	0.041						
.8	0.00178	4.52	25.0	2.46	5.44	1.34	1.25	0.844	0.408	0.481	0.102	1″		1¼″			
1	0.00223	5.65	37.2	3.08	8.28	1.68	1.85	1.06	0.600	0.602	0.155	0.371	0.048			1½″	
2	0.00446	11.29	134.4	6.16	30.1	3.36	6.58	2.11	2.10	1.20	0.526	0.743	0.164	0.429	0.044		
3	0.00668			9.25	64.1	5.04	13.9	3.17	4.33	1.81	1.09	1.114	0.336	0.644	0.090	0.473	0.043
4	0.00891			12.33	111.2	6.72	23.9	4.22	7.42	2.41	1.83	1.49	0.565	0.858	0.150	0.630	0.071
5	0.01114	2″				8.40	36.7	5.28	11.2	3.01	2.75	1.86	0.835	1.073	0.223	0.788	0.104
6	0.01337	0.574	0.044	2½″		10.08	51.9	6.33	15.8	3.61	3.84	2.23	1.17	1.29	0.309	0.946	0.145
8	0.01782	0.765	0.073			13.44	91.1	8.45	27.7	4.81	6.60	2.97	1.99	1.72	0.518	1.26	0.241
10	0.02228	0.956	0.108	0.670	0.046			10.56	42.4	6.02	9.99	3.71	2.99	2.15	0.774	1.58	0.361
15	0.03342	1.43	0.224	1.01	0.094	3″				9.03	21.6	5.57	6.36	3.22	1.63	2.37	0.755
20	0.04456	1.91	0.375	1.34	0.158	0.868	0.056	3½″		12.03	37.8	7.43	10.9	4.29	2.78	3.16	1.28
25	0.05570	2.39	0.561	1.68	0.234	1.09	0.083	0.812	0.041	4″		9.28	16.7	5.37	4.22	3.94	1.93
30	0.06684	2.87	0.786	2.01	0.327	1.30	0.114	0.974	0.056			11.14	23.8	6.44	5.92	4.73	2.72
35	0.07798	3.35	1.05	2.35	0.436	1.52	0.151	1.14	0.704	0.882	0.041	12.99	32.2	7.51	7.90	5.52	3.64
40	0.08912	3.83	1.35	2.68	0.556	1.74	0.192	1.30	0.095	1.01	0.052	14.85	41.5	8.59	10.24	6.30	4.65
45	0.1003	4.30	1.67	3.02	0.668	1.95	0.239	1.46	0.117	1.13	0.064			9.67	12.80	7.09	5.85
50	0.1114	4.78	2.03	3.35	0.839	2.17	0.288	1.62	0.142	1.26	0.076	5″		10.74	15.66	7.88	7.15
① → 60	0.1337	5.74	② 2.87	4.02	1.18	2.60	0.406	1.95	0.204	1.51	0.107			12.89	22.2	9.47	10.21
70	0.1560	6.70	3.84	4.69	1.59	3.04	0.540	2.27	0.261	1.76	0.143	1.12	0.047			11.05	13.71
80	0.1782	7.65	4.97	5.36	2.03	3.47	0.687	2.60	0.334	2.02	0.180	1.28	0.060			12.62	17.59
90	0.2005	8.60	6.20	6.03	2.53	3.91	0.861	2.92	0.416	2.27	0.224	1.44	0.074	6″		14.20	22.0
100	0.2228	9.56	7.59	6.70	3.09	4.34	1.05	3.25	0.509	2.52	0.272	1.60	0.090	1.11	0.036	15.78	26.9
125	0.2785	11.97	11.76	8.38	4.71	5.43	1.61	4.06	0.769	3.15	0.415	2.01	0.135	1.39	0.055	19.72	41.4
150	0.3342	14.36	16.70	10.05	6.69	6.51	2.24	4.87	1.08	3.78	0.580	2.41	0.190	1.67	0.077		
175	0.3899	16.75	22.3	11.73	8.97	7.60	3.00	5.68	1.44	4.41	0.774	2.81	0.253	1.94	0.102		
200	0.4456	19.14	28.8	13.42	11.68	8.68	3.87	6.49	1.85	5.04	0.985	3.21	0.323	2.22	0.130	8″	

FIG. 4-23. Pressure drop of water flowing through Schedule 40 steel pipe. *Courtesy The Crane Co., Technical Paper 410, 1980*

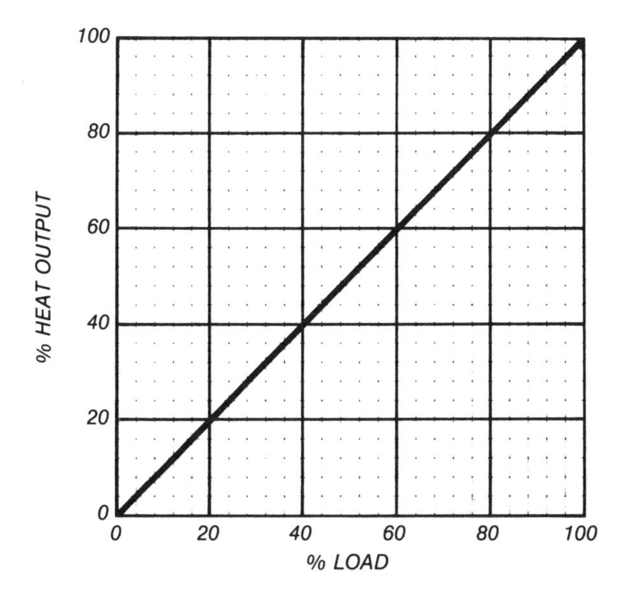

FIG. 4-24. Desired control curve for an air conditioning system. *Courtesy Barber-Colman Co., Technical Paper CA-18.*

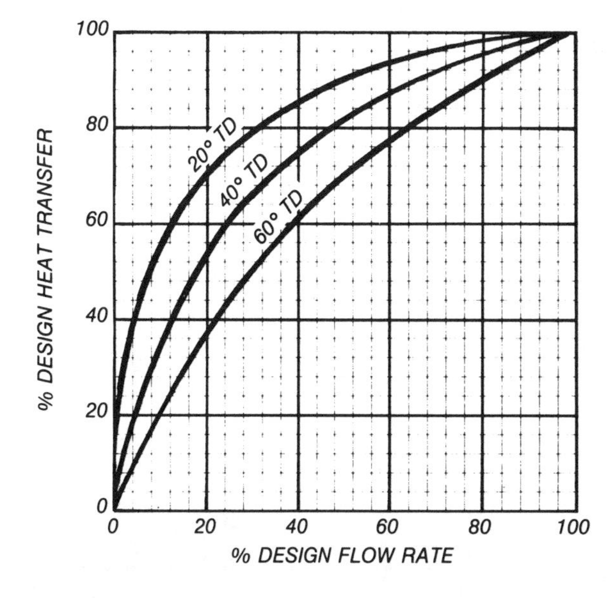

FIG. 4-25 Effects of flow variation on heat transfer for increased design temperature drop. *Courtesy ITT Bell & Gossett, Fluid Handling Div.*

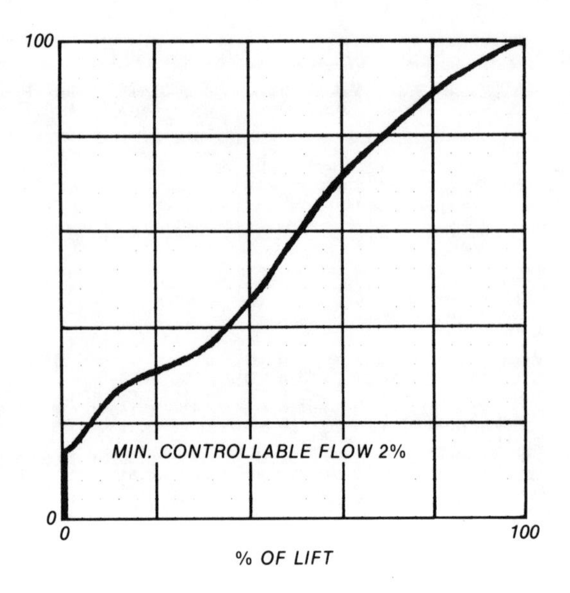

FIG. 4-26. Hot water coil with equal percentage disc contour. *Courtesy Barber-Colman Co., Technical Paper CA 19-1.*

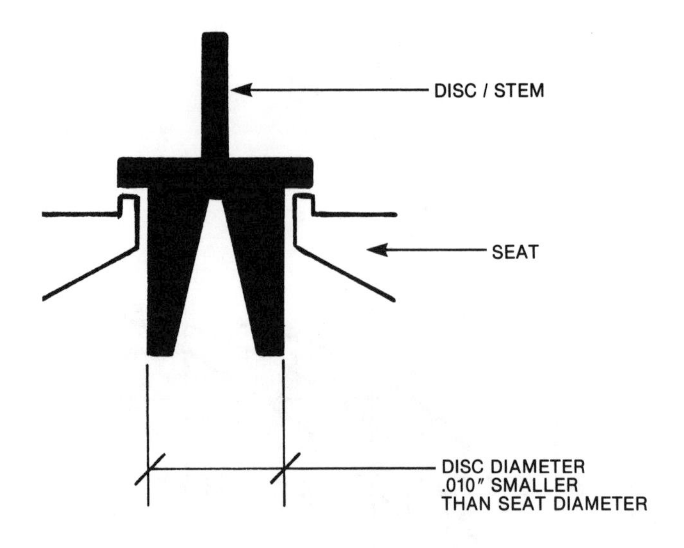

FIG. 4-27. Profile of control valve disc compared to its seat. *Courtesy Barber-Colman Co., Technical Paper CA-18.*

through the coil. The steep curve, for a 20 °F temperature drop, is nowhere near linear. Higher temperature drops, though, do approach being linear. It can be seen, for the 20 °F temperature drop, that only 4% of the flow delivers 40% of the heat, while 10% of the flow delivers 58% of the heat and 20% of the flow delivers 70% of the available heat.

To compensate for such a nonlinear curve, a characterized disc is used. The various disc shapes were shown in Figure 3-19. For a steep heating curve, the *equal percentage* contoured disc is used. The curve of the contoured disc plug and the steep operating curve of the heat transfer device combine for a near linear output, Figure 4-26.

An additional factor in valve construction is clearances. Depending on valve size, clearance between the disc and seat may be .010 to .025 inches. If the clearance area is 5% of the total flow area, this results in 35% of the total heat output of the coil. Finer control cannot be achieved and here is where the problem lies, refer to Figures 4-25 and 4-27.

Minimum flow, due to clearances, is usually given by valve manufacturers as *turndown ratio.* A turndown ratio of 10%, for example, means the flow can be reduced to 10% of its maximum rating. Below that minimum, any movement of the disc towards its seat will have little effect on flow, until final shutoff is reached. It is obvious that the larger the turndown ratio, the better the valve can control flow at the low end of its stroke, but also the tighter the tolerances that are needed in fabrication of the valve.

Clearances are needed to prevent binding and sticking, to take up any thermal expansion or contraction between the various parts, and to allow for possible corrosion. The clearance problem is more critical in the smaller size valves since clearances are a greater percentage of the flow area.

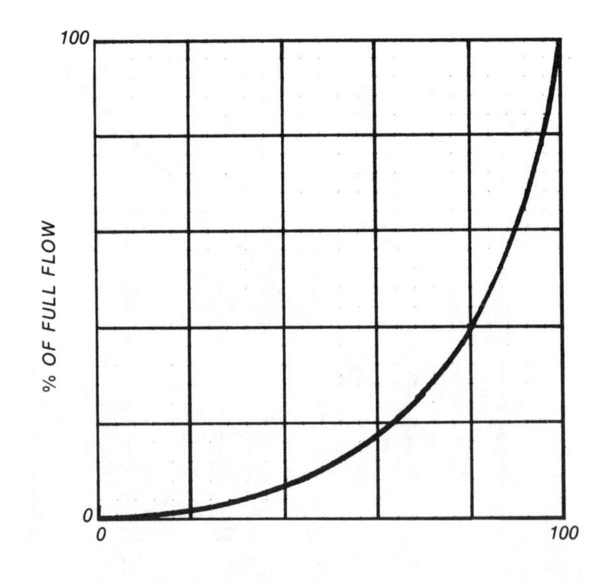

FIG. 4-28. Ideal equal percentage valve with 100:1 turndown ratio. *Courtesy Barber-Colman Co., Technical Paper CA-18.*

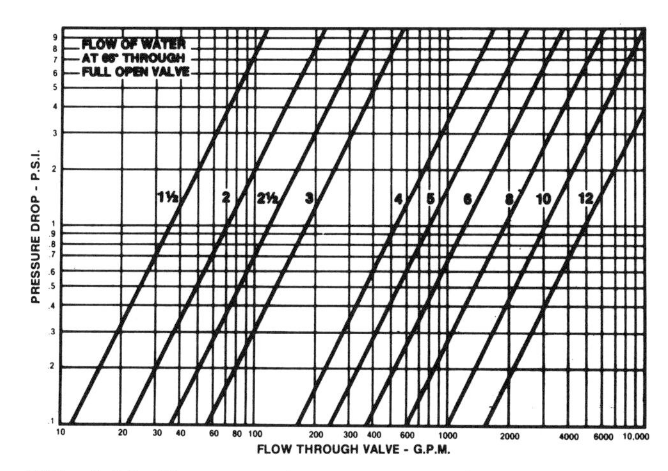

FIG. 4-29. Pressure drop vs. water flow for a wide open butterfly valve. *Courtesy Victaulic Co. of America.*

Good control of hydronic systems can be achieved with proper valve selection:

- The valve must be the major source of pressure drop. The minimum is usually 50% of the total system pressure drop, or higher.
- The proper disc configuration must be selected to eliminate nonlinear output by the thermal device.
- Leakage, due to clearances, should not exceed 1 to 2% of the design flow and the valve should have a high turndown ratio, Figure 4-28

Manufacturers may supply pressure drop data on their line of valves. Figure 4-29 shows pressure drop curves for Victaulic butterfly valves in sizes 1½ through 12-inch and for water flows of 10 to 10,000 gpm. Care must be taken with throttling valves since the disc can be in a number of positions from fully closed to fully open. Figure 4-29 assumes that the disc is fully open, parallel to the flow. In other positions, the pressure drop across the disc will be greater. Determining the pressure drops with other settings will be discussed in the next section on flow coefficient, C_V.

Circuit balancing valves, Figure 4-30, are used to balance hydronic circuits or zones by creating a precise pressure drop within each branch of the system. In hydronic systems utilizing parallel zones or terminals, each circuit must be properly balanced to deliver the required amount of heating or cooling to each zone.

The balancing valve in Figure 4-30 has orifice taps across the valve port so that the pressure drop may be read directly. The handwheel has a vernier calibration and its setting, along with the pressure drop reading, can be coordinated on a chart, Figure 4-31, to find the water flow rate in gpm.

FIG. 4-30. Circuit balancing valve for pressure drop control. Features pressure sensing taps, drain connection, vernier handle for setting flow. *Courtesy Armstrong Pumps Inc.*

The above discussion applies to water systems. Additional care must be taken in selecting valves used with compressible gases such as steam or air. With compressible gases, the pressure drop across the valve should never be greater than 50% of the absolute inlet pressure or choked flow results. Choked flow is a condition where the maximum flow rate of a valve cannot be exceeded even though the downstream pressure is reduced further and further. Choked flow can be accompanied by considerable noise that could be objectionable in non-industrial applications such as domestic systems.

Pressure drops in refrigerant systems are not normally determined by the methods described above. Most refrigerant pipe and valve pressure drops are determined by using charts and tables published in the manuals of the American Society of Heating, Refrigerating and Air-Conditioning Engineers (ASHRAE) and by charts published by valve manufacturers for their specific products.

All components, such as pipe, fittings, solenoids, hand valves, strainers, and driers, must be carefully selected to minimize pressure drops in refrigerant systems. The liquid line is not as critical as the suction and discharge lines. However, too great a pressure drop in the liquid line will cause flashing in the line and reduce the pressure at the expansion valve.

Suction lines are more critical. For every fraction of a pound pressure loss, there will be a corresponding reduction in system capacity. In its *Handbook of Fundamentals*, ASHRAE recommends the following

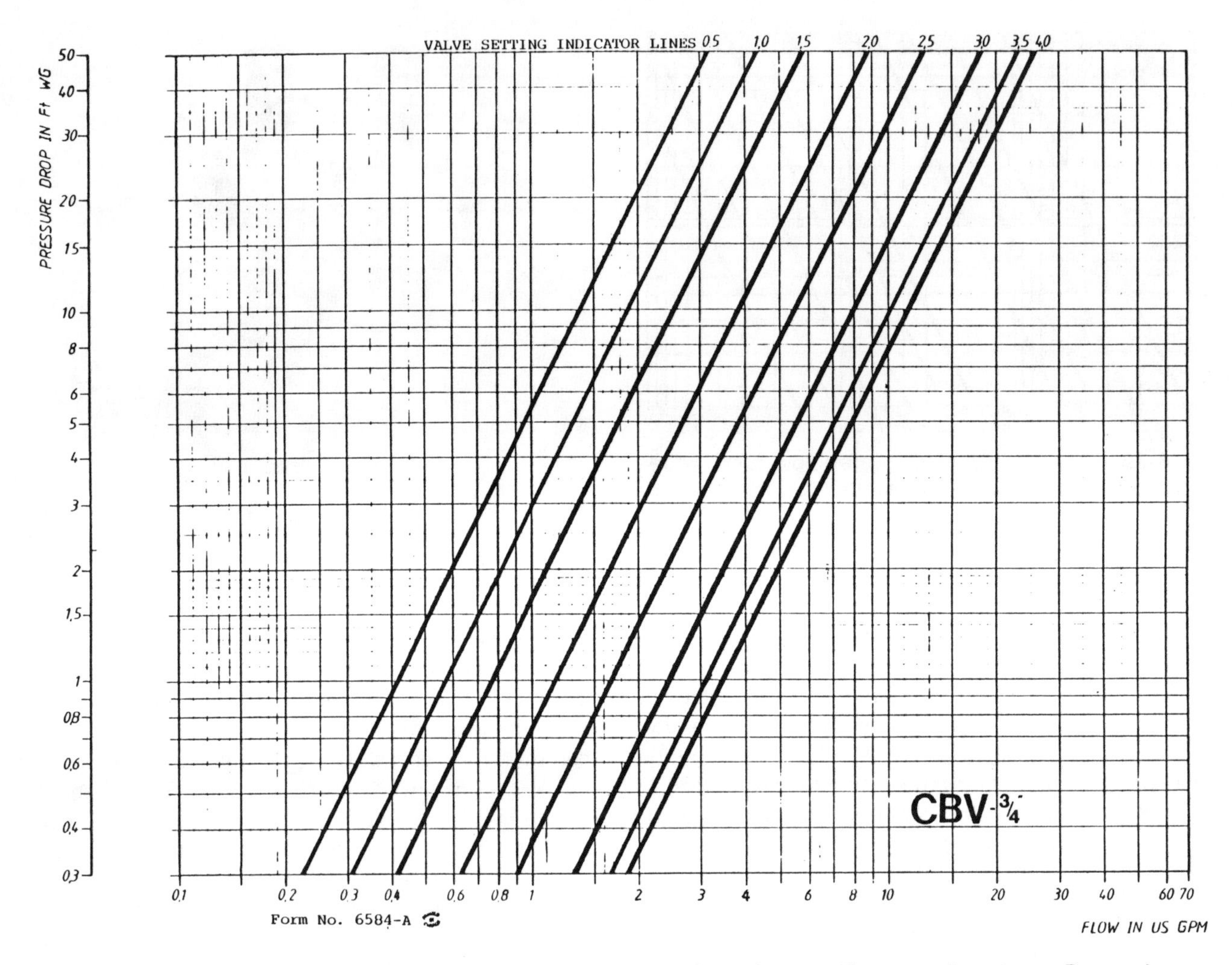

FIG. 4-31. Pressure drop curves for balancing valve – gpm vs. feet of water. *Courtesy Armstrong Pumps Inc.*

pressure losses for suction lines:

Table 4-13. Max. pressure drops in suction lines.

+ 40°F suction	Refrigerant	– 40°F suction
1.81 psi	R-12	.50 psi
2.94 psi	R-22	.75 psi
2.17 psi	R-500	.50 psi
3.14 psi	R-502	.91 psi
1.48 psi	ammonia	.30 psi

It can readily be seen that, at lower suction temperatures, the allowable pressure drop is reduced since it is more critical at these temperatures. Note that there are variations in recommended pressure drops, depending upon the refrigerant.

Expansion valves are designed to provide a pressure drop, from the high side to the low side of a refrigeration system, that may be as high as 200 psi. Chapter 5 discusses these pressure drops and how they are used in sizing the valve.

Flow coefficient, C_V

Most regulators and valves used for throttling are rated by their *flow coefficient factor*, C_V.

The C_V factor is defined as the number of U.S. gallons per minute (gpm) of water, at 60 °F, that will flow through a *wide open* valve at a pressure drop of one pound per square inch (psi) across the valve. Each manufacturer must test his valves in accordance with Instrument Society of America (ISA) standard S39.02.

A valve with a C_V of 10 will handle 10 gpm of 60 °F water at one psi drop across the valve. A valve with a C_V of 500 will handle 500 gpm at one psi pressure drop across the valve.

The flow through a valve or regulator is hardly ever at one pound pressure drop. Figure 4-32 shows the C_V values in the upper left corner. These values are for butterfly valves that are wide open. The

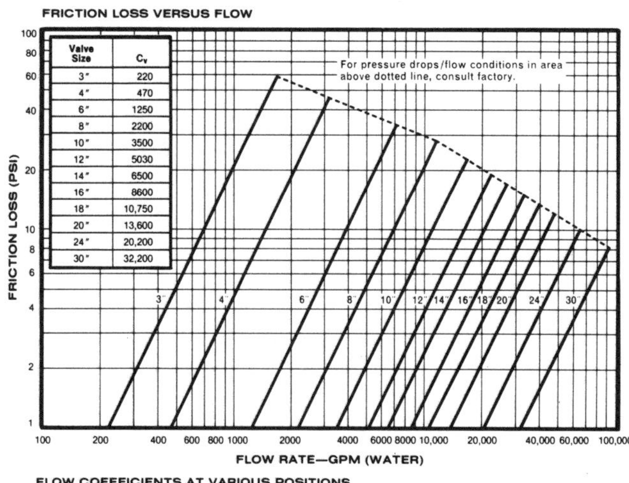

FLOW COEFFICIENTS AT VARIOUS POSITIONS

Disc Position Degrees Open	Valve Size 3	4	6	8	10	12	14	16	18	20	24	30
90	220	470	1250	2200	3500	5030	6500	8600	10,750	13,600	20,000	32,200
80	187	400	1062	1870	2975	4280	5520	7310	9140	11,560	17,000	27,340
70	142	303	806	1420	2260	3240	4190	5550	6933	8770	12,900	20,750
60	103	221	588	1033	1645	2360	3030	4040	5050	6390	9400	15,105
50	71	153	406	715	1139	1635	2115	2800	3490	4420	6500	10,425
40	44	94	250	440	701	1006	1300	1720	2150	2720	4000	6410
30	23	49	131	231	368	528	682	903	1130	1430	2100	3365
20	9.2	20	52	92	147	211	273	361	450	570	840	1345
10	1.6	3.5	9.4	16	26	38	49	65	80	100	150	240

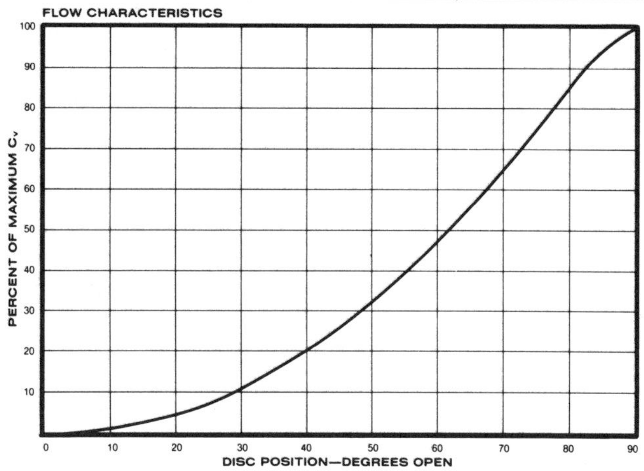

FIG. 4-32. Various methods of presenting Cv values. *Courtesy Rockwell International Flow Control Div.*

slanted curves in the chart show the pressure drops at different flow rates. Looking at the bottom psi line, all the values given there will match the Cv values in the upper left corner since the flow at 1 psi is the definition of Cv. The manufacturer also has included flow coefficients at various disc positions from 10° open to fully open at 90°. Again looking at the fully open position of 90° and the Cv values in the chart, curve in Figure 4-32 will give the Cv values of the butterfly valve, from a disc position of 0° to 90°, in percent of maximum Cv for each size.

Figure 4-33 shows how other manufacturers display the Cv values of their valves. The first manufacturer, Figure 4-33a, b, shows curves of the

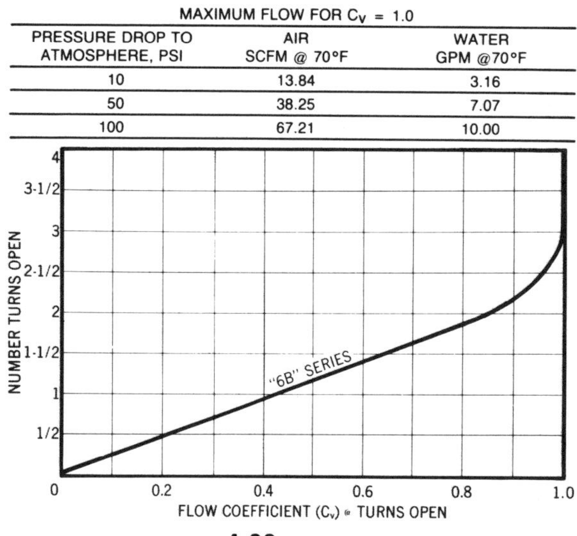

MAXIMUM FLOW FOR C_V = 1.0

PRESSURE DROP TO ATMOSPHERE, PSI	AIR SCFM @ 70°F	WATER GPM @70°F
10	13.84	3.16
50	38.25	7.07
100	67.21	10.00

4-33a

MAXIMUM FLOW FOR C_V = 1.0

PRESSURE DROP TO ATMOSPHERE, PSI	AIR SCFM @ 70°F	WATER GPM @70°F
10	13.84	3.16
50	38.25	7.07
100	67.21	10.00

NUMBER TURNS OPEN

"4BR" SERIES

FLOW COEFFICIENT (Cv) ÷ TURNS OPEN

4-33b

FIG. 4-33. Typical Cv curves. *Courtesy Nupro Co.*

CV VALUES—FLOW IN GPM OF WATER FOR 1 PSI PRESSURE DROP

ANGLE POSIT. SIZE	90°	70°	60°	50°	40°	30°	20°
1½"	36	29	22	15	8	5	2
2"	70	56	43	29	16	9	4
2½"	120	100	75	50	26	15	6
3"	180	145	110	73	40	22	9
4"	520	420	310	210	115	63	26
5"	700	650	480	320	175	90	38
6"	1300	1100	800	520	290	155	65
8"	2000	1600	1250	850	475	260	110
10"	3000	2500	1900	1200	650	375	150
12"	5000	4200	3300	2200	1200	650	280

FIG. 4-33c. Typical Cv table. Courtesy Victaulic Co. of America.

stem travel, from fully closed to fully open, for each disc configuration. Figure 4-33a is a valve with a maximum Cv of 1.0 while that of Figure 4-33b is a valve with maximum Cv of .26. As the disc closes in on the seat, Cv values become smaller and smaller.

Another manufacturer shows the Cv rating of his butterfly valve, Figure 4-33c at various disc positions from 20° open to 90° fully open.

The Cv coefficient is very useful in determining flow rates and pressure drops of other fluids, such as air and steam, and can be used to quickly estimate the required size of any fully opened valve in a piping network.

The equations for determining the pressure drops and flows for water, air and steam are shown in Table 4-14 and are used to determine a range of pressure drops, at varying flows, through a valve with a published Cv. These formulas are basic and many manufacturers have their own methods of determining valve size for a particular application.

Table 4-14. Formulas for Cv, flow, pressure drop*.

Liquid flow

$$Cv = \frac{Q\varrho^{.5}}{7.9(P_1 - P_2)^{.5}}$$

$$Q = \frac{7.9Cv(P_1 - P_2)^{.5}}{\varrho^{.5}}$$

$$\Delta P = \varrho\left(\frac{Q}{7.9Cv}\right)^2$$

Gaseous flow

$$Cv = \frac{.0578W}{\varrho^{.5}(P_2(P_1 - P_2))^{.5}}$$

$$W = \frac{Cv\varrho^{.5}(P_2(P_1 - P_2))^{.5}}{.0578}$$

$$\Delta P = .0033\left(\frac{W}{Cv}\right)^2$$

Steam flow (saturated only)

$$Cv = \frac{W}{3(P_2(P_1 - P_2))^{.5}}$$

$$W = 3Cv(P_2(P_1 - P_2))^{.5}$$

$$\Delta P = \frac{1}{P_2}\left(\frac{W}{3Cv}\right)^2$$

Cv = flow coefficient
Q = flow (gpm)
.5 = square root coefficient
ϱ = density (lb/ft³)
P_2 = downstream pressure, absolute (psig + 14.7)
P_1 = upstream pressure, absolute (psig + 14.7)
W = flow (lb/hr)

* Formulas are for noncritical flow, non-viscous liquids.
Temperature correction for gaseous flow is $Cv_{cor} = Cv \cdot T^{.5}/22.8$ where T = absolute temperature, °R, = (460 + °F)

The formulas for liquids are for those liquids that are near the density and viscosity of water at 60 °F and in turbulent flow. The equations do not apply to liquids with high viscosity or flowing under laminar conditions and the valve manufacturer should be consulted.

In gaseous and steam applications, the formulas are for noncritical flow; that is, for flow where the differential pressure across the valve, $P_1 - P_2 = \Delta P$, is less than 50% of the absolute upstream pressure P_1. For example, if a valve in air service has a differential pressure of 20 psi, where P_1 = 50 psi and P_2 = 30,

$$P_1 - P_2 = \Delta P$$
$$50 - 30 = 20 \text{ psi} = \Delta P$$

The absolute pressure is

$$P_1 + 14.7 = P_1 \text{ absolute}$$
$$50 + 14.7 = 64.7 \text{ psia}$$

50% of the absolute inlet pressure is

$$.50 \times 64.7 = 32.4 \text{ psi}$$

Δ = 20 is less than the critical threshold of 32.4 psia, therefore, this valve will give satisfactory performance.

The reasons for the noncritical flow for the air and steam formula is that as the downstream pressure, P_2, is reduced and the differential ΔP, in turn, is increased towards the 50% value, the air or steam flow is increased accordingly. For differentials above 50%; that is, lower and lower outlet pressures, P_2, beyond the non-critical condition, choked or critical flow develops; that is, a maximum flow condition results. No further increase in flow is achieved even though the outlet pressure P_2 is reduced further. Flow in the critical area can cause extreme noise and possible damage to the valving and piping system.

Figure 4-34 shows schematically the pressure tap locations P_1 and P_2 on the inlet and outlet of a valve. It was shown earlier in Figure 4-20 the pressure reduction of a flowing fluid through a valve and that the outlet pressure P_2 is less than the inlet pressure P_1. The resultant ΔP is that used in the Table 4-14 formulas.

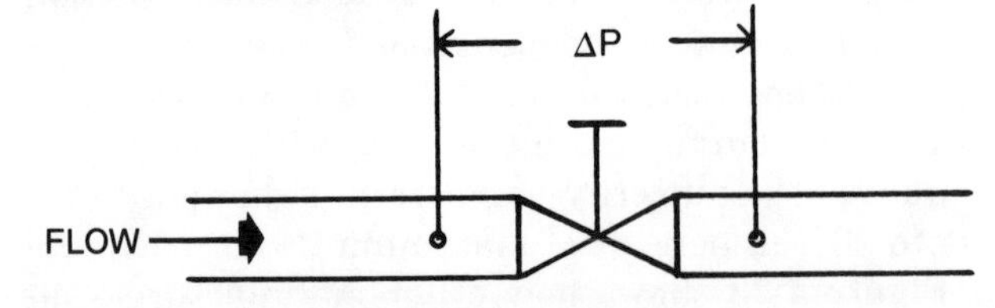

FIG. 4-34. Valve pressure taps P_1 and P_2 across a valve with flow Q for liquids and W for gas and steam.

If the flow and pressure drop are known, say water at 400 gpm and 5 psi drop is allowed across the valve, a valve can be selected that will match these parameters. From Table 4-14 for liquid flow, Cv is calculated to be 178. Figure 4-32 shows that a 3" butterfly valve, 80° open, will handle 187 gpm of water. This valve would be suitable for use in this application.

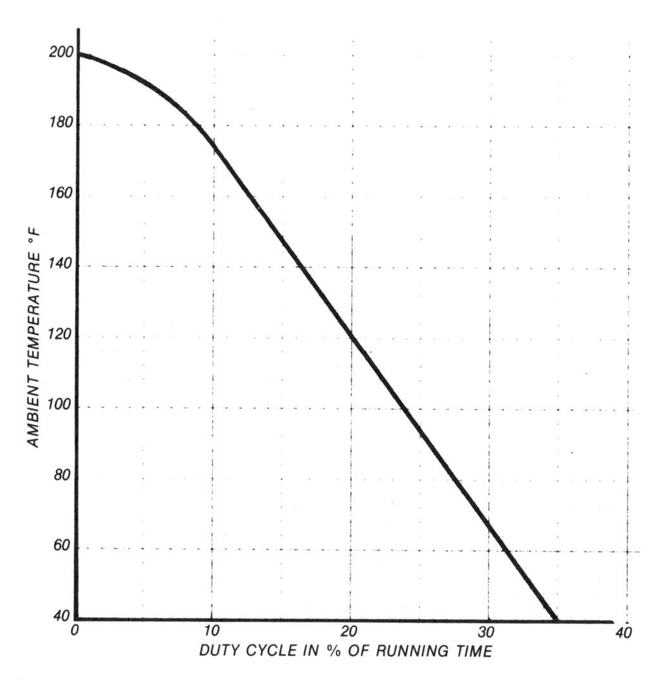

FIG. 4-35. Graphic presentation of the the *duty cycle* on an electric actuator with the maximum allowable duty cycle of at least 20% at ambient temperatures. The higher the ambient temperature, the lower the duty cycle. *Courtesy Jamesbury Corp.*

Service (duty cycle)

Service on a valve is defined as: valve operation and how frequent are the cyclic operations. Opening, closing or throttle adjustments may be made several times an hour, daily, monthly, yearly or only in an emergency. Depending upon the service (duty cycle), a determination can be made on valve style, body and trim materials, manual versus powered actuators and pneumatic, hydraulic or electric power for actuation. Electric actuator duty cycle is shown in Figure 4-35 based on ambient temperature.

Normal position

Normal position is the position in which the valve remains more than 50% of the time.

A valve that is closed or open most of the time may not function when needed, due to corrosion or other factors. Because of the fluids handled, combined with long periods of open or closed operation, the valve stem and packing may be specially selected to assure proper operation immediately upon demand. For example, an external threaded, nonrising stem may be selected over an internal threaded stem because corrosion may attack the stem and nonuse may prevent opening or closing the valve when needed.

Leakage

Leakage, in most cases, is detrimental to valve life. Leakage will erode the disc and seat, further increasing the leakage rate which, in turn, causes more erosion. Leakage through packings and seals can do similar damage. However, small amounts of nontoxic and noncorrosive leakage through packings may keep the packings *alive*, without the drying-out effects of long years of service.

Leakage is common with some valves because of

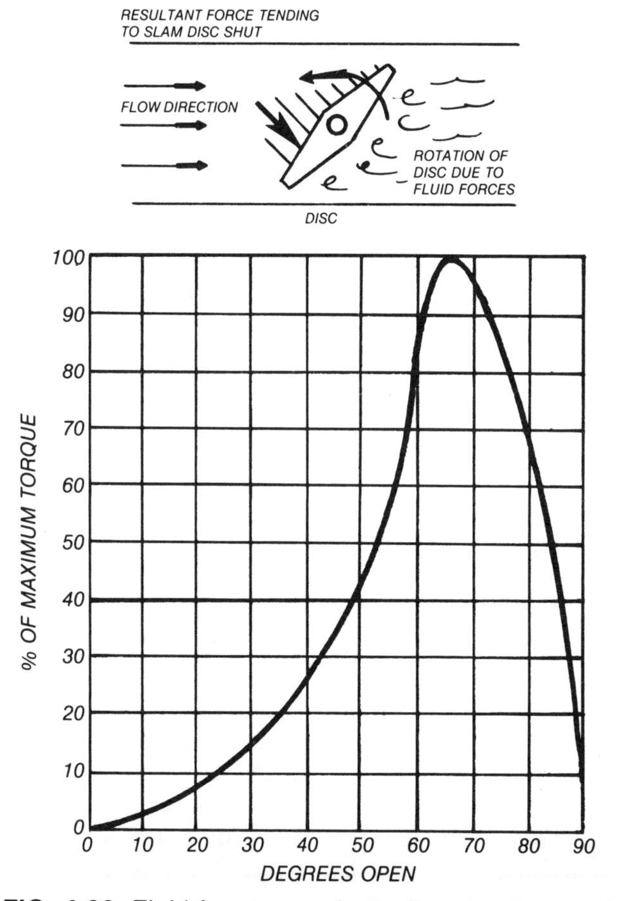

FIG. 4-36. Fluid forces on a butterfly valve that tend to slam disc shut. Maximum torque is developed at 67° open. *(ISA Handbook of Control Valves, 2nd Edition, 1976)*

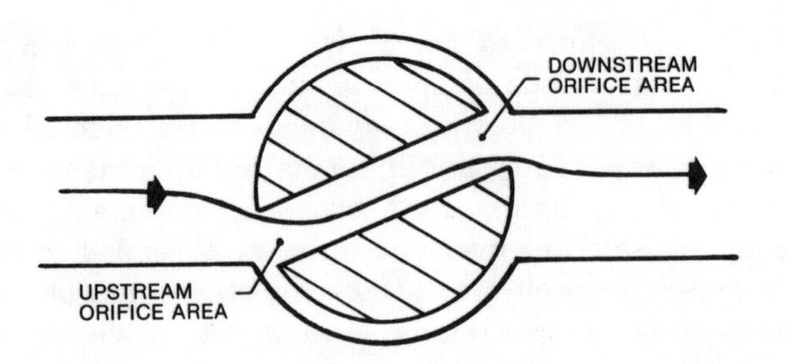

FIG. 4-37. Two orifices or restrictions created by a partially open ball valve. *(ISA Handbook of Control Valves, 2nd Edition, 1976)*

their construction. Three-way, some control valves, and double-ported valves may not have 100% sealing. If 100% blockage is required, a stop valve or solenoid valve should be used in series to stop leakage.

Internal forces

Internal forces, due to flowing fluid, vary with the valve's configuration. Globe valves and regulators based on the globe design have three forces acting on the plug: axial forces parallel to the stem; forces 90°to the stem, trying to bend the stem in the direction of flow; and rotational forces. Flow under the disc tends to promote opening while flow over the disc tends to slam the disc shut. Flow-under increases throttling stability and removes the stem packing as a source of leakage. Flow-over, however, promotes tighter sealing, due to the forces acting in the direction of shutdown.

Butterfly valves have a circular disc placed within the flow stream. In a partially open position, the disc acts like an airplane wing, creating forces that tend to slam it shut, Figure 4-36. When open 67° from the closed position, dynamic forces are at a maximum. Actuator sizing must allow not only for this force, but for the added torque needed to hold the stem steady against fluctuations in flow that would cause the disc to flutter.

When ball valves are used in throttling applications, two orifices are created, Figure 4-37. Forces created by the flow have a tendency to close the valve and any actuator must compensate for this dynamic force. The ball, in most common applications, is of the floating design and is free to move downstream when differential forces are across the ball during partial closure.

Size and location

In most cases, the size of air conditioning and refrigeration valves is of little consequence. The most important consideration is the possiblity that the weight of the valve will stress the piping network and joints, especially if vibration is present. Reciprocating compressors, motors and engines will cause varying degrees of vibration throughout a system and adequate support is required so that vibration will not cause piping or joint failure. Weight may also hamper replacement or make replacement difficult if a valve is located in an inaccessible area.

Valves used for emergency shutdown must be accessible.

Access and space limitations

Opening, closing, repairing or replacing a valve may become a real task if the valve is not within easy reach. If frequent checking or adjustments are needed, then the valve must be located where it is accessible. Electric or mechanical actuators can assist valve operation in inaccessible areas. However, servicing valves in these locations is certainly difficult.

In many instances, piping is so numerous and compact that space limitations require a valve of minimum size and silhouette. Here, ball or butterfly valves (the quarter-turn styles) have a definite advantage over the globe or gate valve versions.

Maintenance

Disassembly of valves for maintenance or replacement is an expensive proposition in today's labor market. Valve repair can be achieved by the use of bolted valves, screw-in seats or discs, or swing-out bodies, Figure 4-38. Larger valves are usually flang-

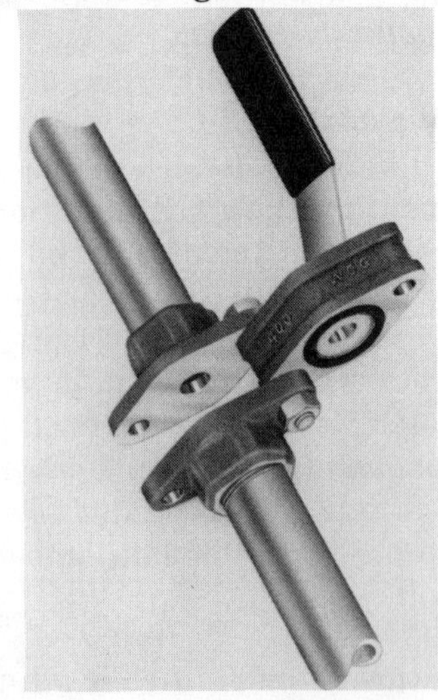

FIG. 4-38. Swing-out type ball valve eliminates pipe unions and allows all parts to be removed for maintenance. *Courtesy The Fairbanks Co.*

ed and are replaced as separate units. Valve maintenance will be discussed further in Chapter 7.

Costs

Valve costs will vary substantially depending upon size, style and materials of construction. Figure 4-39 is a graph showing relative costs for gate and globe valves including installation cost comparisons. Ratings for these valves are 125, 150 and 200-pounds and cost data is based on 1980 prices. If numerous or large-size valves are required, it is obvious that thousands of dollars may be spent quite easily. Material selection becomes very important in order to keep costs at a minimum and still stay within reasonable budgets.

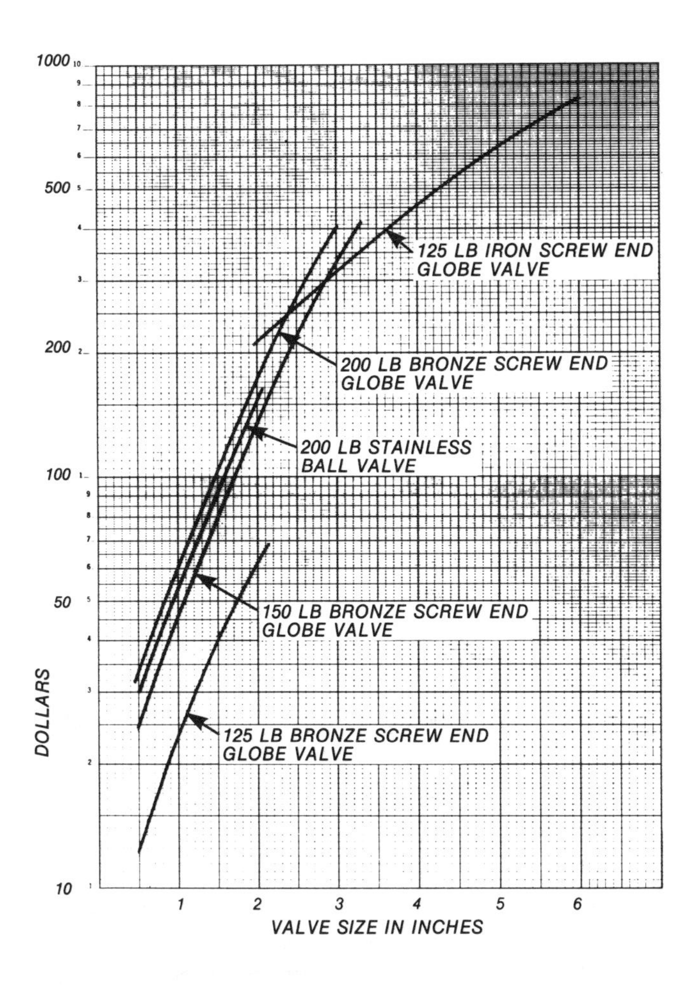

FIG. 4-39. Valve cost comparison, manual valves only.

Safety

Any piping system will contain a fluid under pressure. It may be water, air, volatile liquids, refrigerants or explosive gases. Care must be taken to see that the piping network and valves withstand the pressure/temperature stresses of these fluids including water hammer, overpressure and catastrophic situations involving fire or power failures. Emergency closing or shutdown functions may require automatic controls or mechanical devices and, in many cases, with backfeed to control panels indicating closure position so that personnel will not be exposed to dangerous situations.

Liquid lines in refrigeration systems must not be isolated by closing a pair of stop valves, thus leaving no room for expansion. See Chapter 7 for a further discussion of safety.

Actuators

Most valve actuators found in the air conditioning and refrigeration industry fall into two categories:

- On-off control of quarter-turn valves and solenoid valves
- Control of valves used with heating/cooling terminals in homes, offices and apartments.

Their use and the available designs were covered in Chapter 3.

Most quarter-turn applications will be on ball and butterfly valves used to open or close a fluid piping system. The most commonly used actuators are pneumatic, hydraulic or electric motor driven. Solenoid valves are a special application that utilize the electric features. Gate and globe style actuators are well suited to rising stem applications, using the diaphragm-spring style actuator.

The actuator should be purchased with the valve, as a package, so that all of the application requirements can be met. *Do-it-yourself,* add-on actuators, assembled in the field on existing valves, are not recommended because of mismatched torque requirements and mechanical interface problems that could cause actuator failure.

Actuator selection and sizing is based on torque requirements and are a function of:

- Fluid forces
- Pressure drop across the seat
- Style of seat
- Frictional factors of packings and seals
- Weight of internal parts
- Return spring forces

The fluid forces that were discussed earlier cannot be underestimated, especially in valves larger than

2″ . Discs that remain in the fluid stream, like globe and butterfly valves, will have substantial cross-sectional areas and, because of this, the flowing fluid will transfer forces into the stem, creating actuator torques that could be unstable. High pressure-drop applications, especially in modulating service, will also create large torque requirements.

A butterfly valve disc, for example, if full-ported will have relative area increases as the square of its diameter. If the pressure drop across the valve is of any consequence and the disc in the throttling position, the pressure on the disc will increase proportionally to the area. Actuator sizing must accomodate these forces so that the disc will not drift from its original setting.

Frictional factors can be subdivided into breakaway, running and closing torques. With six inch or larger valves, the weight of the internal parts must be considered. In fail-safe operation, the actuator may be required to open or close when the plant's air or electric power is interrupted. This is usually accomplished by the addition of a spring that will move the valve to a safe setting to shutdown a process or block a fluid line that could be harmful or even start a sprinkler system in case of fire.

If the actuator is undersized, it will not develop enough torque to overcome the above forces. If oversized, it will develop too much torque and could jam internal parts or twist the stem beyond safe limits. Remember that once installed, many valves may not be called on to operate until an emergency arises. This may be after weeks, months or even years, at which time the actuator must function as intended. Most misapplied actuators are usually undersized.

Each valve style has its own unique problems caused by the flowing fluid, the pressure drop across the disc and the torque on the disc. Because the actuator may work well on a 2″ ball valve does not mean it will work satisfactorily on a 2″ butterfly or plug valve.

Torque

Whether manual or powered, actuators on any valve will face similar forces; however, each valve style will offer differing degrees or levels of these forces. Valve torques can generally be classified as:

- Running torque
- Seating torque
- Breakaway torque

Torque, *T*, is defined as the action of a force, *F*, at a distance, *D*, from the center of a rotating device.

$$T = DF$$

The distance is measured from the center of rotation to the applied force at right angles to the applied force, Figure 4-40. Torque is calculated either in inch-pounds or foot-pounds depending on whether the distance *D* is measured in inches or feet.

If a certain amount of torque is required to turn a valve stem, a greater force is required at a smaller distance. At a larger distance, a smaller force is required to do the same job, assuming everything else is the same.

Running torque is defined as the amount of torque needed to move a valve disc between open and closed positions. Running torque determines the force re-

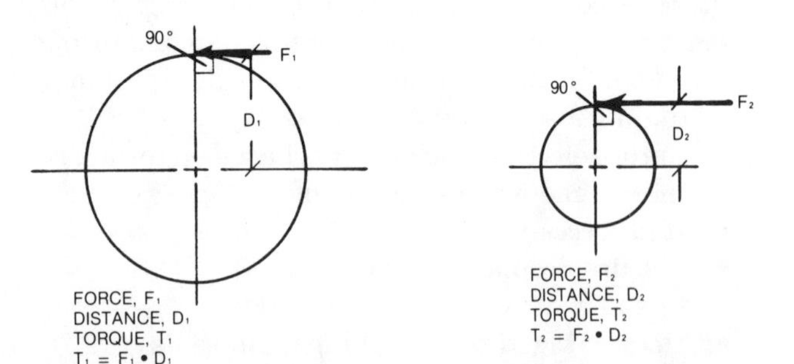

FIG. 4-40. Force on the rim of a hand wheel or a lever will create torque to rotate a valve stem. The farther the force is from the center, D_1 vs. D_2, the smaller the force applied to do the same job.

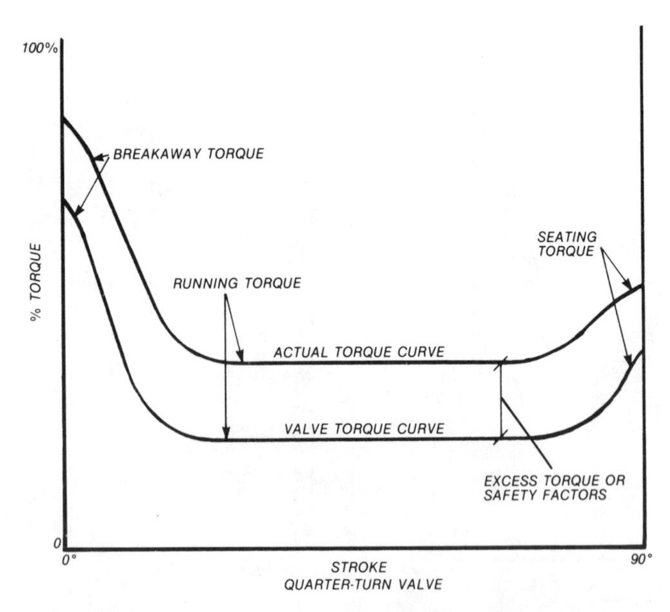

FIG. 4-41. Valve and actuator torque curves show the relationship of breakaway, running, and seating torque. *Courtesy W.R. Hayes Assoc., fluid, pump, and valve consultants.*

quired to operate the disc against existing internal fluid pressure, internal fluid flow, frictional effects of packings and bearings, and the inertia of the moving parts. Flow direction, in relation to the disc, will affect the running torque on globe and butterfly valves.

Seating torque, needed for positive closure, may be 150% of the running torque.

With certain valve designs, breakaway torque can be substantial. Butterfly valves with rubber liners are a good example. Sometimes ten times more torque is needed to unseat a valve than to seat it.

Each valve style, and even the same valve with another seal material, will have unique torque requirements. Tables, charts and curves, that are available from manufacturers, show the torque requirements for each valve style, packing or seal material.

The actuator torque profile must exceed the valve torque requirements profile at all points from breakaway to running and seating. If not, a stall condition results and the valve will not operate.

Figure 4-41 shows both the torque requirements of a hypothetical quarter-turn valve and the actuator torque output. The breakaway torque is highest, running torque is lowest and the seating torque is slightly higher than the running torque. At all points, from 0° to 90° open, the actuator torque output exceeds the valve requirements. About a 10% safety factor is recommended.

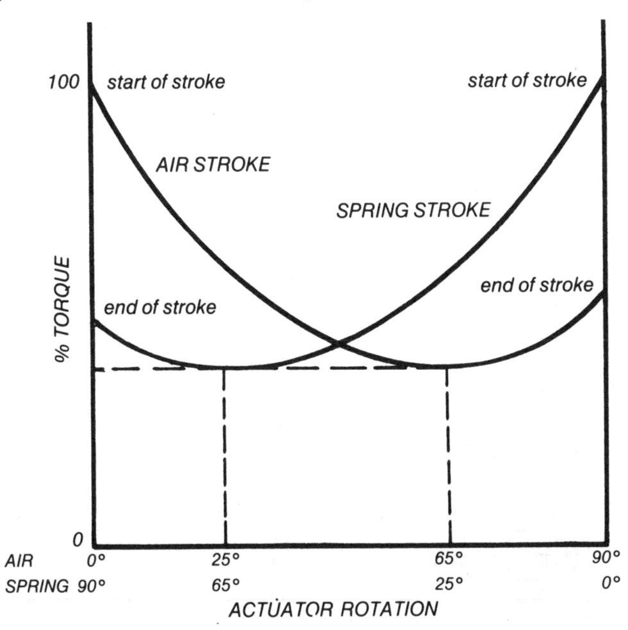

FIG. 4-42. Pneumatic-supply and spring-return actuator torque curves. *Courtesy W.R. Hayes Assoc., fluid, pump, and valve consultants.*

In pneumatic actuator designs, where a spring return is utilized for fail-safe operation, the actuator torque curve, upon application of pneumatic power, will have a different stroke curve than the spring return stroke curve. Figure 4-42 shows the curves for the initial stroke and spring return. The air stroke torque is reduced as the spring is compressed and on the spring stroke, the torque is greatest when the collapsed spring is just beginning to expand.

Opening and closing speed

The operational speed of actuators depends on their design. Electric motor driven designs, working through a set of gears, may take 15 to 30 seconds to open or close a valve. Pneumatic counterparts may take only 2 to 5 seconds for the the same movement. Speed may be required for emergencies, however, too fast a closing may cause water hammer and eventual piping system damage. Each selection must be made on its own merits.

Fail-safe

If power fails and pneumatic or electric energy is lost, valves may be required to go to an open or closed position or even hold their present position. System layout and fluid function will dictate which modes of operation will be needed.

Backup power sources are available in case of main power failure: springs and accumulators for pneumatic systems, AC emergency power or DC battery power for electrical actuators.

Figures 3-73, 3-74 and 3-75 compared three spring diaphragm actuators as spring-to-open/spring-to-close and air-to-open/air-to-close. If pneumatic pressure is lost, the actuator will shift the valve to the open or closed position as designated by the system design. Electrical actuator designs are good for *hold in the last position* if power failure occurs.

Manual Operators

Most valve manufacturers assume that a man can exert 50 pounds of force to operate a valve. Therefore, a *rim* force of 50 pounds is usually designed into valves with handwheels or a 50 pound pull into lever-actuated valves. The 50 pound force is not an industry standard, consequently, operating forces could vary from one manufacturer to another.

Above a certain force, the handwheel or lever is no longer practical because the force required is too much for a typical operator. In this case, a geared or powered actuator must be selected. Various hand-op-

Table 4-15. Torque available with geared actuators.

Gear ratio	Handwheel diameter (inches)	Running torque[1] (ft. lbs)	Seating torque[2] (ft. lbs)	Maximum torque[3] (ft. lbs)
2.92:1	6 12 20	40 80 130	100 330 550	1000
4.11:1	20 24 30	185 225 280	775 935 1170	2500
6.16:1	30	420	1750	6000

Notes:
1. Torque available when handwheel is operated with a spinner handle between fully-open and nearly closed positions.
2. Torque available when two-hand pull is used on handwheel rim.
3. Torque available with hammerblow.

erated actuators were shown in Figures 3-48 and 3-49.

Geared actuators, with ratios of 3:1 to 18:1, reduce the input force required for a specific task. Bevel gear types are the most common.

Table 4-15 shows gear ratios of 2.92:1, 4.11:1 and 6.16:1 for handwheel diameters of 6 to 30 inches. Running, seating and maximum torque values can be compared. Butterfly valves are a special case since they tend to be slammed shut by the moving fluid.

Powered actuators

Three major actuator designs are commonly used in the air conditioning and refrigeration industry:

- The spring/diaphragm type, Figure 3-72, is usually used with a pneumatic air supply of 3 to 15 psi or 6 to 30 psi. These air pressures are readily available in most industrial plants. The advantages of this design are simplicity and few moving parts. The major disadvantage is the limited thrust, due to the spring absorbing a major part of the input force.
- Piston actuators are used with pneumatic pressures ranging from 50 to 150 psi. However, when used with spring-return systems, they have the same limitations as the spring-diaphragm type. A disadvantage of the spring-piston type is the potential for air leakage around the piston, which will allow the valve disc to creep.

The main disadvantage of the straight piston design, Figure 3-71, is the high air supply pressure, the need for a positioner in modulating service, and the lack of a straightforward, fail-safe system since no spring is used for opposing action.

- Electric actuators, consisting of motor and gear train or motor and linkages, can produce a wide range of torques. Electrics are especially handy in remote applications where air is not available. The

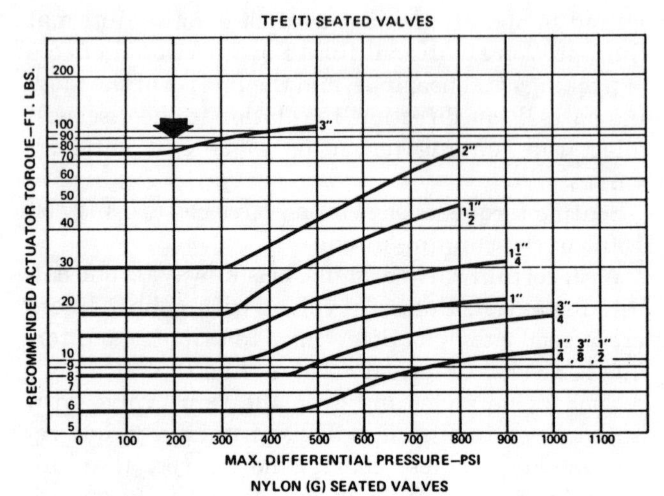

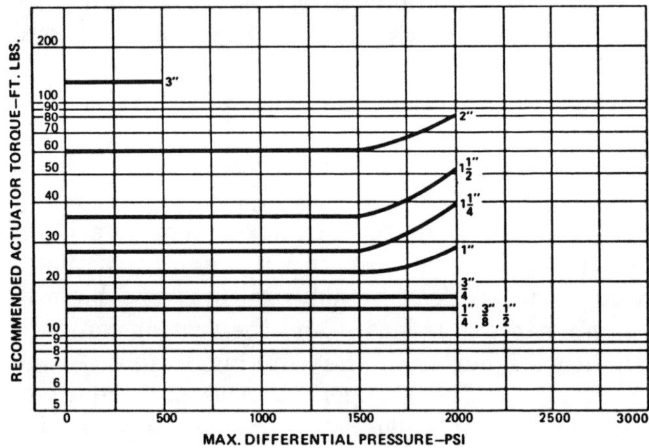

Additional requirements may be imposed by media characteristics, trim, and frequency of valve operation. For clean lubricating fluid service, required torque may be reduced 33% when the valve is equipped with corrosion resistant trim. For difficult services (slurries, semi-solids) increase values by 50%.

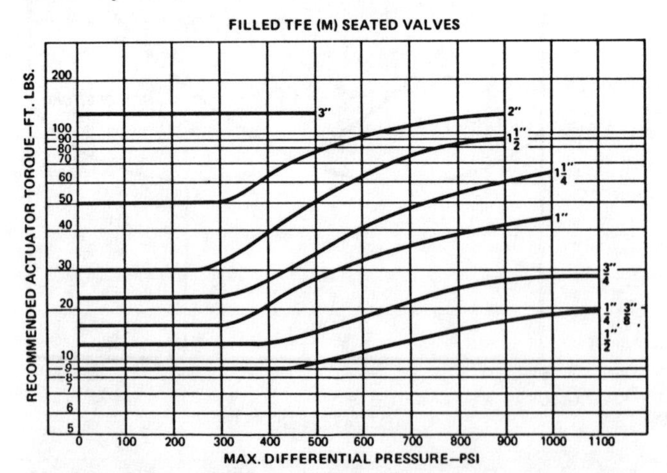

FIG. 4-43. Torque requirements of a double-seal, screwed-end ball valve. *Courtesy Jamesbury Corp.*

DOUBLE-ACTING ACTUATORS

ACTUATOR TYPE	Torque Output — ft. lbs. At minimum supply pressure of:			
	40 psi	60 psi	80 psi	100 psi
SL10	4	6	8	10
ST20	8	12	16	20
ST50	20	30	40	50
ST200	80	120	160	200
ST400	160	240	320	400
ST600	240	360	480	600
ST1200	480	720	960	1200
ST2400	960	1440	1920	2400

SPRING-RETURN ACTUATORS

ACTUATOR TYPE	Torque Output — ft. lbs. At rated supply pressure of:	
	60 psi min.	90 psi min.
ST13MS	13	—
ST60MS	60	—
ST115MS	115	—
ST160MS	160	—
ST290MS	290	—
ST20MS	—	20
ST90MS	—	90
ST175MS	—	175
ST240MS	—	240
ST440MS	—	440

Quadra-Powr actuators

ACTUATOR TYPE	TORQUE OUTPUT — FT. LBS. AT MINIMUM SUPPLY PRESSURE OF:									
	30 psi		40 psi		50 psi		60 psi		65 psi	
	AIR END	SPRING END	AIR END	SPRING END	AIR END	SPRING END	AIR END	SPRING END	AIR END	SPRING END
B30S	12	15	20	15						
B40S			15	20	27	20				
B50S					20	25	32	25		
B60S							25	30	31	30
C30S	30	50	60	50						
C40S			40	65	80	65				
C50S					50	80	90	80		
C60S							60	90	80	90

Type EL line of uni-directional electric actuators

Type EJ and EJX actuators are bi-directional actuators

SPECIFICATIONS GENERAL CHARACTERISTICS						
ACTUATOR	Torque Output (ft. lbs.)*	Motor Drive	Enclosure	Operating Speed (rpm)	Typical Time for 90° Stroke (sec.)	Approx. Weight (lbs.)
EL 8	8	Uni-directional	Watertight	2	8	6
EL 20	20	Uni-directional	Watertight	2	8	9
EJ 20	20	Reversible	Watertight	2	8	8
EJ 50	50	Reversible	Watertight	2	8	10
EJ 90	90	Reversible	Watertight	1	16	16
EJX 20	20	Reversible	Hazardous Location	2	8	10
EJX 50	50	Reversible	Hazardous Location	2	8	12
EJX 90	90	Reversible	Hazardous Location	1	16	18
ERC 160	155	Reversible	Watertight	2	8	35
ERC 250	245	Reversible	Watertight	2	8	35

FIG.4-44. Torque capacities of double-acting pneumatic, spring-return pneumatic, diaphragm-spring return, and electric driven actuators. *Courtesy Jamesbury Corp.*

main disadvantages are the lack of a fail-safe position, limitation of high duty cycles on the motor, and the need to prevent any possible electrical failure from causing an explosion when used in dangerous atmospheres. See Figure 3-70.

Pneumatic actuators require a dry, clean air supply. Because electric power is not required, the pneumatic actuator may be used in hazardous and highly moist environments.

After the valve has been selected to meet all the job requirements, its torque requirements are determined. Figure 4-43 shows the torque curves for a Jamesbury double-seal, screwed-end, ball valve in ¼″ through 3″ sizes. Three charts are shown, one for each seat material available for this valve line. The note added to the charts indicates the torque variations for special trim and for fluids carrying suspended solids. For a double-seal valve with virgin TFE seats (type T), 3″ size at 200 psi differential pressure, the torque requirement is 75 foot-pounds.

Several actuator designs are available: a double-acting actuator (air on both sides of the piston to open or close the valve), a spring-return piston actuator (air on one side only), a spring-diaphragm actuator, or an electric actuator can be selected.

The torque capacity of the actuators is shown in Figure 4-44. With 60 psi air supply pressure, for example, the following actuators, each developing at least 75 foot-pounds of torque, could be selected for use with the valve in Figure 4-43:

ST200, a double-acting piston actuator
ST115, a spring-return piston actuator
C50S, a spring-diaphragm actuator
EJ90, an electric, reversible actuator

Dimensional data for each actuator is contained in the manufacturer's literature and they can be compared for size and speed of operation.

If a butterfly valve, instead of a ball valve, were selected with the same 3″ size and 200 psi differential pressure, the torque requirements are only 32 foot-pounds, less than half that of the ball valve, Figure 4-45.

Figure 4-46 is a torque vs. pressure drop chart for a series of butterfly valves by Rockwell International. The interesting part of this chart is the horizontal line defining the upper limit where manual (hand) operators can no longer be used and powered actuators must be used. For 3″ and 4″ valves, manual operators will suffice. Above 350 pounds differential pressure, a 6″ valve cannot be manually operated in a safe manner. 10″ valves and larger must be power actuated.

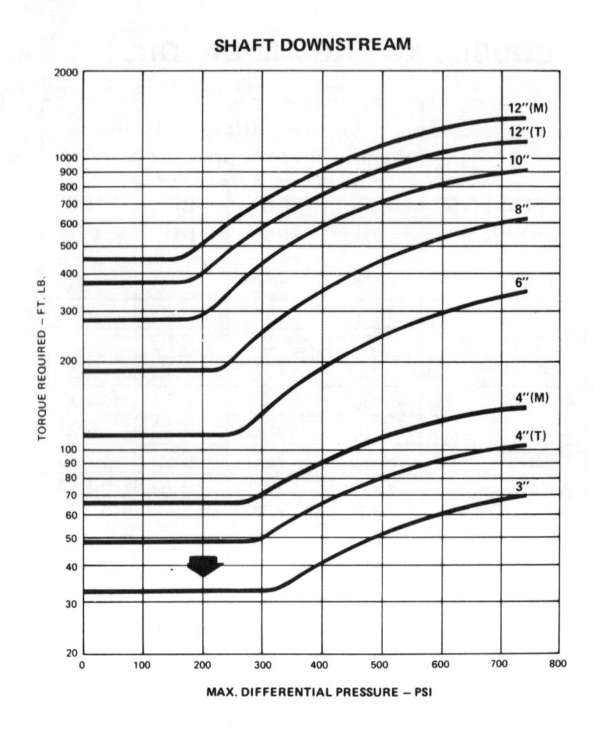

FIG. 4-45. Torque capacities of wafer-sphere butter--fly valves. *Courtesy Jamesbury Corp.*

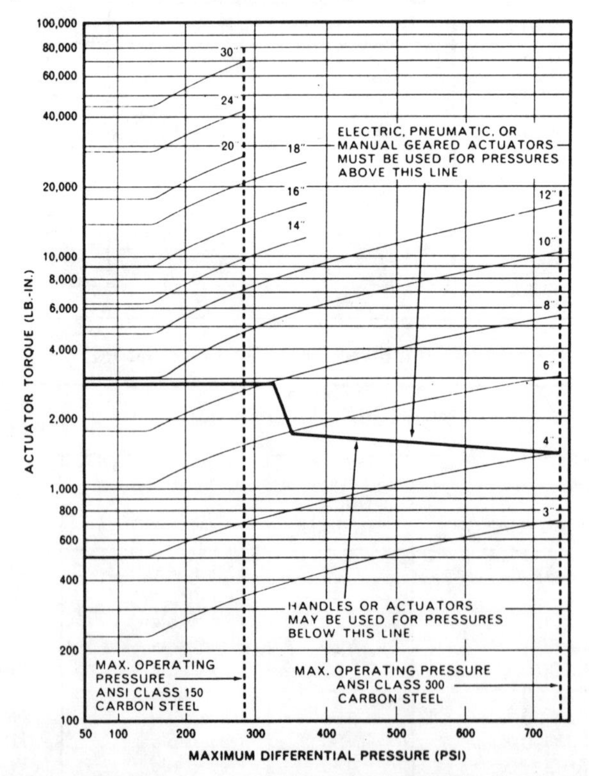

FIG. 4-46. Torque requirements of a series of butterfly valves. Powered actuators must be used when manual limits are exceeded. *Courtesy Rockwell International Flow Control Div.*

Accessories for actuators

Limit switches These electrical switches are usually mounted in pairs: one is actuated when the valve is fully closed, the other when the valve is fully open. Panel lights, in remote control rooms, allow the operator to identify the valve's position at a glance.

Manual override In case of power failure, a manual override is available for most actuators.

Motor brakes Brakes are used to *set* disc position. This is most desirable in valves that tend to creep under the forces of the flowing fluid.

Locking devices To assure that a valve remains in a desired position for a length of time, locks may be provided.

Positioners Positioners are pneumatic relays that are able to rotate a valve to an infinite number of positions. They respond to signals provided by a closed loop automatic control system. An extremely clean air supply is needed since many of the passages and control orifices are very small.

Internal Heaters Many electric actuators are located in outdoor environments and need heaters to prevent moisture condensation within the electrical enclosure.

Potentiometers Electric actuators can utilize a potentiometer to give remote control rooms the exact position of a valve from fully closed to fully open.

Solenoids

Solenoid design was covered in Chapter 3. However, selection of solenoids revolves around a number of basic parameters:

- Maximum operating pressure differential (MOPD)
- Minimum operating pressure differential
- Flow coefficient value, C_V
- Fluid pressure/temperature and material selection
- Normally open (N.O.) or normally closed (N.C.)
- Exterior ambient environment and temperature
- Type of enclosure

MOPD Maximum operating pressure differential is the maximum differential pressure across the valve, $\Delta P = P_1 - P_2$, against which the solenoid coil can operate in raising the valve disc. For direct operating styles, this value will be fairly low. Larger values may be handled by pilot operated versions.

Minimum operating pressure differential Some solenoid valve styles require a minimum pressure differential in order to keep the disc closed or, in the case of three or four-way styles, keep the disc against the proper seat to prevent leakage.

C_V The flow coefficient may be found, as described earlier, for liquids, air and gases, and steam. For direct refrigerant service, though, charts prepared by the various refrigerant valve manufacturers will suffice where the selection is based on the type of refrigerant used, location in the system, allowable pressure drop across the valve, and system capacity.

For liquid, gases and steam, each manufacturer has his own recommended series of calculations tailored to its own valve designs. Table 4-16 shows the Automatic Switch Company's (ASCO) recommended estimating formulas for liquids under 300 SSU (52 centipoise) viscosity. This is equivalent to SAE 10W oil. Correction factors are given for pressure drop (F_g), Figure 4-47, specific gravity (F_{sg}), Figure 4-48, and temperature (F_t), Figure 4-49, for liquids, air and steam. Once the C_V is determined, the orifice size can be found in Figure 4-50 starting

Table 4-16.

Liquid flow	$C_V = \dfrac{Q}{F_g \cdot F_{sg}}$
Gaseous flow	$C_V = \dfrac{W_g}{F_g \cdot F_{sg} \cdot F_t}$
Steam flow	$C_V = \dfrac{W_s}{F_g}$

where
Q = flow (gpm)
F_g = correction factor for pressure drop
F_{sg} = correction factor for specific gravity
W_g = gas flow, standard ft³ per hour
F_t = correction factor for temperature
W_s = steam flow, pounds per hour

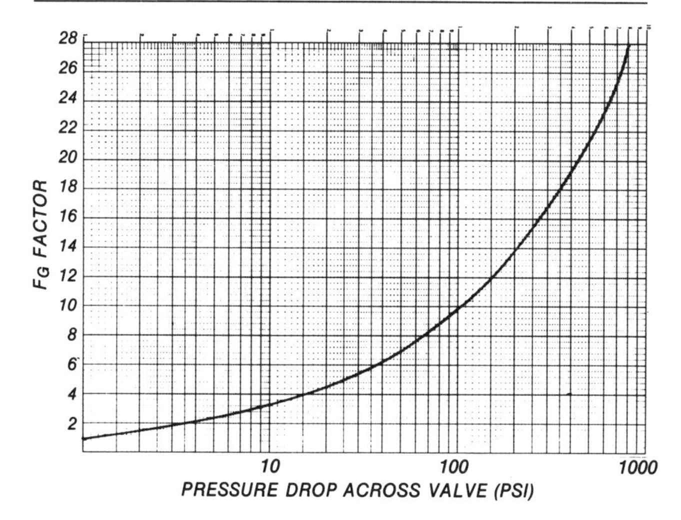

FIG. 4-47. Liquid flow curve used for solenoid valve sizing. *Courtesy Automatic Switch Co.*

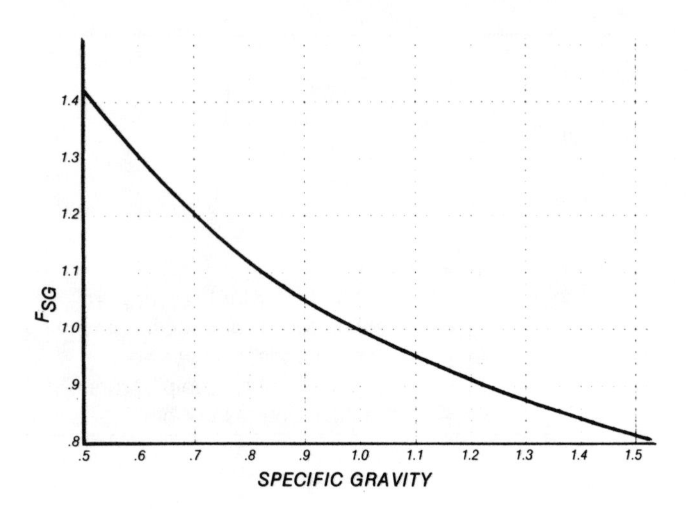

FIG. 4-48. Specific gravity correction. *Courtesy Automatic Switch Co.*

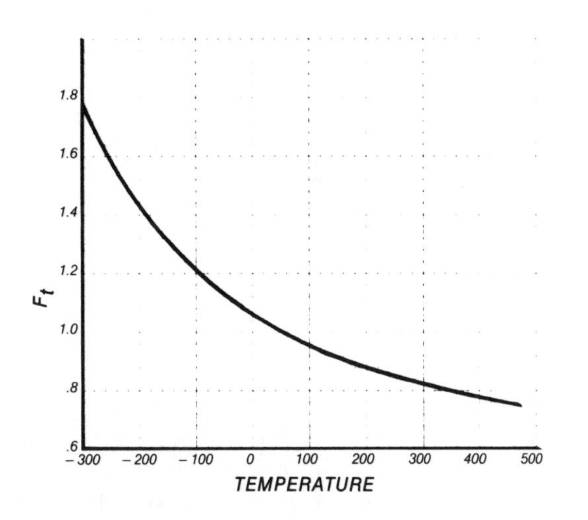

FIG. 4-49. Temperature correction. *Courtesy Automatic Switch Co.*

with the second column from the left.

For example, determine the C_V of 20 gpm of oil with a specific gravity of .9, using a ΔP of 25 psi and a viscosity less than 52 cp. From Table 4-16,

Liquid flow $$C_V = \frac{Q}{F_g \cdot F_{sg}}$$

where

Q = 20 gpm

F_g = 5, from Figure 4-47

F_{sg} = 1.05, from Figure 4-48

$$C_V = \frac{20}{5 \cdot 1.05}$$

$$C_V = 3.81$$

A manufacturer's catalog is reviewed and it is found that a series 8210 valve, and its fittings, are compatible with oil. Solenoid 8210D2, with a 5/8″ orifice and a C_V of 4, is selected. With an AC coil and light oil, a 135 MOPD is recommended, which will suffice.

Figure 4-50 lists a pressure limit of 300 psi and a temperature limit of 180 °F maximum. A normally closed solenoid, that will stay closed until energized, with a NEMA 1 enclosure, complete the specifications. The Appendix describes the various NEMA codes and ratings for electrical devices.

The solenoid coil is rated by *class*, based on temperature tolerance.Table 4-17 shows the temperature limits of four classes of solenoid coils: class A has a limit of 221 °F (105 °C), class B's limit is 266 °F (130 °C), class F's limit is 311 °F (155 °C) and class H has the highest limit, 356 °F (180 °C). Each temperature rating assumes an ambient temperature of 77 °F (25 °C). The higher the class, and the higher the tolerance to operating, fluid or ambient temperature, the tougher the construction .

Most coils are rated by their power consumption in watts.

Actuator Costs

The intial cost of actuators may appear to be high, but the operating cost is low. Table 4-18 is taken from an article in the August 1981 isssue of *In-*

Table 4-17. Temperature limitations of a series of solenoid valve coils.

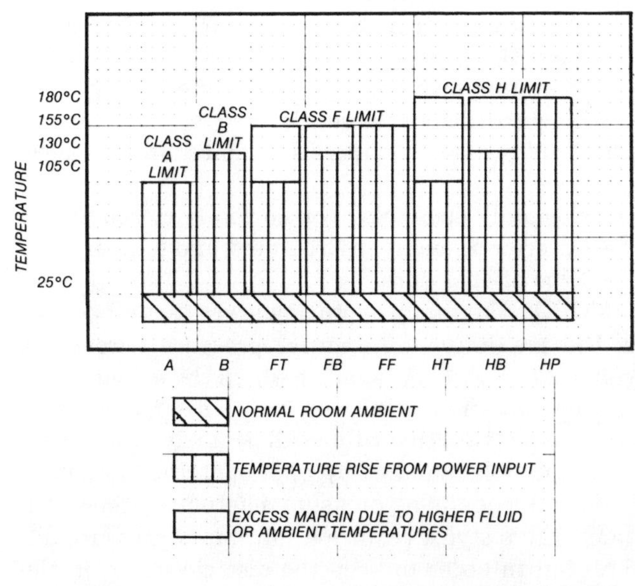

Courtesy Automatic Switch Co.

General Description

This bulletin offers a broad line of 2 way packless solenoid valves for both normally closed and normally open operation. These general service valves are available in various body materials, pipe sizes and operating pressures.

Applications

Completely automatic, these dependable 2 way valves are used in:

- pumps
- spraying
- cooling
- air dryers
- pollution controls
- laundry equipment
- irrigation
- compressors
- dishwashers
- water treatment

Special valves available for: • dry air-gas • continuous cycling • exceptionally long life • heavy-duty operation • clickless and quiet (no A-C hum) operation.

Refer to Long Life Construction on page 68.

Specifications

Operation: Two types are available:

(a) Normally Closed — valves closed when de-energized, open when energized.

(b) Normally Open — valves closed when energized, open when de-energized.

Pipe Sizes: ¼″ to 3″ N.P.T.

Valve Parts in Contact with Media:

Body — Brass, Stainless Steel (Series 300), Nylon or Aluminum, as listed.

Seals and Discs — Buna "N," Teflon or Ethylene Propylene, as listed.

Core Tube — 305 s.s.

Core and Plugnut — 430F s.s.

Springs — 302 s.s.

Shading Coil — Copper (brass, nylon or aluminum body); Silver (stainless steel body).

Solenoid Enclosures: Two types are available:

(a) General Purpose (NEMA 1).

(b) Explosion-Proof and Watertight (NEMA 7C, 7D and 4).

Electrical: Standard Voltages:

24, 120, 240, 480 volts, A-C, 60 Hz (or 50 Hz in 110 volt multiples).

6, 12, 24, 120, 240 volts, D-C.

Other voltages available when required.

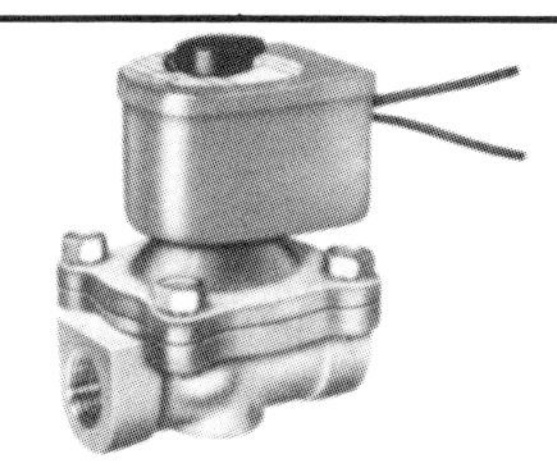

Coil: Continuous Duty Class H and Molded Class A, B and F, as listed.

Temperature: Fluid: To 210°F., as listed. **Ambient:** Nominal Range, 32°F. to 77°F. (104°F. occasionally — see page 6 in Engineering Information Section.)

Installation: Refer to dimension tables on pages 25 and 26 for details. Do note, however, that all Bulletin 8215, D-C valves must be mounted with solenoid vertical and upright.

Approvals: CSA certified. UL listed as indicated below. Refer to page 6 for details and coding explanation.

Important: For shut-off valves and vent valves for fuel gas service, refer to Combustion Section, pages 74-83.

SPECIFICATIONS

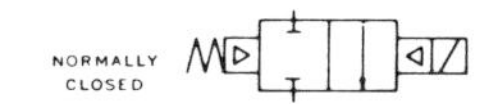

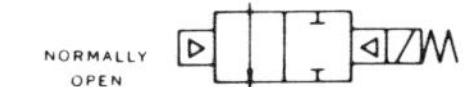

Pipe Size (Ins.)	Orifice Size (Ins.)	Min. Operating Press. Differential (P.S.I.)	Maximum Operating Pressure Differential (P.S.I.) Air-Inert Gas A-C	D-C	Water A-C	D-C	Light Hydraulic Oil @ 300 SSU A-C	D-C	Safe Working Pressure (P.S.I.)	Max.③ A-C Fluid Temp. °F.	Cv Flow Factor	General Purpose Solenoid Enclosure Catalog Number	Optional Feature Ref.	Constr. Ref. ⊖	UL Listing	Explosion-Proof — Watertight Solenoid Enclosure Catalog Number	Optional Feature Ref.	Watt Rating A-C	D-C	Class of Coil Insulation	Approx. Shipping Weight (Lbs.) GP	EP
Normally Closed Operation, Forged Brass Body, Buna "N" or Teflon* Seating for General Service																						
¼	5/16	5	125	—	125	—	—	—	150	180	1.2	8210A20	VII	7D	●	—	—	6.5	—	B	1	—
⅜	¼	5	125	—	125	—	—	—	150	180	1.2	8210A21	VII	7D	●	—	—	6.5	—	B	1	—
	⅜	⑤	125	40	125	40	—	—	300	180	1.5	8210C73Ⓐ	I	5P	●	8211C73Ⓐ	VIII	6	11.2	A	¾	1½
	⅝	0	100	40	100	40	—	—	300	180	3	8210C93	II	9D	O	8211C93	IX	11	11.2	A	2½	3½
	⅝	5	200	125	135	100	135	100	300	180	3	8210D1	I	10D	O	8211D1	VIII	6	11.2	A	2	2¾
	⅝	5	300	—	300	—	300	—	300	175	3	8210C6	IV	9D	O	8211C6	XI	16.7	—	F	2½	3½
½	7/16	⑤	125	40	125	40	—	—	300	180	2.2	8210A15	I	6P	●	8211A15	VIII	6	11.2	A	1	1¾
	⅝	5	200	125	135	100	135	100	300	180	4	8210D2	I	10D	O	8211D2	VIII	6	11.2	A	2	2¾
	⅝	0	100	40	100	40	—	—	300	180	4	8210C94	II	9D	O	8211C94	IX	11	11.2	A	2½	3½
	⅝	5	300	—	300	—	300	—	300	175	4	8210C7	IV	9D	O	8211C7	XI	16.7	—	F	2½	3½
¾	¾	0	100	40	100	40	—	—	300	180	5	8210D95	II	13D	O	8211D95	IX	11	11.2	A	2¾	3¾
	¾	5	125	100	125	90	125	75	300	180	5	8210D9	I	14D	O	8211D9	VIII	6	11.2	A	2	2¾
	¾	0	350	200	300	180	200	180	400	200	6	8210B26*	III	15P	⑪	8211B26*	X	15.4	30.6	⑦	3	3½
	¾	5	250	125	150	125	100	125	300	180	6.5	8210D3	I	16D	O	8211D3	VIII	6	11.2	A	3¼	4½
1	1	5	125	125	125	125	100	125	250	180	13	8210D4	I	18D	O	8211D4	VIII	6	11.2	A	5½	6¼
	1	10	300	225	300	200	300	200	300	200	13.5	8210B78*	IV	19P	●	8211B78*	XI	16.7	16.8	②	7¾	8¾
	1	0	300	—	225	—	115	—	300	200	13.5	8210B27	V	20P	●	8211B27	XII	20	—	F	8¼	9¼
	1	0	125	100	125	100	125	80	250	180	13	8210B54	III	39D	●	8211B54	X	15.4	30.6	⑦	8¾	9¼
1¼	1⅛	5	125	125	125	125	100	125	250	180	15	8210D8	I	22D	O	8211D8	VIII	6	11.2	A	6¼	7
	1⅛	10	300	225	300	200	300	200	300	200	15	8210B80*	IV	23P	—	8211B80*	XI	16.7	16.8	②	7¾	8¾
	1⅛	0	125	100	125	100	125	80	250	180	15	8210B55	III	40D	●	8211B55	X	15.4	30.6	⑦	8¾	9¼
1½	1¼	5	125	125	125	125	100	125	250	180	22.5	8210D22	I	24D	O	8211D22	VIII	6	11.2	A	8½	9¼
	1¼	10	300	225	300	200	300	200	300	200	22.5	8210B82*	IV	25P	—	8211B82*	XI	16.7	16.8	②	12	13
	1¼	0	125	100	125	100	125	80	250	180	22.5	8210B56	III	41D	●	8211B56	X	15.4	30.6	⑦	8¾	9¼
2	1¾	5	125	50	125	50	90	50	300	140	43	8210C46	I	26P	—	8211C46	VIII	6	11.2	A	11¾	12½
2½	1¾	5	125	50	125	50	90	50	300	140	45	8210D49	I	27P	—	8211D49	VIII	6	11.2	A	13¾	14½
3	3	10	250	—	250	—	250	—	300	200	101	8210B51④*	VI	28P	—	8211B51④*	XIII	28	—	H	76	77

FIG. 4-50. Manufacturer's catalog sheet with specifications for 2-way solenoid valves. *Courtesy Automatic Switch Co.*

Table 4-18. Comparative cost of manual vs. actuated 3 in. ball valve.

Manually operated valve	
Labor @ $15.00 an hour	
6 manual cycles per day	$15
2 visual checks per day	1
Daily cost	$16
Total yearly cost (250 days)	$4000

Actuated valve	
Labor @ $20.00 an hour	
4 hours installation	$80
2 hours a year maintenance	40
Materials	
Actuator	500
Other materials	100
Total first year costs	$720
Actuator payback	

$$\text{Payback} = \frac{\$720 \text{ first year actuator cost}}{\$16 \text{ daily manual cost}} = 45 \text{ days}$$

struments & Control Systems titled *Selecting Actuators for Quarter-Turn Valves.* It shows the cost of operating a manual and an actuator-operated valve for one year. It can be seen that the labor required to operate the manual valve quickly exceeds the total cost of the actuator. Multiplying this particular example by maybe hundreds of valves in a single plant can create excessive maintenance costs in little or no time.

General Summary

In order to select and specify valves for a particular application, it is necessary to review the selection parameters:

- Specify fluid: pressure, temperature, flow, cleanliness
- Material selection: body, disc, seat, stem, packings, elastomers
- Valve style: throttling, on-off, pressure drop, C_V
- Configuration: size, end connections, type of stem
- Service
- Normal position
- Leakage tolerance
- Actuator
    - Manual
    - Powered: pneumatic
        - Torque, plant air requirement, type of actuator design, failure mode
        - Electric
        - Torque, electric power AC/DC, frequency, phase, voltage, cycle time, exterior ambient
        - Hazard conditions (explosive environments require NEMA enclosures.)

5
Regulators

The refrigeration, air conditioning, and hydronics industries utilize numerous variations of regulators to control liquid and gas flows by sensing the pressures and/or temperatures of a system.

Regulators should not be confused with control valves. Regulators are self-contained and utilize the working fluid for control. Fluid pressures work against spring forces to create a balanced condition in pressure operated regulators. Temperature regulators contain a closed liquid/vapor system built into the regulator head assembly which, in turn, works against spring forces to control fluid flow.

Control valves, on the other hand, require an *external source* of energy, such as pneumatics, electricity or sophisticated electronic instrumentation to control valve operation. Control valves are usually more sophisticated in design and construction and can control flow to a very fine degree. In larger installations, they may cost upwards of thousands of dollars and extreme care is required in their sizing, selection, and application. A regulator and a control valve are compared in Figures 5-1a and b.

The most common regulators are designed in the globe valve configuration. The globe style shut-off valve utilizes a hand wheel or power actuator to turn the stem. The regulator, on the other hand,

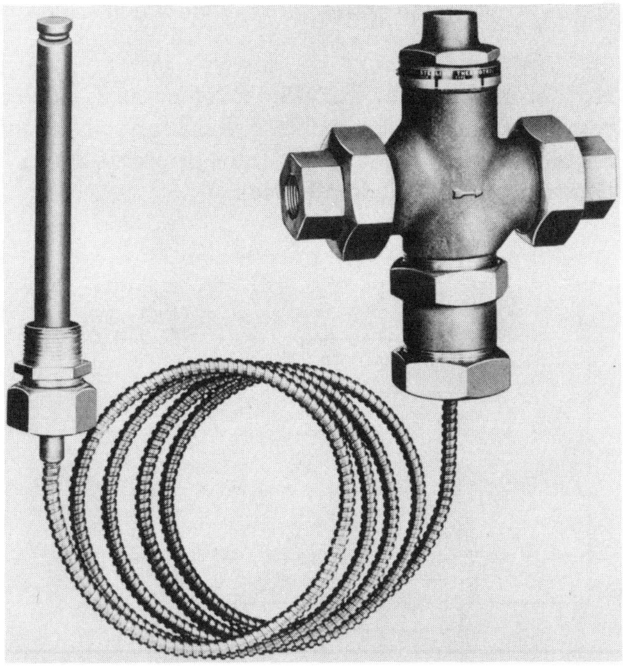

FIG. 5-1a. Temperature control regulator with self-contained thermostatic bulb actuator. *Courtesy Sterling, Inc.*

FIG. 5-1b. Flow control valve with pneumatic actuator, regulator and pneumatic valve positioner. *Courtesy Jamesbury Corp.*

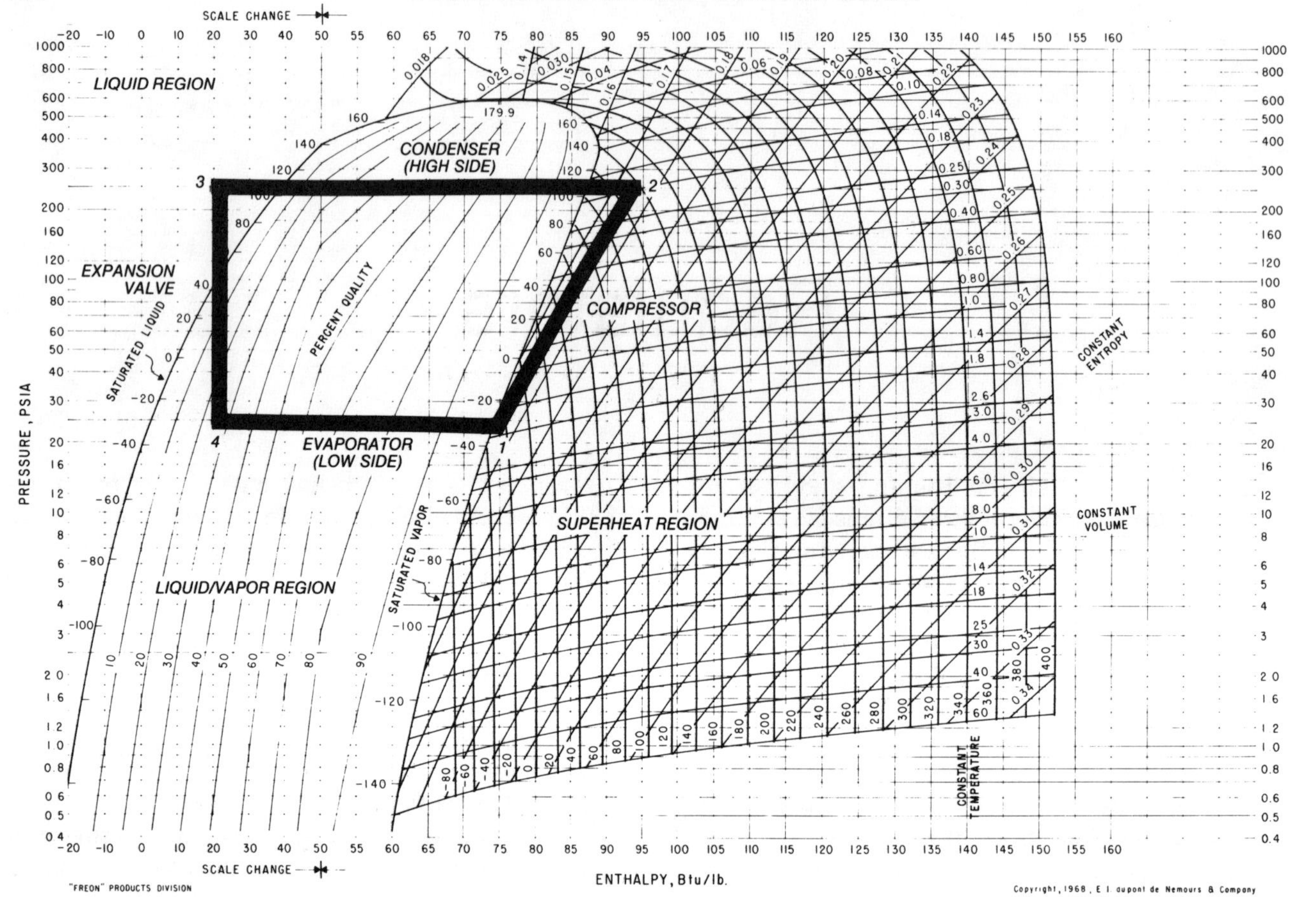

FIG. 5-2. R-502 pressure-enthalpy diagram with refrigeration cycle superimposed. *Diagram courtesy E.I. dupont de Nemours & Co.*

replaces the hand-wheel/stem combination with a spring and diaphragm or spring and bellows. Internal forces regulate the disc in an infinite number of positions from fully open to fully closed.

Refrigeration systems

Refrigeration systems are divided into high and low sides, or condenser and evaporator sides, respectively, and are shown schematically in Figure 5-2. This chart, known as the P-H or pressure-enthalpy chart, has the refrigeration cycle superimposed. A block diagram of the high and low sides is shown in Figure 5-3.

Between points 1 and 2 is the compression process.

Between points 2 and 3 is the condenser and high side piping.

Between points 3 and 4 is the expansion valve and distributor, if used.

Between points 4 and 1 is the evaporator and low side piping.

Figure 5-4 shows the refrigeration cycle and the location of the major components. Head pressure regulators would be installed after point 2, suction regulators just before point 1 and high-to-low side regulators between points 2 and 4 or points 2 and 1, depending upon the application.

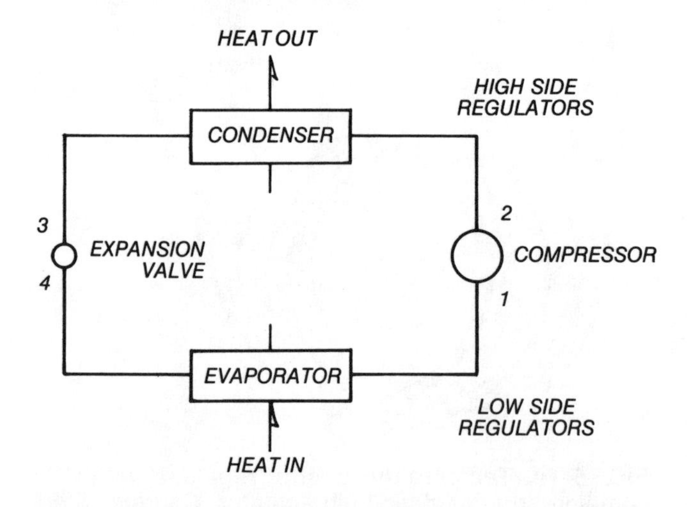

FIG. 5-3. Simple refrigeration system.

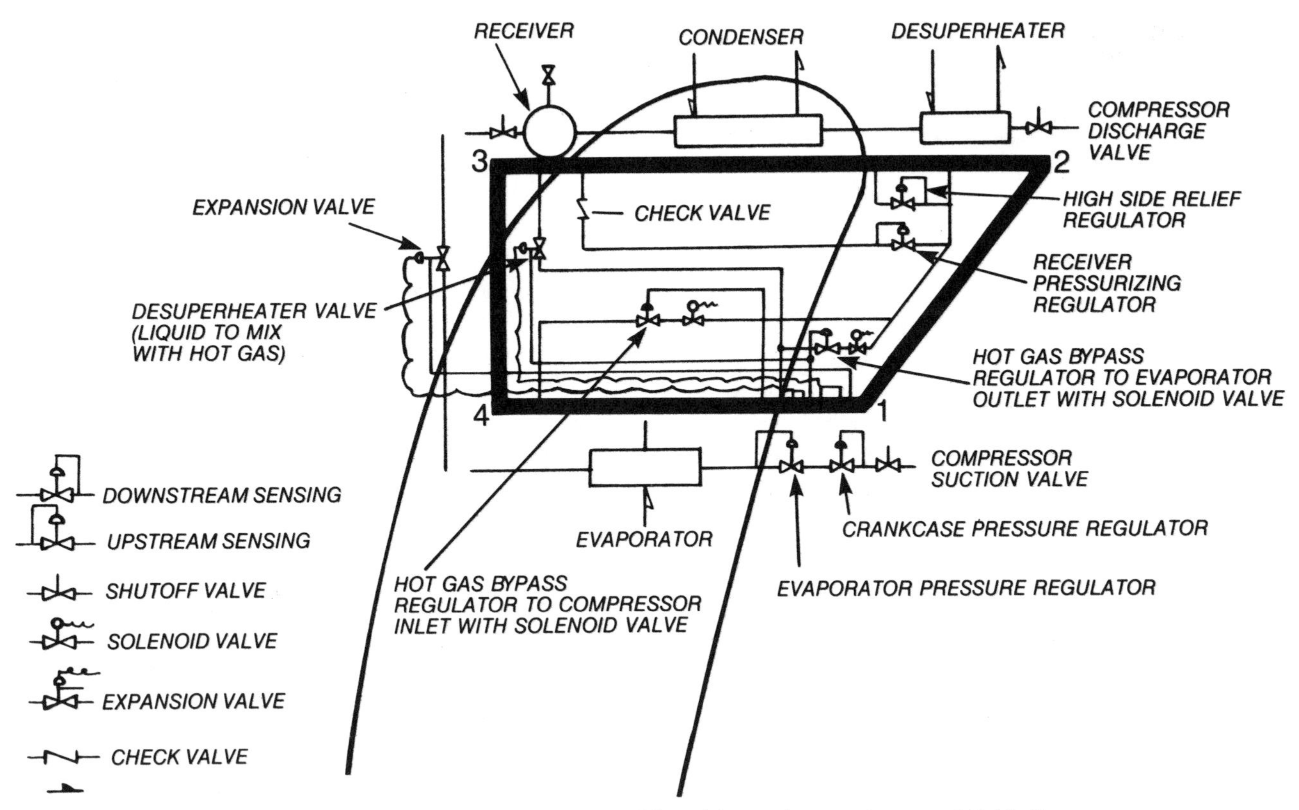

FIG. 5-4. Regulator locations shown with refrigeration system and P-H diagram.

Various regulators will control condenser and evaporator pressures, limit suction pressures to the compressor or divert high temperature-pressure gases from the high side to the low side during hot gas defrost or capacity control. Note that the components dividing the high and low sides are the compressor and the expansion valve.

The P-H diagram, Figure 5-2, shows the three refrigerant regions, in this case for R-502. The three regions are subcooled liquid to the left of the curve, vapor-liquid mixtures within the curve, and superheated vapor to the right of the curve. The left portion of the curve itself indicates saturated liquid conditions while the right portion of the curve indicates saturated vapor conditions. The refrigeration cycle can be plotted on this diagram and specific pressures and temperatures indicate conditions in various parts of the cycle. The functional use of the curve will be shown later.

The saturation curve is an important tool because it demonstrates the relationship between pressure, temperature, and total heat content (enthalpy) during the refrigeration process. For example, at 20 ° F the pressure of saturated R-502 is 67.1 psia and will remain so while the R-502 absorbs heat and changes from liquid to vapor. Raising or lowering the temperature will correspondingly alter the pressure readings in a totally predictable way, as long as the refrigerant is saturated.

To the right of the saturation curve, in the superheat region, there are an infinite number of temperature readings at any pressure. The same holds true in the subcooled region to the left of the saturation envelope.

The refrigerant circuit in Figure 5-2 is a simplified representation and does not account for nor does it show the pressure drops through the various components or lines which may or may not affect component selection and overall performance. It is always best to analyze the system to determine these requirements.

Figure 5-5 shows the refrigeration circuit with both the ideal case and one with the pressure drops of the various components, indicated by dotted lines.

If the pressure drops become excessive, they can cause inefficiencies and reduce performance. Pressure drops cannot be avoided, but excessive ones can and proper sizing and line selections will minimize these losses.

Pressure drops on the high side allow pressure in

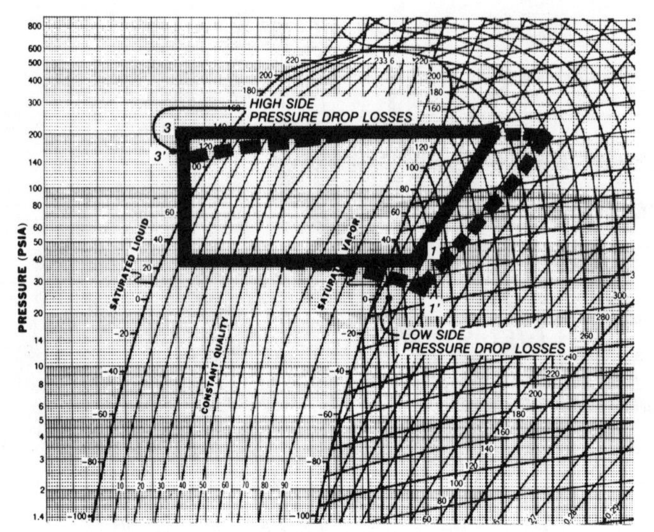

FIG. 5-5. Effect of pressure drops on refrigeration cycle. *R-12 P-H diagram courtesy E.I. dupont de Nemours & Co.*

the liquid line to point #3 to droop. If the pressure drops into the saturated region, the liquid R-502 will begin to vaporize before entering the expansion valve and reduce system performance.

On the low side, a drop in pressure lowers the temperature and increases the volume of vapor per pound of refrigerant. This creates two problems;

- The compressor is a constant volume pump. If the vapor expands, the compressor will pump fewer pounds of refrigerant and system capacity will be reduced.
- Lower pressures into the compressor increases the compression ratio and drives the discharge temperature farther out into the superheat region. This uses more horsepower while making the compressor and circulating oil run at higher and possibly destructive temperatures.

Pocket-sized temperature-pressure tables, Table 5-1, give the saturated conditions for five common refrigerants: R-12, R-22, R-500, R-502 and ammonia or R-717. These are the pressure-temperature relationships that exist within the saturation envelope in Figure 5-2. Subcooled or superheated conditions are not listed. However, the tables are extremely valuable and most refrigeration cycles can be estimated from them.

The two horizontal lines on the P-H chart, representing the condenser and evaporator pressures, can be estimated and drawn by reading the head (high) and suction (low) pressures at the compressor service valves. By subtracting line and valve pressure drops, the actual pressures in the condenser and evaporator may be estimated. Since these are psig values, read from the service gage, Table 5-1 can be read directly. However, when using the P-H diagram, each pressure reading must have 14.7 psi added to obtain absolute pressure values in psia. A 150 psig head pressure and 25 psig suction pressure will

Table 5-1

SPORLAN TEMPERATURE PRESSURE CHART

Vacuum-Inches of Mercury – Italic Figures

Pressure-Pounds Per Square Inch Bold Figures

TEMPERATURE °F.	12-F	22-V	500-D	502-R	717-A
−60	*19.0*	*12.0*	*17.0*	*7.2*	*18.6*
−55	*17.3*	*9.2*	*15.0*	*3.8*	*16.6*
−50	*15.4*	*6.2*	*12.8*	*0.2*	*14.3*
−45	*13.3*	*2.7*	*10.4*	1.9	*11.7*
−40	*11.0*	0.5	*7.6*	4.1	*8.7*
−35	*8.4*	2.6	*4.6*	6.5	*5.4*
−30	*5.5*	4.9	*1.2*	9.2	*1.6*
−25	*2.3*	7.4	1.2	12.1	1.3
−20	0.6	10.1	3.2	15.3	3.6
−18	1.3	11.3	4.1	16.7	4.6
−16	2.0	12.5	5.0	18.1	5.6
−14	2.8	13.8	5.9	19.5	6.7
−12	3.6	15.1	6.8	21.0	7.9
−10	4.5	16.5	7.8	22.6	9.0
− 8	5.4	17.9	8.8	24.2	10.3
− 6	6.3	19.3	9.9	25.8	11.6
− 4	7.2	20.8	11.0	27.5	12.9
− 2	8.2	22.4	12.1	29.3	14.3
0	9.2	24.0	13.3	31.1	15.7
1	9.7	24.8	13.9	32.0	16.5
2	10.2	25.6	14.5	32.9	17.2
3	10.7	26.4	15.1	33.9	18.0
4	11.2	27.3	15.7	34.9	18.8
5	11.8	28.2	16.4	35.8	19.6
6	12.3	29.1	17.0	36.8	20.4
7	12.9	30.0	17.7	37.9	21.2
8	13.5	30.9	18.4	38.9	22.1
9	14.0	31.8	19.0	39.9	22.9
10	14.6	32.8	19.7	41.0	23.8
11	15.2	33.7	20.4	42.1	24.7
12	15.8	34.7	21.2	43.2	25.6
13	16.4	35.7	21.9	44.3	26.5
14	17.1	36.7	22.6	45.4	27.5
15	17.7	37.7	23.4	46.5	28.4
16	18.4	38.7	24.1	47.7	29.4
17	19.0	39.8	24.9	48.8	30.4
18	19.7	40.8	25.7	50.0	31.4
19	20.4	41.9	26.5	51.2	32.5
20	21.0	43.0	27.3	52.4	33.5
21	21.7	44.1	28.1	53.7	34.6
22	22.4	45.3	28.9	54.9	35.7
23	23.2	46.4	29.8	56.2	36.8
24	23.9	47.6	30.6	57.5	37.9
25	24.6	48.8	31.5	58.8	39.0
26	25.4	49.9	32.4	60.1	40.2
27	26.1	51.2	33.2	61.5	41.4
28	26.9	52.4	34.2	62.8	42.6
29	27.7	53.6	35.1	64.2	43.8
30	28.4	54.9	36.0	65.6	45.0
31	29.2	56.2	36.9	67.0	46.3
32	30.1	57.5	37.9	68.4	47.6
33	30.9	58.8	38.9	69.9	48.9
34	31.7	60.1	39.9	71.3	50.2
35	32.6	61.5	40.9	72.8	51.6
36	33.4	62.8	41.9	74.3	52.9
37	34.3	64.2	42.9	75.8	54.3
38	35.2	65.6	43.9	77.4	55.7
39	36.1	67.1	45.0	79.0	57.2
40	37.0	68.5	46.1	80.5	58.6
41	37.9	70.0	47.1	82.1	60.1
42	38.8	71.4	48.2	83.8	61.6
43	39.8	73.0	49.4	85.4	63.1
44	40.7	74.5	50.5	87.0	64.7
45	41.7	76.0	51.6	88.7	66.3
46	42.6	77.6	52.8	90.4	67.9
47	43.6	79.2	54.0	92.1	69.5
48	44.6	80.8	55.1	93.9	71.1
49	45.7	82.4	56.3	95.6	72.8
50	46.7	84.0	57.6	97.4	74.5
55	52.0	92.6	63.9	106.6	83.4
60	57.7	101.6	70.6	116.4	92.9
65	63.8	111.2	77.8	126.7	103.1
70	70.2	121.4	85.4	137.6	114.1
75	77.0	132.2	93.5	149.1	125.8
80	84.2	143.6	102.0	161.2	138.3
85	91.8	155.7	111.0	174.0	151.7
90	99.8	168.4	120.6	187.4	165.9
95	108.2	181.8	130.6	201.4	181.1
100	117.2	195.9	141.2	216.2	197.2
105	126.6	210.8	152.4	231.7	214.2
110	136.4	226.4	164.1	247.9	232.3
115	146.8	242.7	176.5	264.9	251.5
120	157.6	259.9	189.4	282.7	271.7
125	169.1	277.9	203.0	301.4	293.1
130	181.0	296.8	217.2	320.8	—
135	193.5	316.6	232.1	341.2	—
140	206.6	337.2	247.7	362.6	—
145	220.3	358.9	264.0	385.0	—
150	234.6	381.5	281.1	408.4	—
155	249.5	405.1	298.9	432.9	—

Courtesy Sporlan Valve Co.

have 150 + 14.7 = 164.7 psia condensing pressure and 25 + 14.7 = 39.7 psia evaporator pressure. These psia values are then found on the P-H diagram and the cycle plotted.

Various regulators will be discussed and shown in relation to the P-H diagram. These regulators are classified as:

- Pressure sensing:
  - Reducing regulators, control downstream pressures
  - Relief regulators, control upstream pressures
- Combination pressure-temperature sensing:
  - Expansion valves, thermostatic types

Expansion valves

Since the expansion valve is the dividing point between the high and low sides, it will be discussed first.

The expansion valve is designed to meter refrigerant into a heat transfer device such as an evaporator coil, chiller, cold plate, etc. They can be considered high side to low side refrigerant flow regulators. Constant pressure expansion valves respond to pressure changes in the heat transfer device, based on load variations, while thermostatic valves respond to suction superheat which is related to load.

Under normal operating conditions, refrigerant is a high pressure, 100% liquid after it leaves the receiver. As the refrigerant passes through a high to low side throttling device, it enters the evaporator at varying percentages of quality, that is, in a partially liquid, partially gaseous state. In most cases, this is 20% quality, meaning 20% vapor and 80% liquid. The 80% of the refrigerant that is liquid absorbs the heat and provides efficient cooling.

As the refrigerant passes through the evaporator, the liquid boils and becomes a vapor. Leaving the evaporator of a properly sized system, the refrigerant will be 100% vapor and superheated.

Figure 5-6 shows the refrigerant cycle on the P-H chart. Reducing the liquid temperature, prior to its

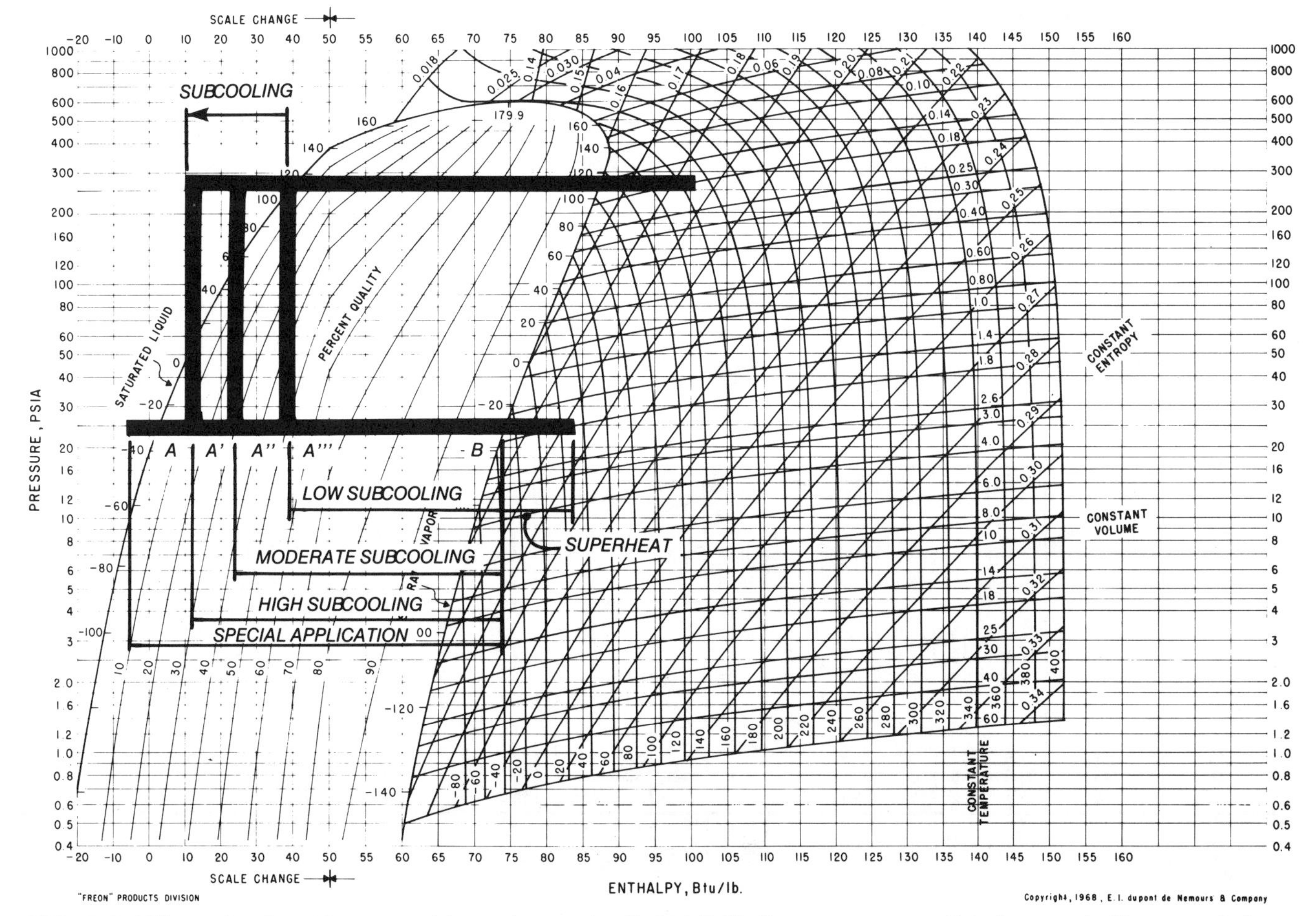

FIG. 5-6. Effect of subcooling on refrigeration cycle. *R-502 P-H diagram courtesy E.I. dupont de Nemours & Co.*

entering the expansion valve, is called *subcooling* and increasing its temperature, beyond the saturated vapor line, is *superheating* the vapor. Saturated liquid, 0% quality, is at point A of Figure 5-6 and saturated vapor, 100% quality, is at point B. As point A moves to the right, there is less subcooling and the amount of heat the evaporator can absorb is reduced. As point A moves to the left, there is greater subcooling, increasing the amount of heat the evaporator can absorb. With subcooling, the capacity of the evaporator is increased. The heat that the evaporator absorbs is known as the *refrigeration effect* and is measured in Btu/lb of refrigerant.

Expansion valves are sized with the aid of tables supplied by the valve manufacturer. Most tables are based on 100 °F liquid to the valve and 7 °F superheat. Variables include evaporator temperatures from 40 °F to −40 °F and varying differential pressures across the valve.

For example: A two ton R-502 system is to operate with a 20 °F (52.4 psig) evaporator and 200 psig (95 °F condensing) head pressure. The liquid entering the valve is subcooled 15 ° and its temperature is reduced to 80 °F by a suction to liquid heat exchanger.

Pressure drop across the valve must be determined in order to select the correct valve. Table 5-2 shows a portion of the selection chart from Sporlan Valve Co. Variables include evaporator temperature and refrigerant pressure drop across the valve.

Pressure drops in the piping, coils, and vertical lift of the liquid, for applications where the coil is above the condensing unit, will lower the actual head pressure at the valve inlet and, if significant, must be accounted for. Actual valve differential pressure is found as follows:

Compressor discharge pressure		200 psig
liquid line pressure drop	1 psig	
valves, solenoids, driers, etc.	3	
condenser coil	1	
static lift of 20 feet	10	
	15 psig	−15
Pressure at valve inlet		185 psig

If the suction pressure is, say 45 psig, then suction losses include:

Compressor suction pressure		45.0 psig
suction line pressure drop	0.5 psig	
valves, strainers	0.5	
evaporator coil	1.5	
distributor	35.0	
	37.5 psig	+37.5
Pressure at valve outlet		82.5 psig

Table 5-2. Thermostatic expansion valve (TXV) capacities in tons. Partial listing, −10° to −40° not shown.

VALVE TYPE	NOMINAL CAPACITY	EVAPORATOR TEMPERATURE DEGREES F.																	
		40°						20°						0°					
		PRESSURE DROP ACROSS VALVE (Pounds Per Square Inch)																	
		75	100	125	150	175	200	75	100	125	150	175	200	75	100	125	150	175	200
G-NI-F	¼	0.22	**0.25**	0.28	0.31	0.33	0.35	0.22	0.25	**0.28**	0.31	0.33	0.35	0.22	0.25	0.28	0.31	0.33	0.35
G-C-S-NI-F	½	0.43	**0.50**	0.56	0.61	0.66	0.71	0.43	0.50	**0.56**	0.61	0.66	0.71	0.43	0.50	0.56	0.61	0.66	0.71
G-C-S-NI-F	1	0.87	**1.00**	1.12	1.22	1.32	1.41	0.87	1.00	**1.12**	1.22	1.32	1.41	0.87	1.00	1.12	1.22	1.32	1.41
G-C-S-H	1½	1.30	**1.50**	1.68	1.84	1.98	2.12	1.21	1.40	**1.56**	1.71	1.85	1.98	1.04	1.20	1.34	1.47	1.59	1.70
G-C-S	2	1.73	**2.00**	2.24	2.45	2.64	2.83	1.64	1.90	**2.12**	2.33	2.51	2.69	1.38	1.60	1.79	1.96	2.12	2.26
C-S	3	2.42	**2.80**	3.13	3.43	3.70	3.96	2.29	2.65	**2.96**	3.24	3.50	3.75	1.73	2.00	2.24	2.45	2.64	2.83
C-S-H	4	3.46	**4.00**	4.47	4.90	5.29	5.66	2.86	3.30	**3.69**	4.04	4.36	4.67	2.42	2.80	3.13	3.43	3.70	3.96
C-S	6	4.76	**5.50**	6.15	6.74	7.28	7.78	3.90	4.50	**5.03**	5.51	5.95	6.36	3.20	3.70	4.14	4.53	4.89	5.23
H	6½	5.62	**6.50**	7.25	7.96	8.61	9.19	4.98	5.75	**6.42**	7.04	7.61	8.13	3.98	4.60	5.14	5.63	6.08	6.50
P-H	9	8.23	**9.50**	10.6	11.6	12.6	13.4	6.06	7.00	**7.83**	8.57	9.26	9.89	4.76	5.50	6.14	6.74	7.27	7.78
P-H	12	11.3	**13.0**	14.5	15.9	17.2	18.4	10.5	12.1	**13.5**	14.8	16.0	17.1	8.40	9.70	10.8	11.9	12.8	13.7
O	6	4.76	**5.50**	6.15	6.74	7.28	7.78	3.90	4.50	**5.03**	5.51	5.95	6.36	3.20	3.70	4.14	4.53	4.89	5.23
O	9	8.23	**9.50**	10.6	11.6	12.6	13.4	6.06	7.00	**7.83**	8.57	9.26	9.90	4.76	5.50	6.15	6.74	7.28	7.78
O	12	11.3	**13.0**	14.5	15.9	17.2	18.4	10.5	12.1	**13.5**	14.8	16.0	17.1	8.40	9.70	10.8	11.9	12.8	13.7
O	30	26.2	**30.2**	33.8	37.0	40.0	42.7	21.7	25.0	**28.0**	30.6	33.1	35.4	17.3	20.0	22.4	24.5	26.5	28.3
O	45	39.0	**45.0**	50.3	55.1	59.5	63.6	34.6	40.0	**44.7**	49.0	52.9	56.6	30.3	35.0	39.1	42.9	46.3	49.5
M	15	13.7	**15.8**	17.7	19.3	20.9	22.3	13.0	15.0	**16.8**	18.4	19.8	21.2	10.8	12.5	14.0	15.3	16.5	17.7
M	20	17.8	**20.6**	23.0	25.2	27.2	29.1	16.5	19.0	**21.2**	23.3	25.1	26.7	12.6	14.5	16.2	17.8	19.2	20.5
M	25	22.2	**25.7**	28.7	31.5	34.0	36.3	19.5	22.5	**25.2**	27.6	29.8	31.8	14.9	17.2	19.2	21.1	22.8	24.3

Courtesy Sporlan Valve Co.

Table 5-2a. TXV correction factors.

REFRIGERANT LIQUID TEMPERATURE CORRECTION FACTORS

Refrigerant Liquid Temperature °F.	40°	50°	60°	70°	80°	90°	100°	110°	120°	130°	140°
Correction Factor	1.52	1.44	1.35	1.26	1.18	1.09	1.00	0.91	0.82	0.73	0.64

Courtesy Sporlan Valve Co.

The final differential pressure across the valve is 185.0 - 82.5 = 102.5 psig. Use 100 psig in the charts.

If the suction pressure at the compressor is unknown, but the evaporator temperature is say 20 °F, then the corresponding saturated pressure is 52.4 psig. When a distributor is used, the valve outlet pressure is 52.4 + 35 psig or 87.4 psig. The saturation pressures are obtained from Table 5-1.

The pressure drops can be shown schematically, Figure 5-7, to obtain the differential pressure across the valve.

The correct valve is selected by looking in the 20 °F, R-502 Table at 100 psig differential. Since this is a 2-ton system, go down the 100 psig column of Table 5-2 to 1.90 tons. The table is rated for 100 °F liquid at the valve inlet but, since the liquid is subcooled to 80 °F, a correction is needed. The valve will have more capacity since subcooling increases valve performance. The correction factor, Table 5-2a, is 1.18 for 80 °F liquid. The 1.9 tons is then multiplied by 1.18, yielding 2.24 tons. A style G, C or S valve may be selected. Double checking, the next smaller size, or 1½ ton valve, yields 1.4 × 1.18 = 1.65 tons. This is not large enough, therefore use the 2-ton valve.

If the evaporator is in a packaged unit or at the same elevation as the compressor, there is no vertical lift. If a simple, low-pressure-drop coil is used and a distributor is not required, those pressure drops are not accounted for. The difference between the compressor discharge and service valve gauge reading will closely indicate the expansion valve differential, but not accurately. In the previous example, this would be 200 − 52.5 = 147.5 psig. Use 150 psig. At this condition, a 1½ ton G, S or H valve would supply 1.71 tons. A 1.18 Correction factor, for liquid subcooling, yields 1.71 × 1.18 = 2.01 tons. A 1½ ton valve could be used in this application, but would not be adequate.

The ordering information for the expansion valve would then include:

- Style of valve
- Tonnage
- Refrigerant type

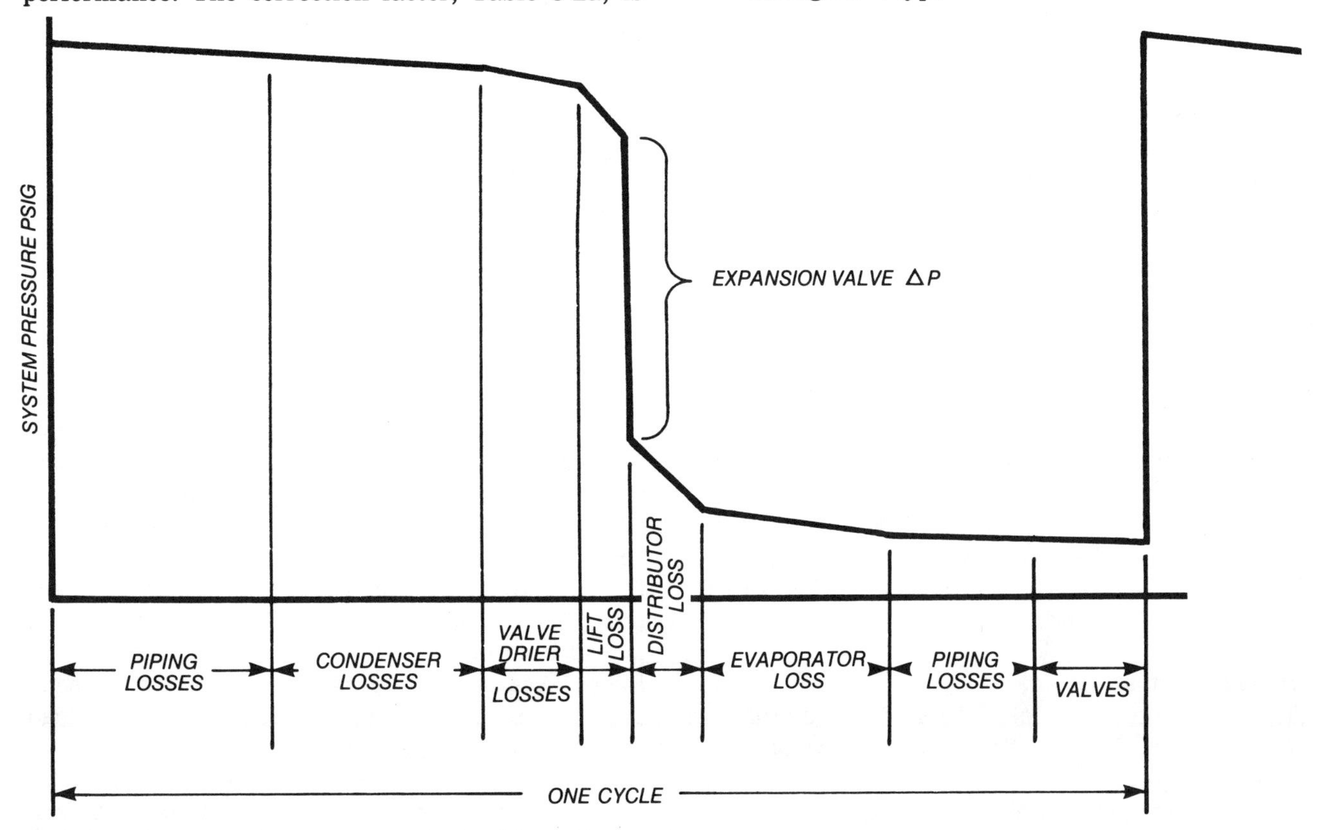

FIG. 5-7. Pressure losses in a typical refrigeration system.

Table 5-3. Static pressure loss due to vertical lift.

REFRIGERANT	VERTICAL LIFT — FEET 20	40	60	80	100
	Static Pressure Loss — psi				
12	11	22	33	44	55
22	10	20	30	40	50
500	10	19	29	39	49
502	10	21	31	41	52
717 (Ammonia)	5	10	15	20	25

Courtesy Sporlan Valve Co.

Table 5-4. Distributor pressure drop at 100% loading.

REFRIGERANT	*AVERAGE PRESSURE DROP ACROSS DISTRIBUTOR
12	25 psi
22	35 psi
500	25 psi
502	35 psi
717 (Ammonia)	40 psi

Courtesy Sporlan Valve Co.

- Internal or external equalizer
- Remote bulb thermostatic charge type
- Capillary tube length
- Line sizes needed

Table 5-3 shows the pressure loss of refrigerants at different vertical lifts. In the R-502 application, a 20 foot lift was encountered. The static pressure loss is found to be 10 psi.

A properly selected distributor, that is loaded at 100% of design, has a pressure drop that varies with the type of refrigerant, Table 5-4. In the example above, the pressure drop was 35 psi for R-502. If the distributor capacity is less than 100%, then the pressure drop varies with the percentage of load, Table 5-5. Note that at 100% loading, the pressure drop across a R-502 distributor is 35 psi and is made up of 25 psi across the nozzle or distributor orifice and 10 psi in the distributor tubes.

For this application, the R-502 expansion valve will have a RC thermostatic charge, Table 5-6.

Low side regulators

There are three types of suction line regulators:

- Crankcase pressure regulators (downstream sensing)
- Evaporator pressure regulators (upstream sensing)

Table 5-5. Distributor pressure drops at various loadings.

Actual Load as a Percent of Published Rating (corrected for Liquid Temperature and/or Tube Length if necessary)	Refrigerant 12		Refrigerants 22 & 502	
	ΔP Nozzle psi	ΔP Tubes and Passages psi	ΔP Nozzle psi	ΔP Tubes and Passages psi
50	4	3	7	3
60	6	4	10	4
70	8	5	13	5
80	10	6	16	6
90	12	8	20	8
100	**15**	**10**	**25**	**10**
110	18	12	30	12
120	20	14	35	14
130	22	16	38	16
140	24	18	40	18
150	27	21	43	21
160	29	23	46	24
170	31	25	49	27
180	33	27	52	30
190	36	29	54	32
200	38	31	57	34

Courtesy Sporlan Valve Co.

Table 5-6. Recommended charges for thermostatic expansion valve remote bulbs are based on operating temperature.

REFRIGERANT	AIR CONDITIONING OR HEAT PUMP	COMMERCIAL REFRIGERATION 40°F. to -10°F.	LOW TEMPERATURE REFRIGERATION 0°F. to -40°F.	EXTREME LOW TEMPERATURE REFRIGERATION -40°F. to -100°F.
12	FCP60	FC	FZ FZP	FX
22	VCP100 and VGA	VC	VZ VZP	VX
502	RCP115	RC	RZ RZP	RX

Courtesy Sporlan Valve Co.

- Stop valves or shut-off valves (solenoid or condenser powered)

Crankcase pressure regulator (CPR) This regulator is installed in the suction line, just before the compressor, to throttle excessive suction line pressures so that the compressor is not overloaded. The CPR senses downstream or compressor pressure. If the pressure to the compressor raises above the set point, the regulator begins to shut down, creating a pressure differential across the valve. This throttling reduces the outlet pressure and protects the compressor from overloading during start up, high load periods or defrost periods.

Table 5-7 shows the capacities of 5 hp high, medium, and low temperature compressors and their operating temperature-pressure ranges for

Table 5-7. Refrigerating capacities of a 5 hp compressor in Btu/hour.

EVAPORATING TEMPERATURE				45°F	35°F	25°F	15°F	5°F	0°F	-5°F	-15°F	-25°F	-35°F	-40°F	-50°F	-60°F	-80°F
SUCTION PRESSURE			R-12	41.7#	32.6#	24.6#	17.7#	11.8#	9.2#	6.7#	2.5#	2.3"	8.4"	11.0"	15.4"	19.0"	24.0"
			R-22	76.0#	61.5#	48.8#	37.7#	28.2#	24.0#	20.0#	13.2#	7.4#	2.5#	0.5#	6.1"	12.0"	20.2"
New Unit Model No.	Old Unit Model No.	Motor Compressor	R-502	88.3#	72.6#	58.7#	46.6#	36.0#	31.2#	26.8#	19.0#	12.3#	6.7#	4.3#	0.0#	7.0"	17.1"
CBAH-0503	CBAH-0504	MRF*-0500	R-12	54000	44900	37000	30200	24000	22000								
CBAM-0503	CBAM-0504	9RA*-0500	R-12			43800	36000	28900	25800	23000							
CBAL-0503	CBAL-0504	9RB*-0500	R-12							29600	22800	16800	12300	10200			
C3AH-0503	C3AH-0504	NRA*-0500	R-22	59500	50750	43000	35500	29400	25000								
C3AM-0503	C3AM-0505	NRM*-0500	R-22			50000	41000	32500	29000	25250							
C3AU-0503	C3AU-0505	9TK*-0500	R-22									21500	17300	15400	12300	9250	5050
C3AM-0503	C3AM-0505	NRM*-0500	R-502			51000	43000	35500	32000	26000							
C3AL-0503	C3AL-0505	MRA*-0500	R-502						35000	31600	25200	19600	15000	12750			
C3AU-0503	C3AU-0505	9TK*-0500	R-502									26200	21000	18750	14875	11000	5550

Courtesy Copeland Corp.

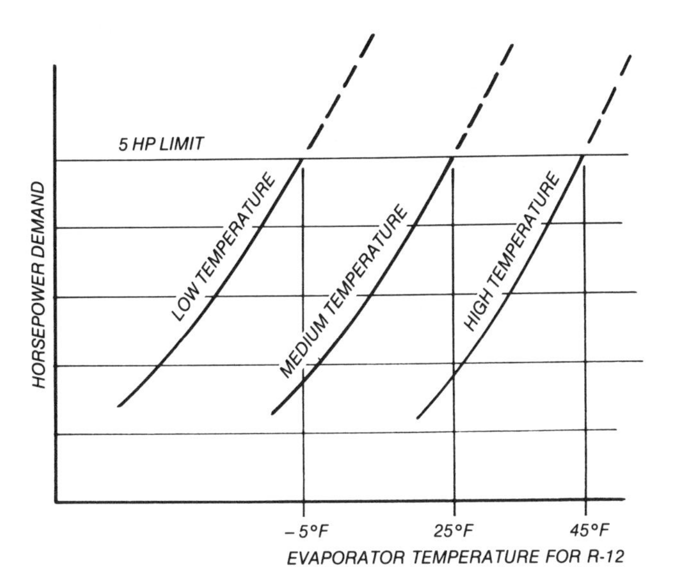

FIG. 5-8. Low, medium, and high compressor capacity vs. horsepower curves, limited by crankcase pressure regulator valve to 5 hp.

R-12, R-22 and R-502.

For an R-12 system, the high temperature model cannot exceed 45 °F (41.7 psig), the medium temperature 25 °F (24.6 psig), and the low temperature version −5 °F (6.7 psig).

For systems with substantial load variations, the compressors could be overloaded if their suction pressures exceed these values. Figure 5-8 shows how the crankcase pressure regulator throttles the pressure to the compressor, keeping the load within range. Excessive suction pressures will cause high horsepower demand on the motor as well as high current draw. If the situation continues, burnout could result.

Here is how a crankcase regulator is selected: A R-502 system has a capacity of 1.4 tons at −30 °F evaporator (9.2 psig saturated). The maximum allowable suction pressure is 30 psig. Table 5-8 shows for a R-502 system and compressor suction pressure

Table 5-8. Crankcase pressure regulator capacities in tons.

VALVE TYPE	DESIGN SUCTION PRESSURE psig	SATURATION TEMPERATURE °F.	½ psi PRESSURE DROP ACROSS VALVE						1 psi PRESSURE DROP ACROSS VALVE						2 psi PRESSURE DROP ACROSS VALVE					
			VALVE SETTING - psig						VALVE SETTING - psig						VALVE SETTING - psig					
			10	20	30	40	50	60	10	20	30	40	50	60	10	20	30	40	50	60
CRO-6 CROT-6	0	–50°	.23	.40	.48	.51	.51	.51	.33	.56	.67	.72	.72	.72	.47	.79	.94	1.00	1.00	1.00
	10	–29°	—	.32	.53	.65	.68	.68	—	.45	.75	.91	.96	.96	—	.64	1.07	1.27	1.33	1.33
	20	–14°	—	—	.40	.67	.81	.85	—	—	.57	.94	1.14	1.19	—	—	.79	1.34	1.58	1.68
	30	–1°	—	—	—	.46	.78	.81	—	—	—	.65	1.10	1.13	—	—	—	.93	1.56	1.85
	40	9°	—	—	—	—	.52	.89	—	—	—	—	.74	1.25	—	—	—	—	1.05	1.76
	50	18°	—	—	—	—	—	.58	—	—	—	—	—	.82	—	—	—	—	—	1.17
CRO-10 CROT-10	0	–50°	.60	1.03	1.03	1.03	1.03	1.03	.85	1.45	1.45	1.45	1.45	1.45	1.17	2.02	2.02	2.02	2.02	2.02
	10	–29°	—	[illegible]	1.39	1.39	1.39	1.39	—	[illegible]	1.96	1.96	1.96	1.96	—	[illegible]	2.74	2.74	2.74	2.74
	20	–14°	—	—	1.02	1.74	1.74	1.74	—	—	1.44	2.45	2.45	2.45	—	—	1.98	3.41	3.41	3.41
	30	–1°	—	—	—	1.19	2.03	2.03	—	—	—	1.67	2.86	2.86	—	—	—	2.31	3.99	3.99
	40	9°	—	—	—	—	1.36	2.31	—	—	—	—	1.91	3.25	—	—	—	—	2.53	4.53
	50	18°	—	—	—	—	—	1.50	—	—	—	—	—	2.11	—	—	—	—	—	2.92

Courtesy Sporlan Valve Co.

Table 5-9. Evaporator pressure regulator capacities in tons.

VALVE TYPE	EVAPORATOR DESIGN TEMPERATURE °F.	SATURATION PRESSURE – psig			REFRIGERANT											
		REFRIGERANT			12				22				502			
					PRESSURE DROP ACROSS VALVE – psi											
		12	22	502	2	5	10	20	2	5	10	20	2	5	10	20
ORIT-6 0/50 or 30/100	40°	37.0	68.5	80.2	.83	1.4	2.0	2.7	1.2	2.0	3.0	4.4	.92	1.5	2.3	3.5
	30°	28.4	54.9	65.4	.75	1.2	1.7	2.2	1.1	1.8	2.7	3.8	.83	1.3	2.0	3.0
	20°	21.0	43.0	52.4	.67	1.1	1.5	1.8	1.0	1.6	2.4	3.3	.75	1.2	1.8	2.5
	10°	14.6	32.8	41.1	.58	.94	1.2	1.5	.87	1.4	2.1	2.7	.62	1.0	1.5	2.1
	0°	9.2	24.0	31.2	.5	.81	1.0	1.2	.79	1.2	1.8	2.2	.54	.90	1.3	1.7
	−10°	4.5	16.5	22.8	.46	.64	.80	—	.67	1.1	1.5	1.8	.46	.77	1.1	1.4
	−20°	0.6	10.1	15.5	.37	.51	.57	—	.58	.9	1.2	1.4	.37	.64	.88	1.1
ORIT-10 0/50 or 30/100	40°	37.0	68.5	80.2	2.0	3.2	4.5	5.8	3.0	4.8	6.9	9.5	2.2	3.5	5.3	7.5
	30°	28.4	54.9	65.4	1.8	2.9	3.9	4.8	2.7	4.3	6.1	8.3	2.0	3.1	4.6	6.5
	20°	21.0	43.0	52.4	1.6	2.5	3.3	3.9	2.4	3.8	5.4	7.2	1.8	2.8	4.0	5.5
	10°	14.6	32.8	41.1	1.4	2.2	2.8	3.2	2.1	3.3	4.7	5.9	1.5	2.4	3.5	4.6
	0°	9.2	24.0	31.2	1.2	1.9	2.3	2.5	1.9	2.9	4.0	4.8	1.3	2.1	3.0	3.7
	−10°	4.5	16.5	22.8	1.1	1.5	1.8	—	1.6	2.5	3.3	3.9	1.1	1.8	2.5	3.0
	−20°	0.6	10.1	15.5	0.9	1.2	1.3	—	1.4	2.1	2.7	3.0	0.9	1.5	2.0	2.3

ALLOWABLE EVAPORATOR PRESSURE CHANGE – psi		2	4	6	8	10	12	14
CAPACITY MULTIPLIER	ORIT-6, 10 - 0/50	.3	.6	.8	1.0	1.2	1.3	1.4
	ORIT-6, 10 - 30/100	—	.2	.6	.7	.9	1.0	1.1

Courtesy Sporlan Valve Co.

of 10 psig and maximum setting of 30 psig and allowable valve pressure drop of ½ psig, a CRO-10 valve will handle 1.39 tons.

If larger capacity systems are employed, then the valve can still be used if a greater pressure drop across the valve can be tolerated. For example, the same valve will handle 1.96 tons at a one psig pressure drop and 2.74 tons at a 2 psig pressure drop. Larger valves are capable of handling upwards of hundreds of tons of refrigeration.

The design conditions needed for selection include:

- Type of refrigerant
- System capacity
- Normal design suction temperature or saturation pressure
- Maximum allowable suction pressure to the compressor
- Allowable pressure drop across the valve

Evaporator pressure regulator (EPR) The evaporator pressure regulator is another suction line regulator, but it regulates upstream pressures rather than downstream pressures like the CPR valve. The EPR will maintain minimum evaporator pressures. Many applications must operate at or just above the 32° freezing point. Under varying and light load conditions, the evaporator pressure will drop and could enter the freezing range. This will frost the evaporator coils or possibly freeze chillers. Both could cause system damage and further capacity reduction.

As the evaporator pressure nears the set point, the evaporator pressure regulator will begin to throttle and will maintain the evaporator pressure at the design minimum. Liquid refrigerant is actually backed up in the evaporator, raising the pressure. As the refrigeration load returns, evaporator pressure will increase and the valve will begin to open. This regulator allows the system to run continuously, rather than in short cycles between starting and stopping, reducing compressor wear.

Overall, EPR regulators have wide application:

- To control saturation pressure in the evaporator for indirect temperature control and freeze-up or frost protection.
- To maintain evaporator pressure during defrost — expediting defrost, conserving power, and reducing flood back.
- To provide safety or pressure relief.

Because evaporator pressure regulators maintain evaporator pressures higher than the common suction line, they are used where a number of evaporators is operating at different temperatures.

Here is how to select an evaporator pressure regulator for a single evaporator, using R-502, with a capacity of 1.8 tons at 35 °F under load (72.8 psig) and a minimum of 28 °F (62.8 psig). Table 5-9 shows

2, 5, 10 and 20 psig valve pressure drops along with R-12, R-22 and R-502 selections at evaporator temperatures of 40 °F to −20 °F. Below the table is a series of capacity multipliers based on pressure change and evaporator pressure.

With a single evaporator or chiller system, an evaporator pressure drop of 2 psig is used. The evaporator pressure change is 35 °F to 28 °F or 72.8 − 62.8 psig = 10.0 psig from loaded to unloaded conditions.

The capacity correction multiplier, from Table 5-9, is .9 at 10 psig. Since 35°F is not listed, an interpolation is required between 40°F and 30°F. 40°F capacity at 2 psig R-502 is 2.2 tons; at 30°F, it is 2.0 tons. Since 35°F is directly in between these two readings, 2.1 tons is used. Multiplying by the correction factor, the final tonnage is 2.1 X .9 =1.9 tons. Therefore, the ORIT-10-30/100, with a 1.218 port and end connections of 1-1/8 or 1-3/8 inches, is satisfactory.

With multiple temperature systems, the common suction line will operate at a pressure lower than that of the warmer evaporators and there is a larger pressure drop across the regulator. From the example above, if the common suction pressure is 62 psig, the pressure drop is then 72.8 − 62.0 = 10.8 psi. With an evaporator capacity of 1.8 tons, a pressure drop of 10.8 psig (use 10 psig column), and again interpolating between 30 °F and 40 °F, select an ORIT-6 regulator with a 2.15 ton rating. Multiplying by the correction factor, the resultant capacity is 2.15 × .9 = 1.94 tons. The ORIT-6-30/100, with a .75 in. port, is suitable for this application.

Selection parameters then include:

- Type of refrigerant
- System and evaporator design capacities in tons
- Evaporator design temperature or pressure
- Minimum evaporator temperature or pressure
- Available pressure drop across the regulator at design capacity

The preceeding series of direct acting, low side regulators is usually reserved for smaller refrigeration systems. On larger systems, where capacities may be in the order of hundreds of tons, regulators may have port diameters as large as 4 to 6 inches. Because of the large port areas and close control required, pilot operated regulators are used. The pilot is a small regulator used to control a larger regula-

Table 5-10. Evaporator pressure regulator capacities.

NOM. PORT SIZE	PORT SIZE CODE	VALVE INLET OR EVAPORATOR PRESSURE AND CORRESPONDING SATURATION TEMPERATURE (SET POINT)															
		37 psig 40° F				28.5 psig 30° F				21.1 psig 20° F				14.7 psig 10° F			
		PRESSURE DROP ACROSS VALVE PSI															
		2	5	10	20	2	5	10	20	2	5	10	20	2	5	10	20
		TONS REFRIGERATION—REFRIGERANT 12															
⅜"	11	.50	.7	1.0	1.1	.4	.6	.8	.9	.4	.6	.7	.7	.3	.5	.6	.6
½"	12	1.25	1.8	2.4	2.8	1.1	1.6	2.0	2.1	1.0	1.4	1.8	1.9	.8	1.1	1.3	1.4
¾"	13	2.2	3.2	4.2	5.0	1.9	2.9	3.7	4.1	1.7	2.5	3.1	3.3	1.5	2.2	2.6	2.7
1"	14	4.0	5.9	7.9	9.0	3.5	5.3	6.6	7.5	3.1	4.6	4.9	6.1	2.7	4.0	4.7	4.9
1¼"	15	6.0	8.8	11.6	13.5	5.2	7.9	10.0	11.2	4.7	6.9	8.5	9.0	4.0	5.9	7.0	7.3
1½"	16	9.0	13.3	17.5	20.3	7.9	11.9	15.1	16.8	7.0	10.4	12.9	13.7	6.0	8.9	10.6	11.0
2"	18	17.0	25.2	33.0	38.2	14.8	22.4	28.5	31.8	13.2	19.6	24.3	25.7	11.3	16.7	20.0	20.7
2½"	110	30.0	42.0	51.0	56.0	26.0	36.0	44.0	48.0	23.0	32.0	39.0	43.0	20.0	28.0	34.0	37.0
3"	112	44.0	61.0	75.0	82.0	38.0	53.0	64.0	71.0	34.0	47.0	57.0	63.0	30.0	42.0	51.0	56.0
4"	116	85.0	119.0	144.0	159.0	76.0	106.0	129.0	142.0	67.0	93.0	113.0	125.0	60.0	84.0	102.0	112.0
5"	120	150.0	210.0	255.0	280.0	130.0	182.0	220.0	243.0	115.0	161.0	195.0	215.0	101.0	141.0	172.0	188.0
6"	124	225.0	315.0	382.0	420.0	192.0	268.0	326.0	359.0	170.0	238.0	289.0	318.0	151.0	211.0	256.0	282.0

Courtesy Alco Controls Div. Emerson Electric Co.

tor or area. The added advantage of the pilots is that they can be driven by pneumatic signal, temperature bulbs with capillaries, electrical signals or through solenoids, allowing multiple settings. Versatility of the pilot operation opens a variety of options.

This type of regulator requires a 2 psi pressure drop across the valve to provide full port opening. The regulator should never be selected for a load or flow rate less than 2 psi pressure drop.

Assume an R-12 system with an EPR regulator that has a setting of 30 °F (28.4 psig) and a load of 5 tons. The compressor will have a 5 ton capacity at a 23.4 psig suction pressure. The pressure differential between the regulator setting and compressor suction, at rated load, is 28.4 − 23.4 or 5 psi. Selection of the EPR regulator is based on this pressure drop and *not* the nominal 2 psi rating which will result in oversizing and poor performance.

Looking at Table 5-10, the evaporator temperature setting is found, that is 30 °F, and the 5 psig differential pressure across the valve column is located. Reading vertically downward until 5 tons is found, a one-inch regulator can be used. If the two psi column had been used instead, a 1¼″ regulator would have been selected and would have been oversized for the job, resulting in poor regulation.

Figure 5-9 shows a suction regulator, with pilot, used as a relief pressure regulator. When the evaporator is isolated during defrost, the pressure will rise to the set point of the regulator, which is the most efficient defrost temperature. A small pressure rise above the set point is required to bring the regulator to the full open position. The refrigerant released to the suction line is usually saturated liquid, so the valve may be considerably smaller than regulators sized for gas flow.

Figure 5-10 shows a suction regulator with pilot remote from the regulator body. This regulator is a combination evaporator pressure regulator and suction stop valve. When the pilot solenoid is energized, the regulator operates as a standard evaporator regulator. When the pilot solenoid is de-energized, the regulator will fully close and acts as a suction solenoid valve.

High side regulators

Most high side regulators are used to control head pressures during low ambient temperatures. Air cooled condensing units are usually sized for 95 °F summer operation. Winter conditions, with low ambient temperatures, cause over-condensing and an overall reduction in system performance. Low head pressure causes a low pressure differential across the expansion valve, thereby reducing capacity. This can be seen in Table 5-2 where, for the same

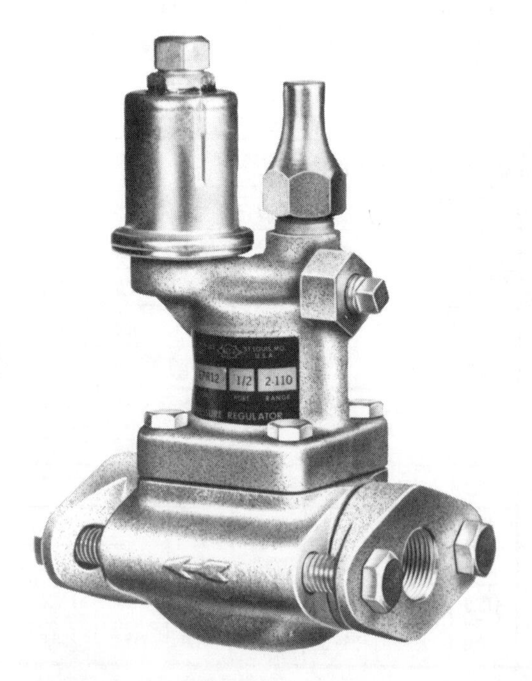

FIG. 5-9. Relief pressure regulator. *Courtesy Alco Controls Div., Emerson Electric Co.*

FIG. 5-10. Evaporator pressure regulator with suction stop combination. *Courtesy Alco Controls Div., Emerson Electric Co.*

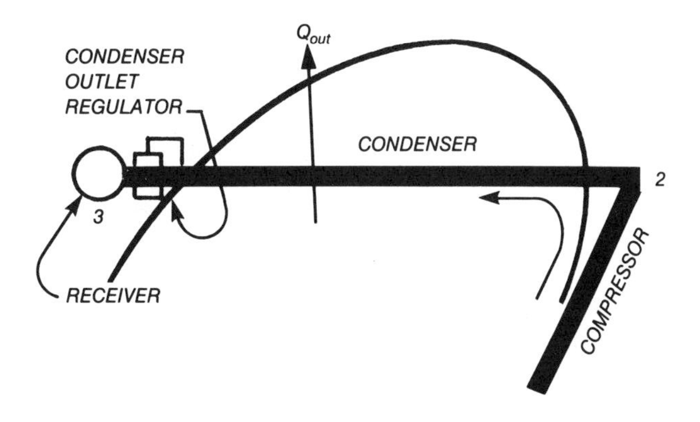

FIG. 5-11. Condenser outlet regulator.

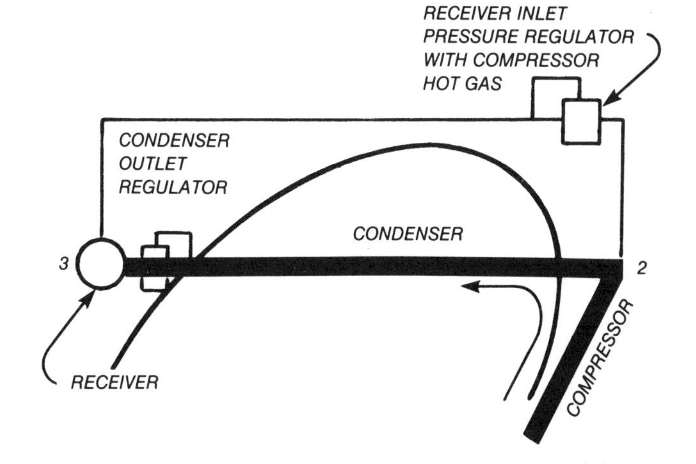

FIG. 5-12. Condenser outlet regulator with hot gas bypass to keep receiver at optimum operating pressure.

valve size and evaporating temperature, lower capacities correspond with lower pressure differentials across the valve. Normal head pressure is needed to maintain adequate inlet pressure to the expansion valve (pressure drop across the valve) and distributor for proper flow characteristics and capacity.

Figure 5-11 shows the placement of the compressor, condenser and receiver on a P-H diagram. The high pressure hot gas leaves the compressor discharge in the superheated condition, flows to the condenser, where it is desuperheated and then condensed to a subcooled liquid by rejecting its heat to the atmosphere. The high pressure subcooled liquid then flows to the receiver.

An upsteam sensing regulator, installed at the condenser outlet, senses the condenser pressure. If the condenser pressure falls below the regulator set point, due to low ambient conditions, the regulator will begin to throttle. When throttling, the regulator will hold back a portion of the condensed liquid and flood the lower section of the condenser to render it inactive. This decreases the condenser's surface area and matches its capacity to the lower ambient temperature.

As the ambient warms up, the condenser pressure increases above the regulator setting and the regulator begins to open. This allows the refrigerant to drain and increases the surface area and capacity of the condenser.

This solves part of the problem. The receiver may also be affected by the low ambient conditions.

With an outlet flow regulator on the condenser, the partially throttled regulator will cause a pressure drop at the receiver. A differential pressure regulator can be employed to deliver compressor discharge gas directly to the receiver, Figure 5-12, to maintain a minimum receiver pressure.

The differential regulator is used in conjunction with the upstream sensing condenser outlet regulator. Both regulators are pressure sensitive types and range from fully open to fully closed.

For water cooled systems, a water regulator is used at the condenser inlet to either throttle or bypass cooling water around the condenser. The regulator senses high side pressures and is actuated by the fluctuating head pressure.

Here is the procedure for sizing the head pressure controls for the above application:

One manufacturer offers two regulators, an ORI-6 with a ¾ inch orifice and an ORI-10 with a 1.218 orifice. Capacities, using R-502, vary from 5 to 10.5 tons and from 10 to 20.5 tons with a 1 to 4 psi pressure drop, Table 5-11.

Hot gas bypass regulator capacities, Table 5-12,

Table 5-11. Condenser outlet regulator liquid flow capacities.

VALVE TYPE	PRESSURE DROP ACROSS VALVE - psi	REFRIGERANT 12	22	502
ADJUSTABLE MODELS				
ORI-6	1	6.5	8.5	5.0
	2	9.5	12.0	7.5
	3	11.0	14.5	9.0
	4	13.0	17.0	10.5
ORI-10	1	12.5	16.5	10.0
	2	18.5	24.0	14.5
	3	22.5	29.0	17.5
	4	26.0	34.0	20.5

Courtesy Sporlan Valve Co.

Table 5-12. Condenser bypass regulator hot gas capacities.

VALVE TYPE	PRESSURE DROP ACROSS VALVE - psi	REFRIGERANT								
		12			22			502		
		MINIMUM AMBIENT DESIGN TEMPERATURES - °F.								
		−20°	0°	20°	−20°	0°	20°	−20°	0°	20°
ORD-4-20	25	15.5	18.0	21.0	22.0	25.0	30.0	12.0	13.5	16.0
	30	18.0	21.0	25.0	25.5	29.0	35.0	14.0	16.0	19.0

Courtesy Sporlan Valve Co.

are shown for both 25 and 30 psi pressure drop at ambients of −20 °F, 0 °F and +20 °F.

If a 10 ton system, with R-502, operates in a northern climate, with a possible −20 °F minimum design temperature, then a ORI-6 regulator is suitable, if a 4 psi pressure drop is tolerable. A ORD-4-20 hot gas regulator, with a 12 ton capacity at −20 °F, is also suitable. If the receiver outlet regulator pressure drop is to be kept to a minimum, then an ORI-10 with a 10 ton capacity at one psi pressure drop should be selected.

The parameters needed to properly select head pressure controls for winterization are:

- System capacity in tons
- Refrigerant type
- Minimum ambient design temperature (winter)
- Allowable pressure drop across each regulator

In this example, the condenser is partially flooded as the ambient temperature and condenser pressure fall.

Use of head pressure controls requires more liquid refrigerant than is required in an unregulated system. There must be enough refrigerant left in the system to keep a full liquid line to the expansion valve. With insufficient liquid, hot gas will enter the liquid line and refrigeration capacity will fall drastically or cease completely. With an increased amount of refrigerant, an extra large receiver is needed to store the excess during high ambient conditions.

Most manufacturers have written procedures for charging systems that employ this type of head pressure control. Information needed includes:

- Volume of condenser coil tubing based on tube I.D., tube length, passes and return bends.
- Minimum expected ambient temperature.
- Type of refrigerant.
- Percentage of compressor unloading, if used.

When charging a system with head pressure controls, especially at low ambients when the high side regulators are in a throttling position, extreme care is needed. If unloaders are part of capacity regulation, it is also necessary to know to what extent the compressor is unloaded during charging.

Another set of high side regulators has to do with heat reclaim from refrigeration systems. High side regulators, both upstream and downstream sensing, can be modified or similar low side regulators such as EPR and CPR valves can be used.

A heat reclaim system with inadequate or no pressure regulation is an inefficient use of heat reclaim equipment, with resultant poor reclaim performance which may even affect the refrigeration capability of the system. Regulators are needed to assure a minimum condensing pressure so that fluid flow to the expansion valve will be adequate for proper valve performance. Receiver pressure must also be maintained at a minimum, usually just below the pressure of the condenser.

Regulator selection for heat reclaim systems must take into account a number of factors:

- Refrigeration is the priority function and should not be compromised in order to recover heat.
- Condensing pressures must not be too high since this will reduce refrigeration capacity and increase power consumption.
- The line leaving the receiver must be pressurized and filled with liquid.
- Condensing pressure must not go below the minimum setting.

Control of the condensing pressure is the key to maximizing the heat that can be reclaimed from any particular system. The higher condensing pressure will raise the temperature level at which heat is available.

If a system has two 10 hp, 4 cylinder, R-502 hermetic compressors, operating at −20 °F (15.3 psig) suction, and is to have heat reclaim added, the following is needed for sizing the regulators:

- Normal condensing is at 90 °F (187 psig)
- Compressor capacity is 112,000 Btu/hr (9.3 tons)
- Heat of rejection is 176,000 Btu/hr
- Electrical consumption is 20.2 kW
- Cooling requirement is 80,000 Btu/hr (6.7 tons)

The 90 °F condensing temperature is too low for effective heat reclaim.

If a 120 ° condensing temperature is selected (283 psig), the compressor capacity is reduced from

112,000 to 82,000 Btu/hr (6.8 tons) and the amount of heat rejected is reduced from 176,000 to 150,000 Btu/hr while the power requirements increase from 20.2 kW to 21.4 kW. The refrigeration requirements are still met since only 80,000 Btu/hr of cooling is needed. It must also be remembered that with higher condensing pressures, in order to achieve higher discharge temperatures for heat reclaim, the discharge temperature, condensing pressure and compression ratio are also increased. The compressor manufacturer should be consulted to be certain that the compressor limits are not exceeded and any other requirements, such as forced air fan cooling or an oil cooler, are not required.

The increase in compressor power is only 1.2 kW or an equivalent of 4098 Btu/hr compared to the 150,000 Btu/hr that is available from reclaim.

Heat reclaim regulators can be used to create a number of system configurations:

- Minimum system without cold weather control
- System with cold weather condenser pressure control
- System with separate higher heat reclaim pressure control
- System with multiple heat reclaim condensers

Figure 5-13 shows a P-H diagram with compressor, condenser, receiver, and heat reclaim heat exchanger with appropriate regulator locations. The heat reclaim system uses regulators to control head pressures in the winter by bypassing both the heat reclaim and condenser coil with a downstream sensing regulator *B*. This regulator senses the receiver pressure and meters hot gas into the receiver to maintain the pressure needed for proper operation of the expansion valve. A second upstream sensing regulator, *D*, will operate during warm weather when *B* will be closed.

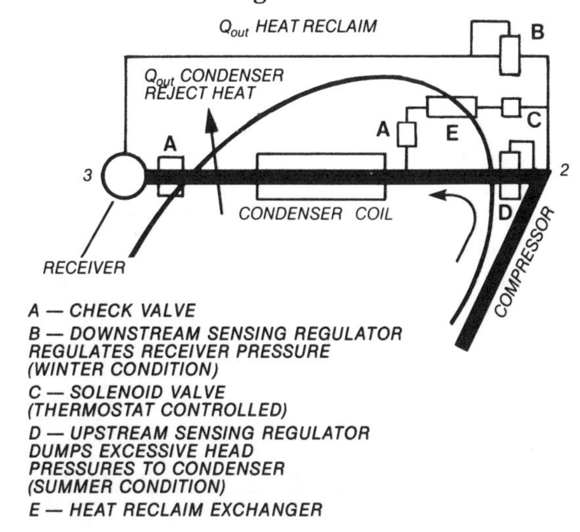

FIG. 5-13. Heat reclaim system in refrigeration circuit. *Courtesy Parker Hannifin Corp., Refrigerating Specialties Div.*

The heat reclaim exchanger is operated by a refrigerant solenoid valve, actuated by a thermostat. Regulator *D* maintains compressor discharge pressure at or above its set point and this pressure/temperature is supplied to the heat reclaim condenser.

If a 6.2 ton refrigeration system, operating with R-502 at 120 °F saturated condensing temperature and −20 °F saturated suction temperature, has 55 °F suction gas entering the compressor, determine the amount of heat to be reclaimed and the size of the regulators.

Table 5-13 shows the factors to be applied to the net refrigeration effect to determine the approximate heat of rejection. For the example above, the factor for −20 °F saturated suction, 120 °F condensing, and a hermetic compressor is 1.82. Therefore, a 6.2 ton system has a heat of rejection of 6.2 × 1.82 × 12,000 Btu/ton = 135,000 Btu/hr.

The regulators and valves for this system are selected by using Tables 5-14, 5-15, 5-16 and 5-17.

From Table 5-14, the outlet sensing regulator B, sized for 6.2 tons and 110ºF condensing-the highest value shown — and R-502 refrigerant, will be an A9 regulator with 5/8" diameter orifice at a 20 psi pressure drop. This regulator has a capacity of 8.1 tons at 0 °F suction. For a −20 °F application, a cor-

Table 5-13. Approximate heat rejection factors for R-12, R-22 and R-502.

Saturated Suction Temperature	Saturated Condensing Temperature				
Hermetic Compressors	90 °F	100 °F	110 °F	120 °F	130 °F
−40 °F	1.85	1.92	2.00	2.13	2.27
−20 °F	1.56	1.64	1.69	1.82	1.96
0 °F	1.41	1.45	1.52	1.61	1.72
+20 °F	1.32	1.35	1.41	1.47	1.56
+40 °F	1.25	1.28	1.33	1.39	1.43
Open Compressors	90 °F	100 °F	110 °F	120 °F	130 °F
−40 °F	1.42	1.47	1.54	1.59	1.67
−20 °F	1.32	1.37	1.41	1.45	1.52
0 °F	1.25	1.28	1.32	1.35	1.41
+20 °F	1.18	1.20	1.23	1.28	1.32
+40 °F	1.11	1.15	1.18	1.20	1.23

Courtesy Parker Hannifin Corp. Refrigerating Specialties Div.

Table 5-14. Condenser bypass valve capacities with R-502.

Size	Type	90°F		110°F		130°F	
		10 psi	20 psi	10 psi	20 psi	10 psi	20 psi
⅝"†	A9	2.2	3.1	1.7	2.4	--	--
⅝"	A9	7.4	10	5.8	8.1	--	--
⅞"	A9	14	20	11	15	--	--
1⅛"	A9	19	26	15	21	--	--
A7-L DIFFERENTIAL REGULATORS "B"							
⅝"*	A7AL	7.4	10	5.8	8.1	--	--
⅝"	A7AL	19	26	14	20	--	--
⅞" & 1⅛"	A7AL	26	37	21	29	--	--
1⅛"	A7A1L	53	74	41	58	--	--
1⅜" & 1⅝"	A7A1L	58	81	45	64	--	--
1⅝"	A72L	95	130	74	100	--	--
2⅛" & 2⅝"	A72L	140	200	110	160	--	--

Courtesy Parker Hannifin Corp. Refrigerating Specialties Div.

Table 5-15. Condenser inlet valve capacities with R-502.

Size	Type	90° to 130°F	
		2 psi	5 psi
INLET REGULATORS "D"			
⅝"*	A7A	0.90	1.5
⅝"	A7A	2.3	3.6
⅞" & 1⅛"	A7A	3.2	5.1
1⅛"	A7A1	6.5	10
1⅜" & 1⅝"	A7A1	7.1	11
1⅝"	A72	12	19
2⅛" & 2⅝"	A72	18	28
SOLENOID VALVES "D"			
3/16"	S6N	0.39	0.61
½"	S8F	1.8	2.8
¾"	S4A	4.7	7.3
1"	S4A	6.5	10
1¼"	S4A	12	18
1⅝"	S4A	22	34
2"	S4A	32	51
2½"	S4A	45	71
3"	S4A	65	100

Courtesy Parker Hannifin Corp. Refrigerating Specialties Div.

rection factor must be used. Table 5-17 shows that the correction factor for −20°F suction is .94. Therefore, the regulator capacity is 8.1 × .94 = 7.6 tons. Use spring B for a 80 to 220 psig pressure range.

The inlet sensing regulator, *D*, is found by using Table 5-15. For the given conditions, a A7A1, with a 1-1/8" orifice and spring range D, is needed. It has a capacity of 6.5 tons at a 2 psi pressure drop and 10.0

Table 5-16. Condenser outlet check valve capacities with R-502.

Size	Type	90°F	110°F	130°F
CHECK VALVES "C"				
½"	CK4A-2	11	9.5	8.3
¾"	CK4A-3	22	12	17
1"	CK4A-4	41	36	32
1¼"	CK4A-6	60	53	46
2"	CK4A-8	140	130	110
2½"	CK4A-9	220	200	170
3"	CK4A-0	350	310	270
4"	CK4-16	660	590	510

Courtesy Parker Hannifin Corp. Refrigerating Specialties Div.

Table 5-17. Correction factors to be applied to values from Tables 5-14, -15, and -16.

Ref't	Evaporator Temperature				
	−40°F	−20°F	0°F	20°F	40°F
R-12	0.90	0.95	1.00	1.04	1.08
R-22	0.94	0.97	1.00	1.03	1.06
R-502	0.88	0.94	1.00	1.05	1.10

Courtesy Parker Hannifin Corp. Refrigerating Specialties Div.

tons at a 5 psi pressure drop. For a −20°F evaporator and R-502, the correction factor is .94. Final capacity is 6.5 × .94 = 6.11 tons and 10.0 × .94 = 9.4 tons respectively.

The heat reclaim solenoid valve, *C*, is sized from Table 5-15. For 120°F condensing and 2 psi pressure drop, the capacity of a S4A 1" solenoid valve is 6.5 tons. At a −20°F suction, the correction factor again is .94. Final capacity is 6.5 × .94 = 6.11 tons.

A CK4A-3 check valve, *A*, with a ¾" orifice, was selected from Table 5-16. It is rated at 12 tons with 110°F condensing. Corrected, this becomes 12.0 × .94 = 11.3 tons and is adequate for both locations.

High side-to-low side regulators

There is another set of refrigeration valves that is quite common: high-to-low side regulators. They are also called hot gas bypass regulators and are used for capacity control.

A portion of the available refrigerant is bypassed

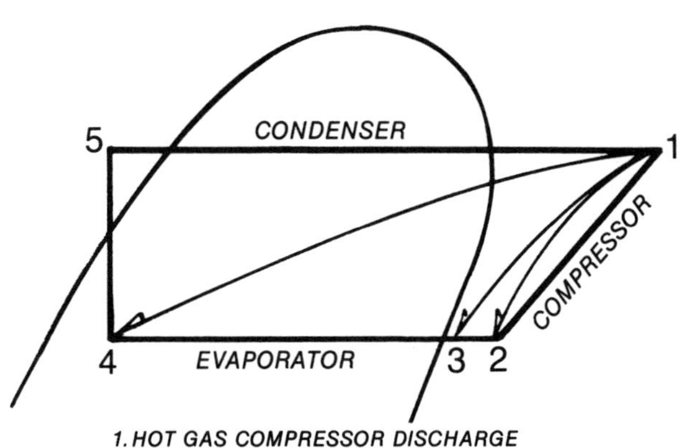

FIG. 5-14. Hot gas bypass for capacity control.

Table 5-18. Probable compressor discharge temperatures.

Refrigerant	Air Conditioning Range °F.	Commercial Range °F.
R-12	120 to 130	150 to 185
R-22	150 to 175	200 to 250
R-502	Not normally used	160 to 210

Courtesy Parker Hannifin Corp. Refrigerating Specialties Div.

Table 5-19. Hot gas bypass regulator ratings.

R-12

Condensing Temperature °F.	*Compressor Discharge Temperature °F.	A9E or A9 Hot Gas Bypass Regulator Size			A4AOE or A4AO Hot Gas Bypass Regulator Port Size					
		⅝	⅞	1⅛	¾	1	1¼	1⅝	2	2½
86°F. 93.3 psig	120	2.8	6.4	9.9	17	28	49	79	140	190
	140	3.3	6.7	10.	18	29	51	84	150	200
	160	3.5	7.1	11.	19	31	54	88	160	210
	180	3.7	7.5	12.	20	33	57	92	170	220
100°F. 117.2 psig	120	3.6	7.4	12.	20	32	56	90	160	220
	140	3.8	7.8	12.	21	34	59	95	170	230
	160	4.0	8.2	13.	22	36	62	100	180	240
	180	4.2	8.6	14.	23	38	65	110	190	250
	200	4.4	9.0	14.	24	40	68	110	200	260
110°F. 136.4 psig	140	4.2	8.6	13.	23	38	64	100	190	240
	160	4.5	9.1	14.	24	40	68	110	200	260
	180	4.8	9.6	15.	25	42	72	120	210	280
	200	5.0	10.	16.	27	45	76	120	220	290
	220	5.3	11.	17.	28	47	80	130	240	310
120°F. 157.6 psig	140	4.6	9.4	14.	24	40	70	110	210	270
	160	4.9	10.	15.	26	43	74	120	220	290
	180	5.2	11.	16.	28	46	78	130	230	310
	200	5.5	11.	17.	29	48	83	130	250	320
	220	5.8	12.	18.	31	51	87	140	260	340

R-22

Condensing Temperature °F.	*Compressor Discharge Temperature °F.	A9E or A9 Hot Gas Bypass Regulator Size			A4AOE or A4AO Hot Gas Bypass Regulator Port Size					
		⅝	⅞	1⅛	¾	1	1¼	1⅝	2	2½
86°.F 158.2 psig	140	5.9	12	18	32	53	91	150	270	350
	160	6.1	12	19	34	56	95	150	280	360
	180	6.4	13	20	35	58	99	160	290	380
	200	6.7	14	21	36	60	100	170	300	400
100°F. 195.9 psig	140	6.8	14	22	38	62	100	170	300	400
	160	7.1	14	23	39	64	110	180	320	420
	180	7.4	15	24	41	67	110	190	330	440
	200	7.7	16	25	43	70	110	200	340	460
	220	8.0	16	26	44	72	120	210	360	480
110°F. 226.3 psig	140	7.4	15	23	40	66	120	180	330	430
	160	7.8	16	25	42	69	120	190	350	460
	180	8.2	17	26	44	73	130	200	370	480
	200	8.6	18	27	46	77	140	210	390	500
	220	9.0	19	29	48	80	140	220	410	530
120°F. 259.9 psig	140	8.1	16	25	43	71	130	200	360	480
	160	8.6	17	27	46	75	130	210	380	500
	180	9.0	18	28	48	79	140	220	400	530
	200	9.4	19	29	50	83	150	230	420	560
	220	9.9	20	31	53	87	150	240	440	580

Courtesy Parker Hannifin Corp. Refrigerating Specialties Div.

around the condenser and expansion valve, consequently, it is not available for cooling purposes and the net refrigeration capacity is reduced, Figure 5-14.

This method of capacity control prevents the compressor from operating below normal suction pressures at low loads which could cause oil pumping and compressor short cycling. If the pressure falls below the saturation pressure equivalent to 32 °F for the refrigerant used, then freezing could take place. Frost on an evaporator coil or ice in a chiller will act as an insulator causing a further reduction in capacity.

Three methods are used to bypass with hot gas regulators:

- Hot gas to evaporator entrance
- Hot gas to evaporator exit
- Hot gas to compressor suction inlet

Each application requires additional components for liquid and temperature control of the system.

Figure 5-14 shows the three bypass routes on a P-H diagram.

Since hot gas bypass is a method of capacity control, hot gas is needed winter and summer. Some method of winterization is needed to maintain adequate levels of hot gas or system performance will be reduced.

Hot gas bypass systems open on a fall in suction pressure. Because of this, a refrigeration system with *pumpdown* cannot be employed unless the hot gas regulator is prevented from feeding hot gas during pumpdown. A hand shutoff valve, solenoid valve or a hot gas regulator, with a pilot electric shutoff, may be employed to shut down the hot gas leg during the pumpdown mode. Any one of these methods can shut down the system for pumpdown, negating the regulator bypass feature.

Tables 5-18 and 19 show gas temperature and capacity tables for one particular regulator. For example, assume that the hot gas enters the evaporator just downstream of the expansion valve, at the evaporator entrance, and that a 75 ton, R-22, direct expansion air conditioning system is to operate at varying capacities down to 0 tons at an evaporator temperature of 32 °F. The compressor is capable of being unloaded, using internal cylinder unloaders, by 66 or 33%. This leaves 25 tons left to unload.

An air-cooled, winterized condenser maintains 100 °F throughout the year. An average discharge gas temperature, from Table 5-18, is 162.5 °F, that is, (175-150)/2 = 12.5. This is added to 150 °F to give the 162.5 value. For these conditions — R-22, 100 °F condensing, and 160 °F discharge temperature — select regulator A9E, with 23 ton capacity and 1-1/8″ end connections, from Table 5-19. The regulator is set to be fully open at 57.8 psig (32 °F for R-22). For hot gas runs of less than 20 feet and a single globe-style valve upstream, a line of the same size as the A9E connections is usually satisfactory.

If, however, the line is of considerable length, which may be the case because the entrance is into the evaporator inlet, the line size must be increased to keep the pressure drop to no more than 10 psi. If the line is of considerable length, 100 to 200 feet long with valves and elbows, usually two line sizes larger is selected. In this case, the line size would be increased to 1-5/8″ while the regulator has 1-1/8″ fittings.

If load variations are large and sudden, it may be wise to add an accumulator in the suction line just ahead of the compressor. Large amounts of hot gas will *dump* liquid from the evaporator and flush it into the suction line to the compressor. Compressor damage may result.

When hot gas is bypassed to the evaporator exit, the hot gas line is connected to the suction line between the evaporator exit and the expansion valve remote bulb location.

Another example of regulator application is a 5-ton R-12 system that is required to operate at 0% of its capacity while the evaporator is receiving cold inlet air. The system is to run continuously, using automatic hot gas bypass. From Table 5-18, the discharge temperature is estimated to be 160 °F. The evaporator is 30 °F and the saturated condensing temperature is given as 86 °F. From Table 5-19, a 7/8″ regulator, style A9E, is chosen. With a capacity of 7.1 tons, its size is ample. If the line length is less than 20 feet, then a 7/8″ line is adequate. If the line length approaches 100 to 200 feet, with numerous valves and elbows, then a line two sizes greater than the regulator size is required, in this case, 1-3/8″.

The third method bypasses hot gas directly to the compressor suction inlet, Figure 5-14. Because the hot gas enters the suction at the compressor, it must be cooled or it will overheat the compressor. Therefore, a separate thermal expansion valve, called a desuperheat valve or liquid injection valve, is used. A line is run from the liquid receiver, through the desuperheater valve, and is connected to the outlet of the hot gas bypass valve. The common line, carrying liquid from the desuperheater expansion valve and hot gas regulator, must be of proper length and configuration to insure thorough mixing of the two fluids so that no hot spots are created and the suction gas is desuperheated for complete compressor protection.

Still another example, is a 30 ton commercial R-12 system that is required to operate at varying capacities down to 0 tons at an evaporator temperature of 20 °F. The compressor is capable of being unloaded to 33%. The condenser is winterized to maintain 100 °F condensing and is air cooled.

Table 5-18 shows a discharge temperature estimated to be 180 °F for the R-12 gas from the compressor. From Table 5-19, for R-12 with a 100 °F condensing and 180 °F discharge temperature, an A9E regulator is selected with end connections of 1-1/8″.

The liquid injection valve is sized using Tables 5-20 and 21. For the A9E regulator, a liquid injection valve will have an uncorrected capacity of 11 tons per Table 5-20 for R-12, +20 °F evaporator, 100 °F condensing and 180 °F discharge temperature. The correction factor, from Table 5-21 for the

Table 5-20. Capacities of TXV valves used for liquid injection.

R-12

Condensing Temperature °F.	Compressor Discharge Temperature °F.	Suction Temperature (Saturation)			
		−40°F.	0°F.	+20°F.	+40°F.
86°F. Condensing	120	7.6	5.8	4.7	3.8
	140	8.9	7.1	6.0	5.2
	160	11.	8.8	7.7	6.8
	180	12.	10.	9.3	8.4
100°F. Condensing	120	8.7	6.5	5.1	4.0
	140	11.	8.4	7.1	6.0
	160	13.	10.	9.1	8.0
	180	16.	12.	→ 11.	9.9
	200	17.	14.	13.	12.
110°F. Condensing	140	12.	9.4	7.8	6.5
	160	14.	12.	10.	8.7
	180	16.	14.	12.	11.
	200	19.	16.	15.	13.
	220	21.	18.	17.	16.
120°F. Condensing	140	13.	10.	8.2	6.8
	160	16.	13.	11.	9.3
	180	18.	15.	14.	12.
	200	21.	18.	16.	14.
	220	24.	20.	19.	17.

Courtesy Parker Hannifin Corp. Refrigerating Specialties Div.

Table 5-21. Liquid injection TXV valve correction factors.

A9E or A9			A4AOE or A4AO					
5/8″	7/8″	1 1/8″	3/4″	1″	1 1/4″	1 5/8″	2″	2 1/2″
0.11	0.23	0.36	0.61	1.0	1.7	2.8	4.9	6.6

Courtesy Parker Hannifin Corp. Refrigerating Specialties Div.

A9E regulator and 1-1/8″ size, is .36. The final size is then 11 × .36 = 4.0 tons with 100 °F condensing and 20 °F evaporator.

Since hot gas regulators supply hot gas suction pressure control, some designs are externally equalized and therefore maintain constant pressure at the equalizer location.

Capacity of the regulator depends upon the amount of heat required to maintain pressure at the equalizer. As loads decrease, the hot gas will put a false load on the system to maintain the desired low side pressure.

The following information must be available when ordering or specifying hot gas regulators.

- Load in Btu/hr (or tons)
- Type of refrigerant
- Hot gas entrance location
- Minimum saturation pressure with regulator delivering full capacity
- Regulator sensitivity

The regulator is a modulating device. It will begin to open at the set point and begin to pass hot gas. As the suction pressure decreases, the regulator will continue to open until it is fully open. The range of pressures, from start to fully open, is known as the *sensitivity* of the regulator.

If, for example, the regulator is to maintain a minimum evaporator pressure of 35 psig and has a sensitivity of 8 psi, it should be set to start opening at 43 psig and be wide open at 35 psig, operating at full load. If it is incorrectly set to open at 35 psig, it will not be fully open and operating at full capacity until 27 psig and the evaporator coil will probably have icing problems.

Specialty refrigeration control valves

The high-to-low side regulators are manually adjusted for a specific application. If the setting is to be changed, the mechanic must install gages on the regulator and make the necessary adjustment. This is usually time consuming and costly and not easily accomplished if located in equipment rooms or in large industrial complexes where long piping runs are common.

FIG. 5-15. Modulating electromagnetic regulator for refrigerant service. *Courtesy Staefa Control Systems Inc.*

A control valve with remote controllers and sensors, Figure 5-15, can be used to bypass hot gas to control rooms or processes to a ±1 °F tolerance with a rangeability of 500:1, meaning that flow can be regulated down to 1/500th part of that of a fully open valve.

Conventional regulators may take as long as several minutes to react and cause overshooting of the control set point. The control valve in Figure 5-15 can position the disc within one second of controller demand. A two wire, 0 to 20 volts D.C. signal is required for valve actuation with the position of the disc varying as a function of the applied voltage.

The refrigerant version is hermetically sealed to prevent leakage. The controller can be set up to react to temperature, humidity or pressure signals and is ideal for close temperature and humidity control.

It was stated earlier that control valves would not be discussed since this particular chapter is on regulators. However, since this valve is used in refrigeration systems and is used in place of many high-to-low-side regulators, it will be covered here.

The control valve, Fig. 5-15, may be used, in a three-way version, to divert hot gas from condenser to evaporator, for capacity control and possible heating, and with heat recovery systems.

A two-way version of the refrigerant control valve in Figure 5-15 can control refrigerant system capacity three ways:

- Throttling refrigerant vapor into the compressor suction.
- Bypassing high side hot gas to either the evaporator inlet or outlet.
- Bypassing high side hot gas to the compressor inlet with additional liquid injection for desuperheating.

For hot gas bypass applications, the selection parameters include the capacity in Btu/hr, evaporator and condenser temperatures and refrigerant type. For example:

- Capacity is 145,000 Btu/hr
- Evaporator temperature is 40 °F
- Condensing temperature is 120 °F
- Refrigerant is R-22

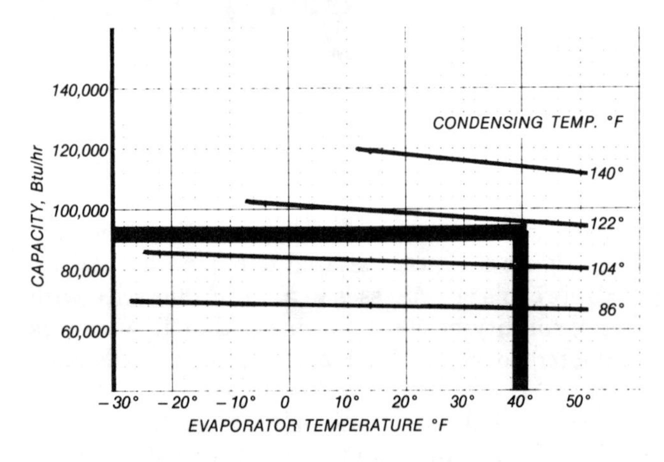

FIG. 5-16a. Capacities of modulating valve size 08 in hot gas bypass application with R-22. *Courtesy Staefa Control Systems Inc.*

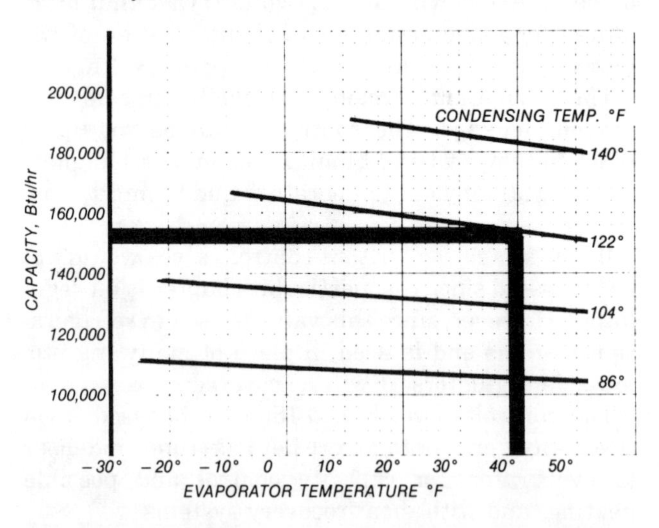

FIG. 5-16b. Capacities of modulating valve size 10 in hot gas bypass application with R-22. *Courtesy Staefa Control Systems Inc.*

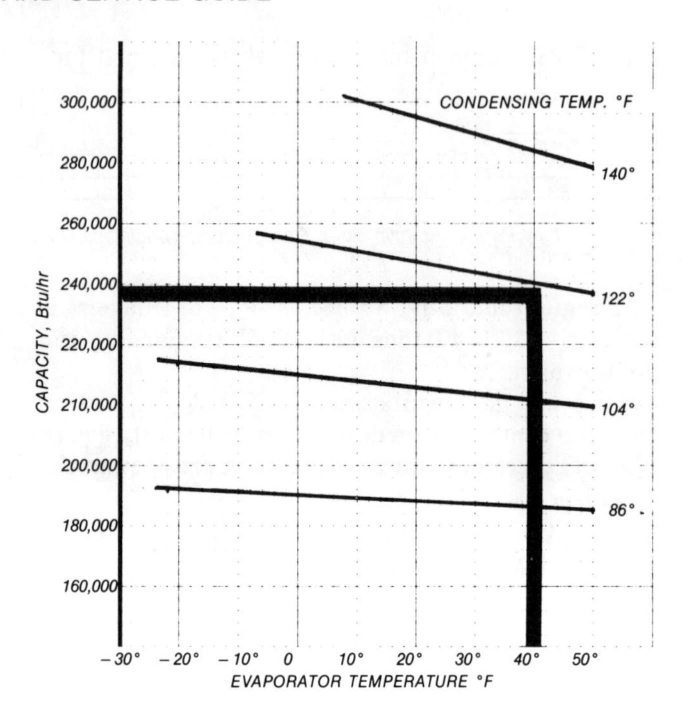

FIG. 5-16c. Capacities of modulating valve size 15 in hot gas bypass application with R-22. *Courtesy Staefa Control Systems Inc.*

The curves in Figures 5-16a, b and c show the capacity of the valves available in sizes 08, 10 and 15. The charts include the evaporator and condensing temperatures. These are for hot gas application only. At 145,000 Btu/hr and 40 °F evaporator and 120 °F condensing, a size 10 valve meets the above requirements. Checking each curve, the size 08 is too small and the size 15 is too large as seen in Figures 5-16a and c.

In selecting the modulating control valve for suction gas service, the required parameters are the evaporator capacity and temperature, the condensing temperature, refrigerant type, and pressure drop through the valve. An example:

- Evaporator capacity is 20,000 Btu/hr
- Evaporator temperature is 50 °F
- Condensing temperature is 120 °F
- Pressure drop is 5.5 psi (equivalent to approximately 5.4 °F)
- Refrigerant is R-12
- Suction correction factor is 1.2, Figure 5-17b.

The capacity curves in Figure 5-17a are for pressure drops corresponding to a drop of 3.6 °F. For deviations other than the 3.6 °F drop, a correction factor must be applied.

The capacity curves in Figure 5-17a show the mod-

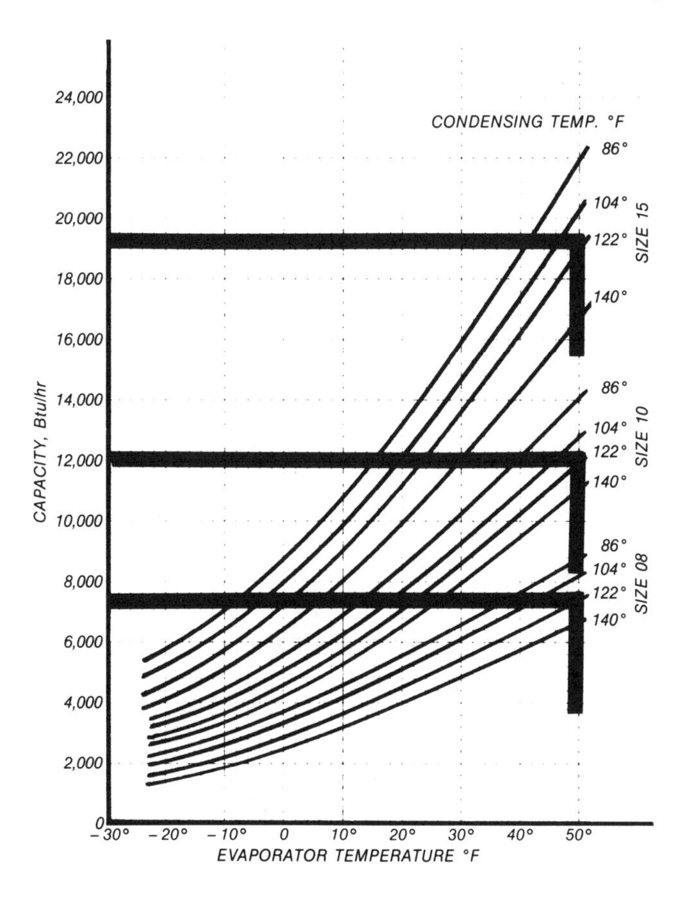

FIG. 5-17a. Suction gas applications using various size modulating valves and R-12. *Courtesy Staefa Control Systems Inc.*

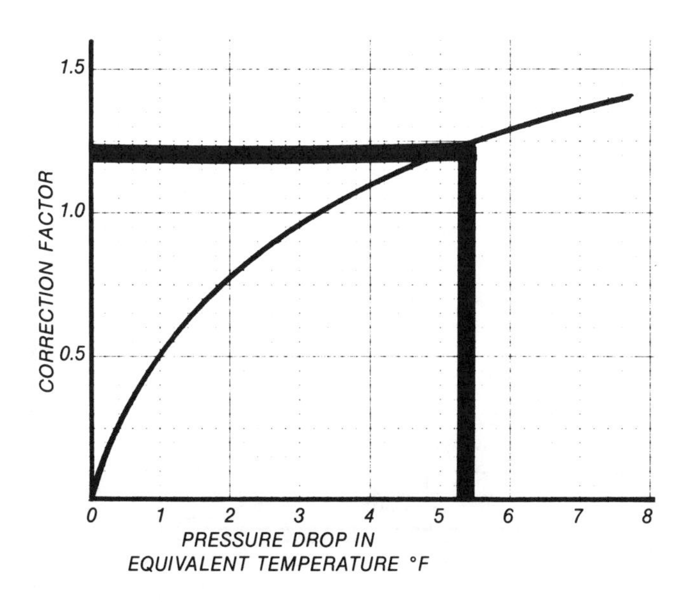

FIG. 5-17b. Correction factor vs. equivalent temperature. *Courtesy Staefa Control Systems Inc.*

el 08, 10 and 15. Each has its own capacity at different condensing and evaporator temperatures. For 20,000 Btu/hr and 50 °F evaporator and 120 °F condensing, a size 15 is required. With the correction factor of 1.2, the capacity is found to be 19,100 Btu/hr × 1.2 = 22,900 Btu/hr, meeting the 20,000 Btu/hr requirement.

Regulators for steam, water and air

Regulators for fluids other than refrigerants function similar to refrigerant regulators. They are classified as pressure reducing (downstream sensing), relief regulators (upstream sensing), and differential regulators. These regulators are not hermetically sealed even though they provide the same service as refrigerant regulators. Control accuracy may be only 10% to 20% which is sufficient for many purposes. In some cases, accuracy of 2% to 5% can be achieved by utilizing pilot operated versions for closer control and accuracy when it is needed.

Figure 5-18 shows *droop* curves for regulators and shows how set point accuracy falls off or *droops* as flow increases. The droop in direct-acting regulators is caused by three factors:

- Spring rate
- Diaphragm area
- Length of stroke

The curve in Figure 5-18, however, has a linear portion and, between the minimum and maximum limits, will have a reasonably linear output.

To compensate for droop, many regulators will have two, three or four sets of springs and diaphragms, each covering a segment of the regulator range. These spring and diaphragm sets allow an in-

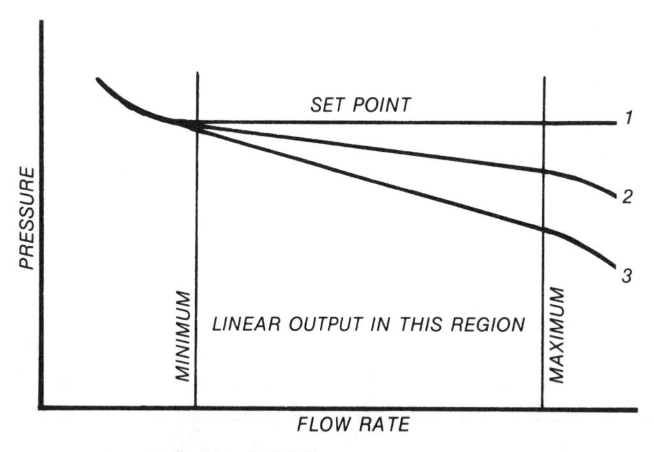

FIG. 5-18. Droop curves for regulators.

Table 5-22a. Sizing table for saturated steam, flow in pounds per hour.

Inlet Pressure (psig)	Outlet Pressure (psig & inches mercury vacuum)	Flow Coefficient — Cv .84	1.6	2.5	4.4	5.0	6.4	9.5	15	25	30	35	45	50	55	65	70	85	115	130	200	395
5 (227°F)	3	16	30	47	81	100	118	174	272	451	539	627	803	889	974	1151	1234	1499	2018	2282	3510	6933
	1	21	39	61	106	120	153	223	346	569	678	785	1001	1103	1202	1420	1515	1840	2463	2785	4284	8461
	5" Hg. Vac.	25	47	72	123	139	176	252	380	617	730	831	1062	1157	1255	1486	1575	1911	2545	2876	4425	8740
	10" Hg. Vac.	26	48	74	125	139	176	252	380	617	730	831	1062	1157	1255	1486	1575	1911	2545	2876	4425	8740
10 (240°F)	8	18	34	52	92	104	133	196	308	510	611	710	911	1009	1107	1308	1404	1705	2298	2598	3997	7894
	5	26	49	76	131	149	189	277	429	706	840	973	1241	1367	1490	1761	1878	2281	3054	3453	5312	10490
	2	30	56	87	149	168	213	308	470	765	905	1041	1318	1437	1558	1844	1953	2372	3134	3543	5450	10765
	1" Hg. Vac.	32	59	91	155	174	220	308	470	765	905	1041	1318	1437	1558	1844	1953	2372	3134	3543	5450	10765
15 (250°F)	12	23	44	69	121	137	175	258	404	668	800	930	1191	1318	1444	1707	1831	2223	2994	3385	5207	10284
	8	33	61	95	165	187	237	346	534	876	1042	1205	1533	1686	1834	2168	2307	2801	3743	4231	6509	12855
	4	37	69	107	183	206	261	374	567	917	1085	1244	1578	1726	1861	2209	2340	2836	3779	4272	6572	12980
	1" Hg. Vac.	39	72	110	186	208	263	375	567	917	1085	1244	1578	1726	1861	2209	2340	2836	3779	4272	6572	12980
25 (266°F)	20	34	64	99	173	196	250	367	574	950	1135	1318	1687	1866	2042	2413	2584	3138	4222	4772	7342	14501
	15	44	82	127	220	248	316	460	709	1160	1379	1592	2025	2224	2417	2857	3037	3688	4921	5563	8558	16903
	10	49	91	140	239	269	341	489	740	1198	1417	1629	2062	2255	2437	2886	3058	3710	4941	5586	8593	16971
	5	50	92	141	239	269	341	489	740	1198	1417	1629	2062	2255	2437	2886	3058	3710	4941	5586	8593	16971
50 (298°F)	40	59	112	174	303	343	438	643	1002	1653	1973	2289	2926	3231	3531	4173	4464	5420	7280	8229	12660	25004
	30	75	141	224	375	423	537	776	1188	1934	2292	2638	3343	3657	3959	4679	4958	6016	8011	9056	13932	27516
	25	79	148	227	389	438	554	794	1201	1944	2297	2641	3344	3657	3959	4679	4958	6016	8011	9056	13932	27516
	15	83	152	232	393	441	557	794	1201	1944	2297	2641	3344	3657	3959	4679	4958	6016	8011	9056	13932	27516
75 (320°F)	60	84	158	245	427	484	617	905	1408	2322	2770	3212	4103	4528	4946	5845	6247	7586	10181	11509	17706	34970
	50	100	188	290	475	566	719	1043	1604	2619	3108	3584	4551	4990	5413	6397	6787	8241	10973	12405	19084	37691
	40	109	203	312	534	600	760	1087	1646	2664	3149	3618	4580	5008	5421	6409	6791	8246	10973	12405	19084	37691
	30	161	209	318	539	604	763	1088	1646	2664	3149	3618	4580	5008	5421	6409	6791	8246	10973	12405	19084	37691
100 (338°F)	80	108	205	318	553	626	798	1170	1819	2998	3576	4145	5293	5840	6375	7536	8049	9775	13113	14824	22805	45041
	70	125	235	363	629	710	903	1312	2021	3306	[illegible]	4533	5762	6325	6869	8118	8624	10473	14169	16017	24641	48666
	60	142	263	403	686	770	973	1387	2098	3396	[illegible]	4612	5839	6386	6913	8171	8657	10512	13989	15814	24329	48050
	40	144	266	405	686	770	973	1387	2098	3396	4013	4612	5839	6386	6913	8171	8657	10512	13989	15814	24329	48050
125 (353°F)	100	136	250	389	676	766	976	1429	2222	3661	4365	5059	6460	7125	7776	9190	9816	11919	15986	18071	27802	54908
	90	149	281	435	753	851	1082	1574	2428	3976	4725	5457	6941	7625	8287	9794	10413	12644	16874	19075	29346	57958
	75	165	309	475	815	918	1163	1672	2541	4114	4864	5588	7075	7738	8375	9900	10490	12737	16950	19161	29479	58221
	50	175	322	491	832	933	1179	1680	2542	4114	4864	5588	7075	7738	8375	9900	10490	12737	16950	19161	29479	58221
150 (366°F)	125	146	276	430	749	849	1083	1589	2477	4090	4882	5665	7242	7998	8742	10331	11052	13420	18027	20379	31352	61920
	110	174	327	507	878	922	1262	1838	2837	4649	5527	6386	8126	8931	9711	11477	12208	14824	19796	22378	34428	67994
	90	195	364	560	961	1082	1371	1969	2990	4840	5721	6574	8322	9101	9854	11646	12339	14981	19939	22539	34676	68485
	60	206	379	578	979	1098	1387	1977	2991	4840	5721	6574	8322	9101	9854	11646	12339	14981	19939	22539	34676	68485

Courtesy Jordan Valve Div. of Richard Ind., Inc.

crease in regulator accuracy over the entire range of the regulator's operating pressures by dividing the regulating range into smaller segments.

Droop can be reduced in self-operated regulators by increasing the diaphragm area, decreasing the spring rate and by decreasing the length of stroke. It is, therefore, necessary to select spring and diaphragm sets covering the desired operating range.

For example, Table 5-25 for Jordan Series 60 and 61 regulators shows various ranges of operation for different regulator sizes. If, for example, the outlet pressure of a particular system requires 70 psig for a one-inch size regulator, then a range of 20-80 is selected. The 40-115 and 40-175, even though usable, are not wise selections since their sensitivity is lower and the resultant droop will be greater. Always choose the lowest rated spring available that covers the set pressure. As the range of the spring becomes lower, the sensitivity at the regulator becomes higher.

For regulator sizing, the following is required for downstream sensing applications:

- Determine flow rate required for the process
- Inlet pressure
- Outlet pressure
- Specific fluid data

Flow rate The flow rate is determined by either heat transfer calculations or by the requirements set by the respective equipment manufacturer. For example, if 3100 pounds/hour of saturated steam is required for a process at 80 psig with 100 psig steam available, a regulator with a C_V of 25 is required as shown in Table 5-22a.

TABLE 5-22b

When superheated steam is used, an adjustment must be made using this formula:

$$N = \frac{W}{F_s}$$ where

W = Flow, in pounds per hour, of superheated steam.
N = Capacity of saturated steam flow, from steam sizing chart.
F_s = Superheat factor.

Superheat factor

50°	.96	100°	.94
150°	.90	200°	.87

Courtesy Jordan Valve Div. of Richard Ind., Inc.

Table 5-22c. Flow coefficients, C_v.

VALVE MARK	50-51-52-53 54-55-56 60-61-62-63 64-65-66	57-67-500-600	
SEAT MATERIAL	Stainless Steel Jordanite Jordanic	Stainless Steel Jordanic	Jordanite
VALVE SIZE	STANDARD CV *		
¼	.84	–	–
3/8	1.6	–	–
½	2.5	5.0	5.0
¾	4.4	9.5	9.5
1	6.4	15	15
1¼	9.5	25	25
1½	15	30	25
2	25 or 30**	50	35
2½	55 +	–	55
3	115 +	–	115
4	200 +	–	200
6	395 +	–	395

Courtesy Jordan Valve Div. of Richard Ind., Inc.

If 750 pounds of superheated steam per hour at 395 °F is used for a process with an inlet pressure of 75 psig and outlet pressure of 30 psig, what C_V value is required? Saturated steam at 75 psig is 320 °F (steam tables), the superheat is 395 - 320 = 75° F The superheat factor must be obtained from Table 5-22b. For 75 psig, this is .95, by interpolation. The saturated steam flow or adjusted steam flow is:

$$N = \frac{W}{Fs} = \frac{\text{Superheated steam flow}}{.95} = \frac{750}{.95} = 790 \text{ lb per hour}$$

Using Table 5-22a, find the 75 psig inlet and 30 to 0 psig outlet row and move to the right until a value of 790 or more is given. In this case, 1132 pounds per hour corresponds to a C_V of 9.5.

Table 5-22c shows that 1¼″ regulator in one series and a ¾″ size in another series have the required C_V. A Series 60 regulator is selected with a control range, for spring and diaphragm, of 10 to 30 psig.

It must be pointed out that, when substituting or exchanging regulators, it is not always possible to exchange brands, line size for line size, since different manufacturers may have varying C_V values for their comparable designs. Each manufacturer's tables should be checked for their C_V ratings.

It is also bad practice to let the pipe size govern the regulator size. Doing so will usually oversize the regulator and produce poor regulation.

Regulators are usually smaller than the piping adjacent to the regulator. The maximum pressure drop downstream of the regulator, including pipe fittings, piping and valves, should not exceed 10%, of the static pressure at the regulator outlet, per 100 feet of pipe.

If, for example, a regulator reduces 100 psig water to 20 psig at 50 gal/min, a regulator with a C_V of 6.4 would be selected from Table 5-23a. Per Table 5-22c, this would be a 1″ regulator. If the above pressure drops are adhered to, the maximum downstream pressure is 10% of the outlet pressure or 10% of 20 psig = 2 psig per 100 feet of pipe.

Figure 4-23 showed that, with water at 50 gal/min, a 2″ schedule 40 pipe has 2.0 psi pressure drop per 100 feet of pipe. A 1½″ pipe will have 7.2 psi drop and a 1¼″ pipe will have a 15.7 psi drop per 100 feet of pipe. A 1″ pipe size, the same as the regulator, is not even shown in the chart because the drop is too great. Therefore, the 1″ regulator will have a 2″ outlet. If the piping is over 100 feet long or includes many valves and fittings, a 2½″ schedule 40 pipe may be necessary to meet the 10% rule.

Water and other liquid applications

The following must be known in order to select a regulator for liquid service:

- Pressure drop across the regulator
- Fluid flow rate in gallons per minute
- Fluid viscosity, if other than water

If water is flowing at 150 gal/min, with the inlet pressure at 120 psig and outlet at 45 psig, the differential pressure is 120 − 45 = 75 psig. Using Table 5-23a, enter the 75 psig differential pressure row and move right until 150 gal/min or more is shown. This is 215 gal/min. Reading directly upward, a C_V of 25 is found. Select a regulator with a C_V of 25 for this application.

Table 5-23a. Sizing table for water, flow in gallons per minute.

PRESSURE DROP (△p) PSI	.84 Cv	1.6 Cv	2.5 Cv	4.4 Cv	5.0 Cv	6.4 Cv	9.5 Cv	15 Cv	25 Cv	30 Cv	35 Cv	50 Cv	55 Cv	115 Cv	200 Cv	395 Cv
3	1.4	2.8	4.3	7.6	8.6	11	16	26	43	52	60	86	95	200	346	684
5	1.9	3.6	5.6	9.8	11	13	21	32	55	67	78	112	122	256	446	881
10	2.6	5	7.9	13.9	16	20	30	47	79	94	110	158	173	363	631	1250
15	3.3	6.3	9.8	17	19	25	37	57	96	115	135	193	212	442	771	1520
25	4.3	8	12.5	22	25	32	47	75	124	149	174	249	274	572	991	1960
50	5.9	11.3	18	31	35	45	67	106	176	212	247	353	389	811	1410	2120
75	7.3	14	22	37	43	55	82	130	215	259	302	432	476	995	1730	3420
100	8.4	16	25	43	50	63	95	150	250	300	350	500	550	1150	2000	3950
125	9.4	18	27	48	56	72	106	167	256	336	392	560	616	1282	2240	4417
150	10	19	30	53	61	77	116	182	305	366	427	610	671	1400	2440	4815
175	11	21	32	57	66	85	126	197	330	396	462	660	726	1517	2640	5220
200	12	22	35	62	71	90	133	212	355	426	497	710	780	1630	2840	5600
250	13	25	38	70	79	101	150	237	395	474	551	790	869	1810	3160	6220
300	14	27	42	76	85	111	165	260	432	519	605	865	952	1990	3460	6820

Courtesy Jordan Valve Div. of Richard Ind., Inc.

Table 5-23b

With viscous fluids, such as #6 fuel oil or syrups, an adjustment must be made using this formula:

$$N = \frac{V_V}{F_V}$$

where

V_V = Flow, in gallons per minute, of viscous fluid.
N = Water flow capacity, in gallons per minute, from water sizing chart.
F_V = Viscosity correction factor.

Viscosity factor, F_V

SSU	F_V	SSU	F_V
50	.86	500	.62
100	.78	1000	.56
200	.71	2000	.51

Courtesy Jordan Valve Div. of Richard Ind., Inc.

High temperature water can cause problems if the pressure drop across the regulator causes the water to flash into vapor. At low temperatures, water may vaporize, if the pressure is reduced to very low values, and cause serious piping problems. It is recommended that the regulator manufacturer be consulted, if this is the case.

Fluid flow ratings must be adjusted to compensate for fluids with viscosities greater or less than water. For example, to determine the regulator for 7 gal/min of oil, with a viscosity of 400 SSU and a 15 psig pressure drop, use Tables 5-23a and 5-23b. A correction is needed to compensate for the greater viscosity of oil since the tables are based on the viscosity of water. Table 5-23b, by interpolation, yields a viscosity correction factor of .65 for 400 SSU. Therefore, the new flow is 7 gal/min divided by .65 or 10.7 gal/min. This is the equivalent flow of water, so Table 5-23a may now be used. Select the row for a pressure drop of 15 psig and go to the right until 10.7 gal/min or greater is obtained. 17 gal/min is the equivalent flow, with a C_V of 4.4.

Air and gases

There are a number of factors to be considered when selecting a regulator for gaseous service:

- Most charts are based on standard air at 60 °F
- Density factors come into play because gas densities vary with pressure.
- The specific gravity of the gas is used rather than that of air.
- Inlet and outlet pressures must be specified.
- Flow rate, in *standard cubic feet per minute*, SCFM, is required.

Like most selection tables, Table 5-24 is based on

Table 5-24. Sizing table for air, flow in SCFM.

Inlet Pressure (psig)	Outlet Pressure (psig & inches mercury vacuum)	Flow coefficient – Cv															
		0.84	1.6	2.5	4.4	5.0	6.4	9.5	15	25	30	35	50	55	115	200	395
5	3	5	9	14	25	29	36	54	86	142	176	200	282	311	636	1092	2193
	1	6	12	19	33	37	48	71	111	184	220	255	361	396	805	1372	2660
	10"-28" Hg Vac.	8	15	23	39	44	56	83	127	212	245	288	400	441	877	1459	2838
10	8	5	10	16	28	32	41	54	97	161	194	230	323	355	731	1253	2479
	5	8	15	23	41	46	59	88	138	230	275	319	452	495	1009	1719	3333
	4"-28" Hg Vac.	10	19	28	49	55	70	104	159	259	309	361	501	554	1095	1820	3555
15	12	7	14	22	38	43	55	81	129	213	256	301	425	469	961	1650	3304
	8	10	21	29	52	58	75	111	174	285	343	400	563	617	1257	2132	4115
	2-0	12	22	34	59	65	84	123	191	311	369	434	602	661	1323	2103	4276
25	20	10	20	32	56	63	80	119	188	312	374	435	621	678	1400	2407	4737
	15	14	26	41	71	80	102	152	237	389	467	545	759	841	1704	2875	5577
	7-0	16	32	45	79	89	112	166	255	415	496	580	805	876	1761	2929	5670
50	40	19	35	56	98	109	141	209	326	546	655	759	1085	1182	2439	4152	8179
	30	24	46	70	122	136	176	260	405	663	795	917	1288	1421	2834	4751	9306
	20-0	26	51	74	130	145	184	271	416	676	807	944	1311	1436	2873	4770	9306
75	60	27	53	82	143	160	205	303	475	790	944	1099	1566	1705	3528	5978	11622
	50	33	62	96	166	186	240	352	552	903	1083	1266	1767	1950	3920	6566	12790
	30-0	36	70	104	181	197	256	378	579	941	1123	1313	1822	2054	3994	6570	12790
100	80	36	67	105	184	205	262	391	613	1020	1220	1417	2002	2200	4550	7700	14940
	70	41	78	120	209	235	300	442	695	1137	1370	1587	2250	2452	4982	8315	16150
	50-0	46	88	131	227	257	326	476	740	1210	1432	1670	2325	2567	5072	8432	16500
125	100	44	83	129	228	254	322	481	751	1256	1509	1747	2470	2708	5581	9455	18300
	90	50	95	144	253	283	360	532	838	1372	1646	1921	2717	2961	6025	10140	19700
	60-0	56	109	161	281	314	396	588	901	1464	1738	2013	2810	3093	6161	10280	20130
150	125	49	92	144	252	283	361	532	837	1404	1688	1944	2772	3024	6246	10610	20660
	110	58	111	170	295	331	423	628	978	1616	1940	2260	3185	3480	7050	11870	23050
	75-0	69	129	190	331	370	469	693	1063	1723	2063	2412	3348	3704	7337	12180	23760
175	150	52	100	157	273	307	393	583	872	1525	1827	2150	3022	3312	6831	11710	23060
	125	68	131	199	348	388	499	733	1156	1893	2278	2646	3726	4071	8240	13800	26900
	90-0	77	148	219	380	424	538	795	1223	1987	2359	2773	3850	4205	8437	14000	27320
200	170	61	117	182	317	356	455	674	1044	1771	2112	2459	3493	3830	7860	13760	26300
	150	74	141	217	380	424	542	804	1324	2073	2485	2901	4081	4488	9112	14730	29810
	100-0	87	168	248	431	482	608	902	1382	2246	2686	3135	4352	4797	9453	15820	30890

Courtesy Jordan Valve Div. of Richard Ind., Inc.

free air at 60 °F in cubic feet per minute, ft^3/min. Any temperature other than 60 °F must be corrected to convert to *standard air*, using the following formula:

$$N = SCFM = ACFM \times \frac{psia \times 520}{14.7\ (°F + 460)}$$

where
SCFM = CFM of standard air
ACFM = CFM of actual air
psia = Absolute air pressure
520°F = Standard absolute air temperature (460°F + 60°F)
°F = Actual air temperature

For gases at 60 °F, but with a different specific gravity than air, adjust for flow using this equation and the exponent .5 as the equivalent of the square root radical:

$$Q = \frac{N}{G_g^{.5}}$$

where
Q = Gas flow in SCFM
N = Air capacity, in SCFM, from table
G_g = Specific gravity of gas

Now, find a valve C_V to handle air under the following conditions: 40 ACFM • 100 psig inlet air reduced to 50 psig outlet air • 200 °F temperature.

The chart value N must be found first:

$$N = SCFM = ACFM \times \frac{psia \times 520}{14.7\ (°F + 460)}$$

$$= \frac{40 \times (100 + 14.7) \times 520}{14.7 \times (200 + 460)}$$

$$N = 246\ SCFM$$

This value is found in Table 5-24.

First look for the row that has a 100 psig inlet pressure and a 50 psig outlet pressure. Move to the right until a value of 246 or just larger is found. This will yield a C_V of 5.0.

If ammonia is used, under the following conditions: 200 psig inlet pressure • 30 psig outlet pressure • 5 ACFM • 150 °F, then the selection process is:

$$N = SCFM = ACFM \times \frac{psia \times 520}{14.7\ (°F + 460)}$$

$$= \frac{5 \times (200 + 14.7) \times 520}{14.7 \times (150 + 460)}$$

$$N = 24.9\ SCFM$$

An adjustment for the specific gravity difference between ammonia and air is made, yielding a new SCFM to be used in the air charts. Ammonia has a specific gravity of .596, therefore:

$$N = Q \times Gg^{.5} = 24.9 \times .596^{.5} = 24.9 \times .78$$

$$N = 19.4\ SCFM$$

From the charts for inlet and outlet pressures and the corrected flow, C_V = .84 per Table 5-24.

The flow rate in the charts, at the above conditions, is 87 SCFM, more than required. A special orifice may be needed to obtain good regulated flow with a C_V less than the .84 shown in the charts. Contact the manufacturer, a C_V as low as .0008 can be obtained with this type regulator.

Regulators can be mounted in pairs, in either parallel or series. Parallel service is usually reserved for situations where flow varies widely. The regulators are sized to have the smaller one handle normal flows while the larger regulator will begin to open only after the capacity of the smaller regulator is exceeded.

Regulators are installed in series to handle large pressure drops. If only one regulator is utilized for high pressure reduction, sensitivity is poor, Table 5-25. An intermediate pressure and regulator is selected to yield maximum output through both regulators. With steam or gases, the second regulator is usually larger than the first to handle the larger volume of lower pressure gas. The second regulator should be installed not less than 10 pipe diameters downstream of the first.

To select regulators for non-refrigerant service, that is steam, liquids and gases, the following information is needed:

- Capacity
- Lbs/hr, steam, and superheat or
- Gal/min, liquids, and viscosity or
- SCFM standard air, gases, specific gravity, temperature
- Inlet to outlet pressure drop

Regulator characteristics, such as sensitivity, droop, rangeability and tightness of shutoff, should

Table 5-25. Control ranges of spring-diaphragm sets for two styles of regulators.

REDUCED PRESSURE CONTROL RANGES							
Size in Inches	RANGES IN PSIG						
MARK 60							
¼-⅜	2-10	4-30	15-60	25-75	40-105	40-170	75-200
½-¾	2-20	15-35	20-60	30-85	40-145	40-170	50-175	75-220
1-1¼	1-5	2-12	2-25	10-30	20-55	20-80	40-115	40-175
1½-2	1-5	2-6	2-16	10-25	20-50	20-70	40-100	30-160
MARK 61							
¼-¾	1-5 2-12 2-25 10-35 20-55 20-80 25-160 30-180 40-115						

◄*Better sensitivity* *Poorer sensitivity*►

Courtesy Jordan Valve Div. of Richard Ind., Inc.

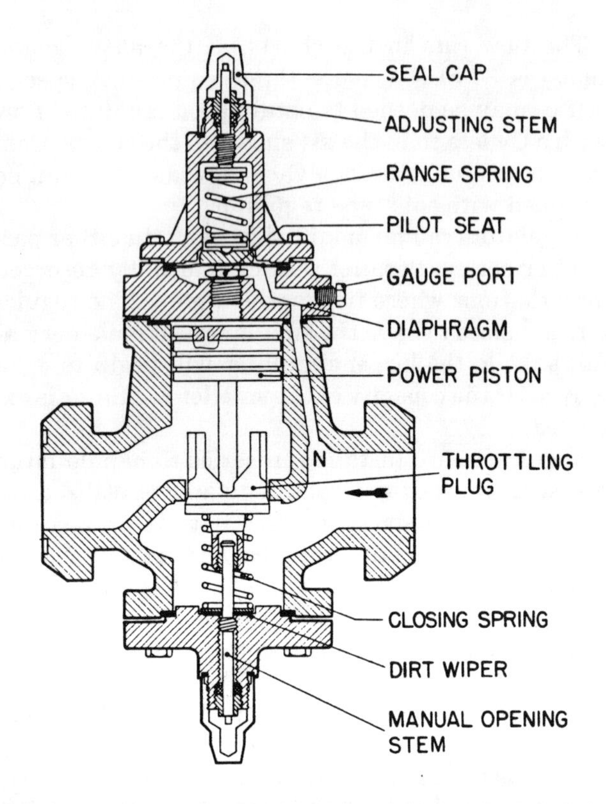

FIG. 5-19. Cross-sectional view of an Adaptomode® modular inlet pressure regulator. *Courtesy Parker Hannifin Corp., Refrigerating Specialties Div.*

also be evaluated for each application and are available from the manufacturer.

Modular style regulators

In industrial plants with numerous refrigeration applications and where regulators of many types are used, modular or built-up regulators can save substantially in repair, maintenance and inventory costs. An example of modular construction is the *Adaptomode*® pressure regulator, Figures 5-19 and -20, sold by Refrigerating Specialties Co. Many control variations are possible through the use of adapter kits, pilot solenoids and pilot regulators.

The basic body is utilized for inlet, outlet pressure sensing regulators; temperature sensing regulators; pilot solenoid operated regulators; and pneumatic and electrically compensated regulators. The regulator is converted from one function to another by a series of pilot adapters called *Modudapters*® and *Moduplates*® . These devices mount on top of the regulator body and direct fluids to various internal flow passages. The flow paths, both within and external to the adapter, connect the diaphragm, inlet fluid, outlet fluid, and top of the piston to modular pilot solenoid valves and/or pressure/pilot regulators. Both the solenoid valve and pressure pilot regulator act as *piggy-back* devices to the main regulator

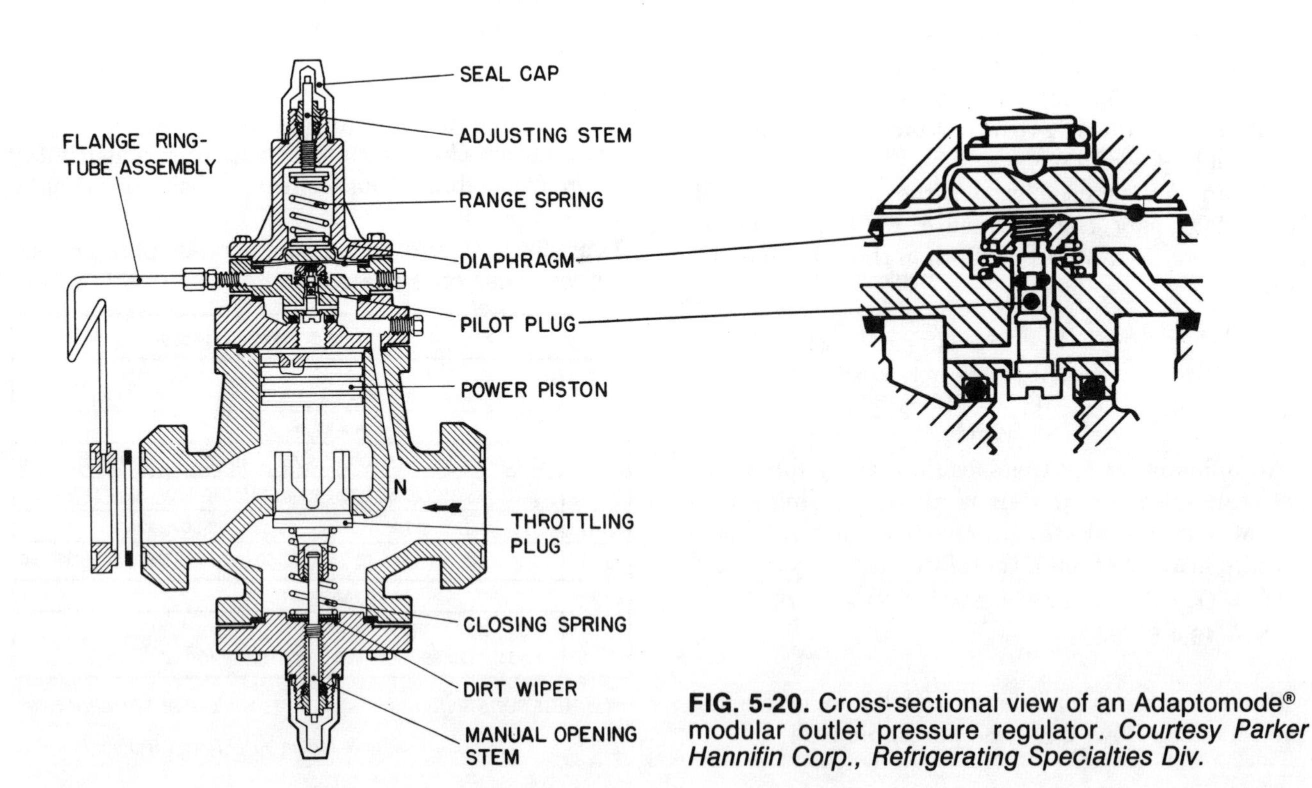

FIG. 5-20. Cross-sectional view of an Adaptomode® modular outlet pressure regulator. *Courtesy Parker Hannifin Corp., Refrigerating Specialties Div.*

and control pressure or block fluid in the internal flow passages to achieve the desired control.

Inlet sensing series

The inlet sensing regulator series has eight variations including:

- Inlet pressure regulator
- Inlet pressure regulator with remote sensing
- Relief regulator
- Differential regulator
- Pneumatically compensated inlet regulator
- Inlet pressure regulator with electric shutoff or wide open variation
- Dual pressure inlet regulator
- Electrically compensated inlet pressure regulator

Most pilot operated inlet sensing regulators are used as upstream or back pressure controls to maintain constant evaporator pressures and temperatures even though load variations may be considerable. The regulator will open when the evaporator pressure rises above the regulator setting and throttle down when pressure begins to fall below the setting.

These designs are available with line sizes of ¾″ to 12″ and from one ton to 3,000 tons capacity with −20 °F to 40 °F refrigerant temperatures. Refrigerants include R-12, R-22, R-502 and ammonia.

These regulators are used to control refrigerant flow to water chillers, to prevent freezing; control air cooling evaporators, to prevent excessive dehumidification; control air conditioning coils, to prevent icing; maintain different temperatures in multiple evaporator refrigeration systems; and maintain a high evaporator temperature during hot gas defrost as well as being used as a blocking device to entirely stop flow.

A pilot seat is utilized to control a power piston which, in turn, drives a throttling plug which has a characteristic V-port design for precise control, Figures 5-19 and 5-20. The pressure is set by the turning the adjusting stem to increase or decrease the pressure on the diaphragm through the range spring. Several spring ranges are available for each regulator. Pressure sensing under the diaphragm is through a passage from the inlet port, sensing upstream pressure. This is passage N in Figures 5-19 and 5-20.

Remote sensing can be achieved by rotating the adapter on top of the regulator body, by one bolt hole, to block flow (pressure) from the internal sensing passage. The sensing pressure then enters the adapter, via the gage port, from any desired control point.

The basic inlet sensing regulator may be used as a relief regulator. The buyer determines the setting needed to meet his requirements for upstream pressure relief and notifies the manufacturer. The regulator is adjusted and set at the factory, a seal cap is wired to the bonnet, and the relief setting is marked on the seal.

The relief regulator does not close 100% and a slight leakage is always present. Therefore, this device cannot be used to relieve to atmosphere. Rather, it relieves to a lower pressure portion of the system.

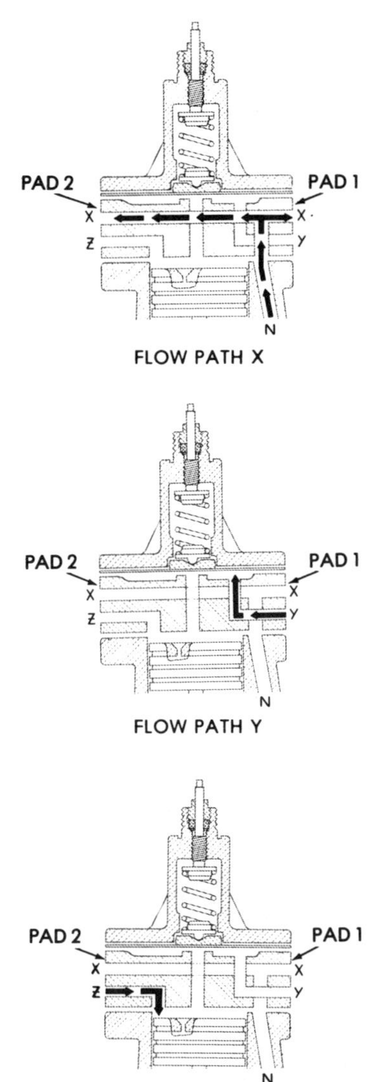

FIG. 5-21. Flow paths in regulator sensing passages. *Courtesy Parker Hannifin Corp., Refrigerating Specialties Div.*

Differential regulators are used sparingly, but can be employed in liquid recirculation pumping systems. Inlet pressure is applied under the diaphragm through the normal internal flow passage. Downstream pressure is applied to the top of the diaphragm from the outlet of the regulator via an external sensing tube. The diaphragm, therefore, senses the difference between the inlet and outlet pressures and modulates accordingly on the pilot seat.

By equipping the regulator with an adapter called *Modudapter®*, and *Moduplates®*, the pilot passages can be routed to externally-mounted, *piggy-back* pilot solenoids or regulators. Figure 5-21 shows a sectional schematic of the Modudapter® with pads 1 and 2 for mounting the Moduplates®. A Moduplate® and pilot solenoid are shown, Figure 5-22, mounted on the pads for an inlet pressure regulator, electric shut-off version. The two ported pads, 1 and 2, are used to mount the Moduplate®, varying the regulator functions. The *Moduplate®* may be reversed, front-to-back, to again vary the regulator's function. When a normally closed solenoid is used as a pilot, and mounted on pad 1 of the *Modudapter®*, the regulator may be used to block flow. The *piggy-back* solenoid, regulator and pads are shown in Figure 5-23.

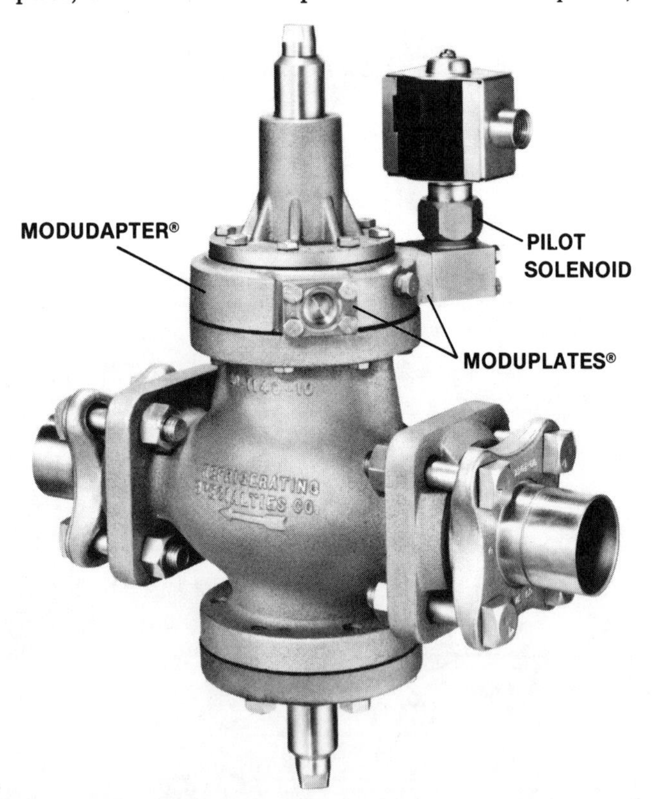

FIG. 5-22. Moduplates mounted on modudapter and pilot solenoid to stop (block) flow in sensing passages. *Courtesy Parker Hannifin Corp., Refrigerating Specialties Div.*

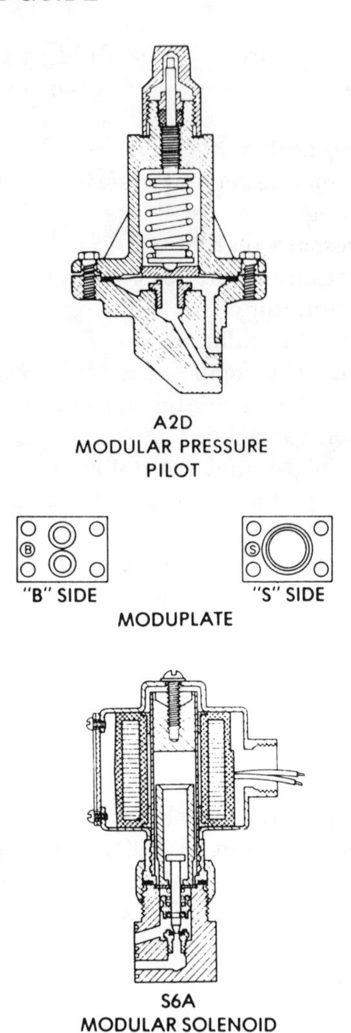

FIG. 5-23. Modular components: pilot regulator, moduplates, and pilot solenoid. *Courtesy Parker Hannifin Corp., Refrigerating Specialties Div.*

When the pilot solenoid (N.C.) is energized via a thermostat, the inlet sensing pressure passage is open to the diaphragm underside as well as the pilot port. The regulator then acts like an inlet pressure sensing regulator. However, when the *piggy-back* pilot solenoid is de-energized, the pilot passage is completely blocked and the regulator shuts down tight, Figure 5-24.

This feature is used with flooded evaporators or large pipe-coil banks to halt refrigeration immediately upon reaching the control temperature and to stop possible overcooling if the large quantity of refrigerant is pumped out.

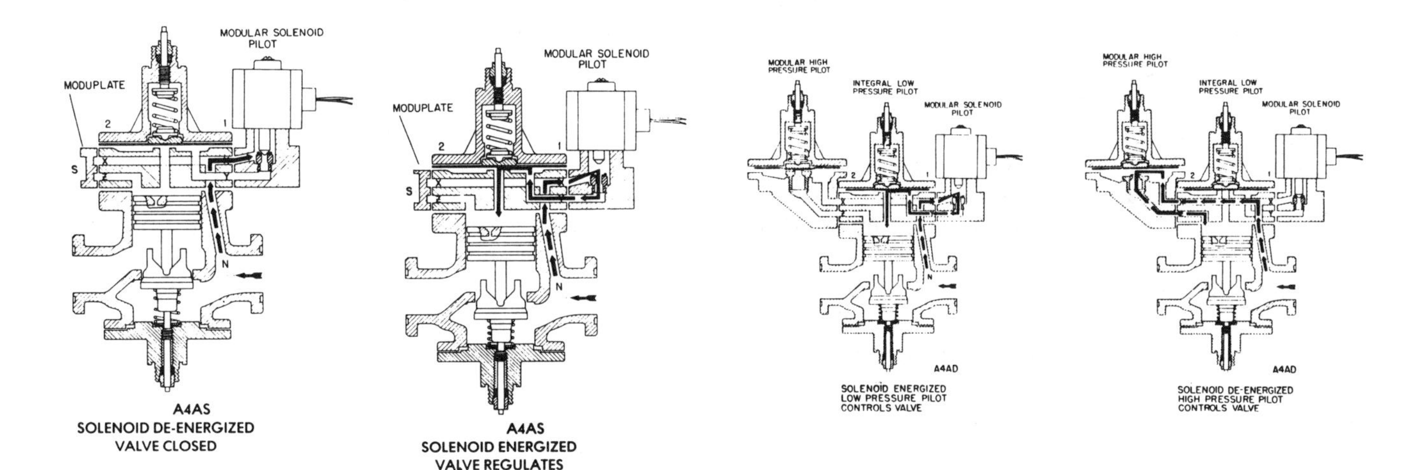

FIG. 5-24. Cross-sectional view showing pilot solenoid blocking flow when de-energized and passing flow when energized. *Courtesy Parker Hannifin Corp., Refrigerating Specialties Div.*

FIG. 5-25. Pilot solenoid and pilot regulator mounted on modudapter. Pilot solenoid is diverting flow. *Courtesy Parker Hannifin Corp., Refrigerating Specialties Div.*

Using the second port on the *Modudapter®* with a *piggy-back* solenoid, creates a wide opening inlet regulator. When the pilot solenoid is de-energized, flow is under the diaphragm and the inlet regulator controls inlet pressure. However, when the solenoid is energized, the ported passage is opened to the top of the piston. The piston then drives the throttling plug off its seat, to a wide open position, allowing full flow through the regulator.

This version can be used to raise evaporator pressure and promote rapid defrost or, in the normal refrigeration mode, it can be wide open, for maximum cooling. It may also be used in batch freezing systems where rapid chilling is required initially and a controlled evaporator pressure is necessary afterwards.

If both pads of the Modudapter are used, that is, pad 1 for a pilot solenoid and pad 2 for a regulating pressure pilot, a dual pressure regulating function may be achieved, Figure 5-25. The pilot solenoid alters the flow in the control passages by diverting and blocking the controlling fluid. When energized, the pilot solenoid allows the regulator to pass controlling flow to the main diaphragm and the regulator controls evaporator pressure at one setting. When de-energized, the pilot solenoid allows the flow to bypass the main diaphragm and diverts it to the pilot regulator which allows the main regulator to control upstream pressures at another setting.

Two levels of cooling may be maintained with this regulator. One temperature is used when loading

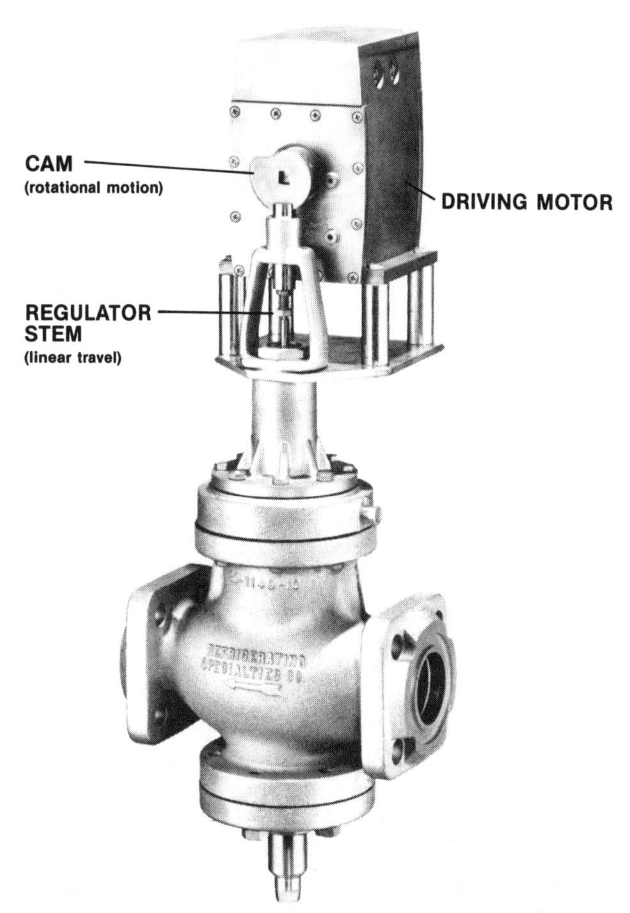

FIG. 5-26. Electric compensated pressure regulator. *Courtesy Parker Hannifin Corp., Refrigerating Specialties Div.*

warm product into a cooler, and another for holding the load. Humidity may also be controlled by running an evaporator at higher temperatures.

Temperature, pressure or humidity in refrigeration systems may be precisely controlled by the addition of externally compensated devices that modulate the regulator input pressures to meet load requirements. Compensation controllers may be electric or pneumatic and are driven by either electric or pneumatic thermostats.

The electric compensated version, Fig. 5-26, is an electric motor with a cam that translates rotational motion into linear motion to position the valve stem. The motor will respond to an electric signal from a potentiometer type thermostat. Rather than maintaining a constant evaporator pressure, like a mechanically driven device, the electrically compensated regulator will lower refrigerant temperature during heavy loads and increase refrigerant temperature during light loads. This adjusts the temperature difference (TD) between the refrigerant and product to increase or decrease system capacity.

The Modudapter and Moduplates can also be used with the compensated style of regulator to create dual functions within a single regulator.

Outlet sensing series

For outlet sensing regulators, the basic valve body is used, as with the inlet regulator. However, an external downstream sensing tube assembly is connected to an adapter ring sandwiched between the top of the valve body and the bonnet containing the range spring and diaphragm.

The outlet series of regulators has four versions:

- Outlet sensing pressure regulator
- Outlet sensing pressure regulator with remote sensing connection
- Outlet pressure regulator with electric shutoff
- Outlet pressure regulator with electric compensation

The outlet series is used to control compressor inlet pressure, to protect the motor from overloading; as a hot gas bypass for capacity control; to limit liquid line pressures in special applications; to prevent pressure rise in evaporators in multiple evaporator systems; and to control hot gas pressures in defrost systems.

Outlet sensing regulators have the same modular concept as the inlet sensing regulators and use the Moduadapter and Moduplates to alter regulator operation.

The basic inlet or outlet sensing regulator may be converted to a temperature sensing regulator by the addition of a bellows-actuated thermal bulb system. A special temperature pilot replaces the standard pilot. A drop in temperature causes the regulator to open, a rise causes it to close. This is a reverse acting or *heating* regulator. Two temperature ranges are available: 60 °F to 140 °F and −20 °F to 80 °F. An electric pilot solenoid may be added to override the regulator and shut it off regardless of the temperature setting or pressure in the regulator.

A direct acting, cooling regulator is also available. A rise in temperature will cause the regulator to open, a drop will cause it to close.

Electronics

Several years ago, Singer produced the thermo-electric expansion valve and electric suction throttling regulator, Figures 3-62, 3-63, and 3-64. A thermister is used to sense temperature and to vary the voltage through the valve head. In the suction throttling regulator, this is done through an electronic controller as the thermister senses higher or lower temperature about the set point.

The electronic controller increases or decreases the voltage output until the temperature at the thermister agrees with the set point temperature. If the temperature is above the set point, there is a change in the sensor's resistance and that increases the voltage input to the regulator controlling head. The suction regulator moves to a more open position, reducing the evaporator's pressure and temperature.

Alco and Sporlan *Electronic Temperature Control Systems* are capable of controlling refrigeration and air conditioning systems to ±1 °F. Figures 5-27,

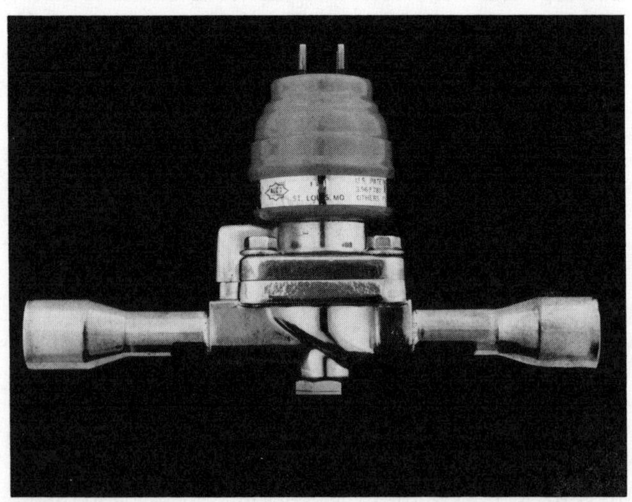

FIG. 5-27. Solid state, electrolled suction regulator. *Courtesy Alco Controls Div., Emerson Electric Co.*

5-28, 5-29, 5-30, and 5-31 show such systems. Even though these two designs vary, they both utilize solid state electronics mounted in a controller. The controller receives a signal from a remote temperature-sensitive thermister (sensor) and converts the change in resistance to a modulating electrical output to the regulator motor or pilot solenoid.

In the Alco version, the probe is installed where it can sense representative return or discharge air. A continuous signal is transmitted to the control circuit where it is compared to the reference signal for the set point temperature. Increasing power tends to close the valve, increasing evaporator temperature. Reduced power opens the valve, decreasing evaporator temperature. The Alco design has a feedback sensor, located in the valve heat motor, that constantly signals valve position to the controller. The valve position signal is processed by the controller, along with temperature probe and set point signals, to provide accurate control that is unaffected by changes in voltage supply or by refrigerant or ambient temperatures.

During the initial *pull-down* period, when the space is warm, the regulator is wide open and allows full flow of refrigerant from the evaporator. As the controlled space reaches its preset operating temperature, the sensing probe signals the control circuit to start closing the regulator.

FIG. 5-28. Regulator electronic controller processes temperature signals to maintain set point. Controller takes signals from remote sensors and defrost clock and controls regulator function. *Courtesy Alco Controls Div., Emerson Electric Co.*

FIG. 5-29. Diagnostic board used to check operation during start up and service. *Courtesy Sporlan Valve Co.*

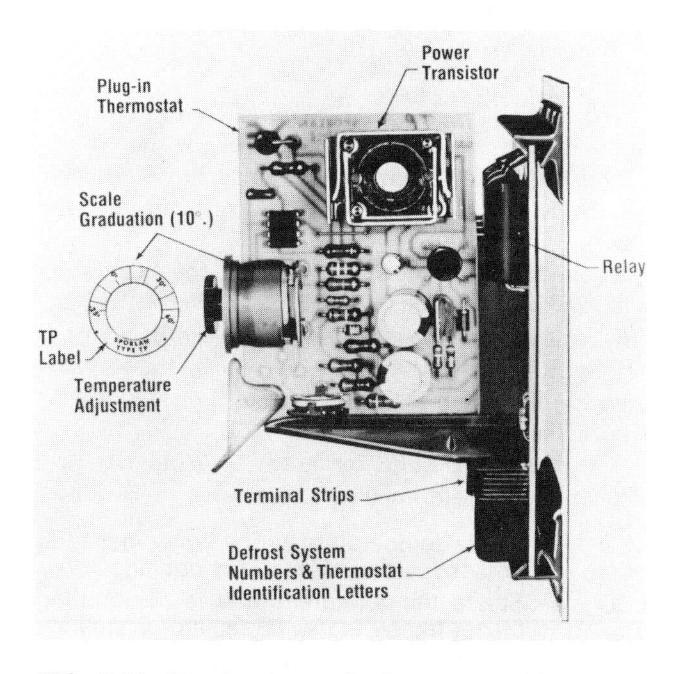

FIG. 5-30. Plug-in electronic thermostat with temperature adjustments and electronic components. *Courtesy Sporlan Valve Co.*

FIG. 5-31. Temperature sensor thermister with protective shield and neoprene covered lead wires. *Courtesy Sporlan Valve Co.*

Accordingly, a green LED comes on and power is applied to the regulator. A heater inside the regulator head is thus energized, warming a bimetallic motor in the regulator that drives it toward the closed position.

As the regulator slowly modulates, the exact stem position is continuously fed back to the control circuit. Once the space temperature is stabilized at the set point, power to the regulator heat motor is cycled on and off to maintain the stem of the regulator at the correct position.

The green LED indicator on the control board provides a visual reference for the valve's operation.

LED off	Space temperature probe is warmer than preset value. *Regulator is opening.
LED on	Space temperature probe is colder than preset temperature. *Regulator is closing.
LED blinking	Space temperature probe is near set point. *Regulator is holding position.

*Suction line nominal capacities are based on 2 psi pressure drop across the regulator with a 40° evaporator.

As a fail safe feature, the regulator will automatically position itself in the full-open position in the event of a power failure.

Both models include sequencing of the regulator during defrost. Upon defrost initiation, a signal from the controller shifts the regulator to a closed position and converts it to a suction stop valve. After defrost is complete, the controller shifts back to the normal refrigeration mode controlled by the sensor input.

In the Alco version, a red LED lights, indicating that the regulator should close. The green LED will light and, when the regulator temperature feedback diode senses that the regulator is fully closed, 24 volt AC is supplied to a pair of terminals on the control module. This voltage is used to energize a customer-furnished defrost contactor in the condensing unit control panel and the heating phase of defrost begins.

The heating phase of defrost is terminated by a customer furnished close-on-rise thermostat connected to the control module.

The regulator remains closed and the red LED stays lit until the defrost timer contacts close, thus providing runoff time. When the timer contacts close, the red and green LEDs will go out and the regulator will begin to open. When the space temperature has reached the set point, the green LED will begin to flash.

Electronic controllers

Figure 5-32 shows several solid state electronic controllers that can be programmed for .5% measur-

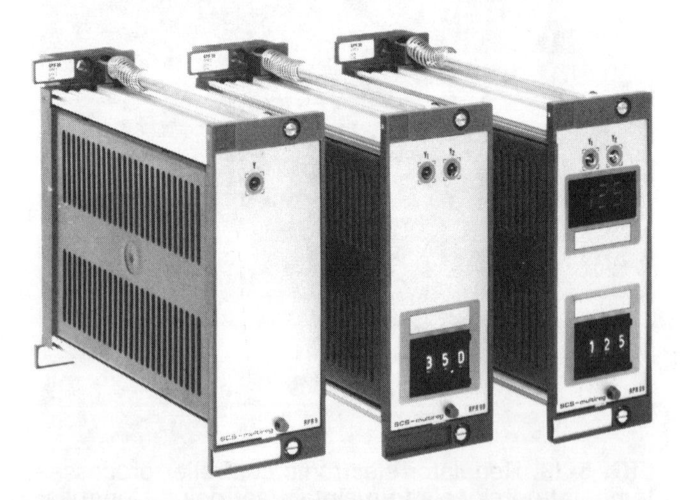

FIG. 5-32. Various electronic controllers for heating, cooling, and humidity control. *Courtesy Staefa Control Systems Inc.*

ing accuracy over their range. When connected to solid state, contactless sensors and modulating control valves, they will deliver precise heating, cooling and humidity control for:

- Environmental rooms in medical research facilities
- Laboratories
- Fermentation processes
- Dairy and food plants
- Industrial environmental rooms for electronic production and assembly
- Paper and textile plants

Controllers can monitor, control and indicate temperature, humidity, pressure, wind, solar radiation,and air volume and compare indoor requirements to outdoor conditions. Once the loads and demands are determined, proportionally controlled heating and cooling systems may be sequenced with regulators to produce the selected conditions.

Three-way control valve applications

Chapter 3 and Fig. 3-28 covered the construction and uses of three-way valves. A number of application and selection details will be covered here.

A variety of terms were discussed in Chapter 4 and will be reviewed and expanded. This review includes:

Cv Cv is the water flow in gallons per minute, GPM, at 60 °F through a valve in the full open position with a one psi pressure differential across the valve. Valves that are used in applications other than water, that is, steam, refrigerants, gases, etc., still have Cv ratings given in GPM of water at one psi.

Each manufacturer will have charts, tables or curves of the Cv values for each of his valve designs and sizes. This was seen in Tables 5-22 for steam, 5-23 for water and 5-24 for air. If tables are not readily available, estimates based on calculations can be made similar to those shown in Table 4-14.

Rangeability Rangeability is the ratio of maximum controllable flow at constant pressure drop.

$$R = \frac{W_{max}\text{ (controlled flow)}}{W_{min}\text{ (controlled flow)}}$$

Turndown Ratio Turndown ratio is the ratio of maximum usable flow to minimum controllable flow at constant pressure drop.

$$R = \frac{W_{max}\text{ (useable flow)}}{W_{min}\text{ (controlled flow)}}$$

If a regulator has a maximum controlled flow of 50 GPM, maximum usable flow of 30 GPM and a minimum controllable flow of one GPM, then the rangeability is 50:1 and the turndown ratio is 30:1.

Critical Pressure Ratio (CPR) When regulators are employed in gaseous service, that is with steam, air, or gaseous refrigerants, the flow is measured either in pounds per hour or standard cubic feet per minute (SCFM) and is directly proportional to the pressure differential across the regulator. This is only true up to a point, however, since increasing the pressure differential above a certain value will not increase the fluid flow further. The point at which this happens is the *critical pressure ratio.*

$$CPR = \frac{\text{Absolute pressure downstream}}{\text{Absolute pressure upstream}} = \frac{P_2 + 14.7}{P_1 + 14.7}$$

For steam, the CPR value is .546. (See Figure 4-34.) In control valves or regulators, the CPR is assumed to be .5, therefore, the flow does not increase even if the absolute outlet pressure P_2 is less than .5 of the absolute inlet pressure P_1.

Wire Drawing Wire drawing is a destructive condition where a fluid has eroded a *slit* in the regulator disc or seat. This occurs when a regulator is continuously allowed to operate *near* its closed position. When nearly closed, the fluid velocities are enormous, causing severe turbulence. This usually happens when regulators are *oversized* and is one important reason to make sure that regulators are properly sized and a *pipe size-to-pipe size* selection is not casually made.

Fluid Flashing Flashing of a liquid can have destructive effects on regulators, pipe and pipe fittings. Noise and reduced capacity can also result. It occurs when the pressure on a fluid is reduced below that required to keep the fluid in its liquid state. A car radiator may allow the coolant to circulate at 250 °F at 15 psig pressure, but when the cap is opened, the fluid immediately flashes to steam as the pressure is released.

The same thing can happen if the pressure drop across a regulator's port area is great enough to allow the downstream pressure, P_2, to fall below the fluid's saturation pressure at the operating temperature.

Figure 3-19 showed the various regulator plug designs and their flow characteristics as a percentage of lift of the disc in relation to its seat.

The *quick opening* disc is designed mainly for on-off control and can attain a major portion of its flow with little disc travel.

Linear disc designs were reserved for double seat,

balanced regulators where pressure differentials in a single seat design would require extensive operational power to open and close the regulator. This design does not close 100% and a slight leakage is always present.

Many refrigeration regulators use the V-port or *modified linear* disc configuration where rangeability is important. At low flows, a large disc movement produces little flow change. This characteristic is essential for heat exchanger application and good temperature control in heating systems.

Maximum rangeability is achieved with the *equal percentage disc.* A high degree of control accuracy is achieved over a wide range of pressure, flow and load conditions.

A more recent control valve design has a rangeability of 500:1, Figure 5-33. This regulator style control valve is used in hot and cold water mixing service. It achieves high rangeabilities in two ways:

- Precise stem control (disc movement)
- Quickness of response to a signal from an electronic controller.

Three-way valves may be used as proportional control valves to vary flow of heating or cooling liquids to match varying load requirements. They may also be used as two-position valves to switch heating or cooling liquids on demand. The three-way valve may also be used in two-pipe, three-pipe and four-pipe dual temperature water systems. Some of these configurations will be seen later.

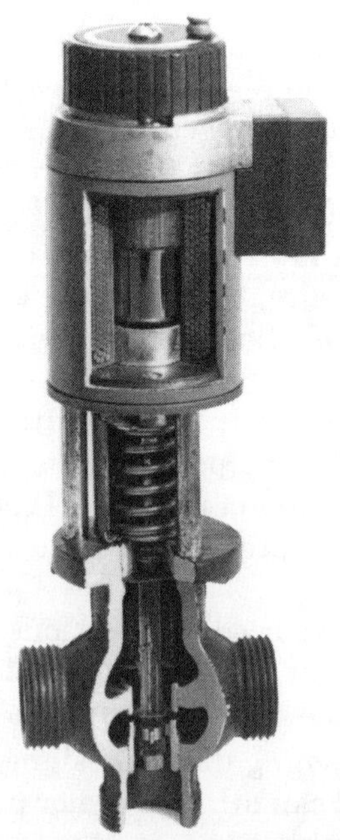

FIG. 5-33. Cutaway view of modulation electromagnetic valve. *Courtesy Staefa Control Systems Inc.*

When three-way valves are used in the two position application, they are usually sized with minimum pressure drop since the valve is used primarily as a switch-over device from heating to cooling, that is, boiler to chiller. The lower the fluid resistance, the lower the pump head required to move a given quantity of fluid. When used as a control device, to proportionally control fluid flow to a heat exchanger, the pressure drop in the valve is critical for good temperature control.

Application

Chilled or hot water flow to a heat exchanger is commonly used to cool or heat offices, homes, apartments, and industrial plants. Control of the fluid has been studied over the years and many different approaches are used to create a comfortable and controlled condition.

The major problem with heating and cooling terminals is that they are designed to handle maximum demand. This requirement, however, only lasts for a few days a year. The rest of the time, the equipment must operate at partial load and operate efficiently; therefore, the equipment must operate over a range of conditions from low to full design load.

Figure 5-34a shows a two-way valve, in a *throttle circuit,* that can open or close at the demand of a thermostat. The major drawback of this design is that, as the valve closes to restrict flow, the pump *head* changes and, near valve closeoff, the pump will approach a *dead head* condition. At near closeoff, the disc could be subjected to wire drawing. In addition, fluid flashing could occur, resulting in damaged equipment as well as noise and loss of further control.

If the throttle valve is used in parallel installations, Figure 5-34b, the major drawback is that, as one or more of the valves begins to throttle, more

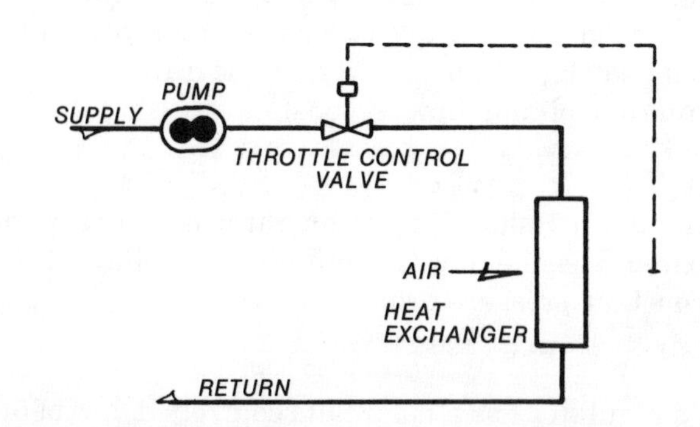

FIG. 5-34a. Throttle circuit.

water is forced through the open valves and heat exchangers. This imbalance causes a loss of thermal control, noise, wire drawing and flashing. The pump outlet pressure could increase to the point where the open valves will have a pressure differential that exceeds their rating. Consequently, the valves may not be able to close when needed. It is important that the maximum rated shutoff differential pressure of a control valve should exceed the pump's *dead head* pressure. If the *dead head* of the pump is 30 psig, then the valve differential pressure should exceed 30 psi, say 40 psi.

Other possibilities include three-way valves that divert flow around the heating/cooling coil, Figure 5-35. The major problem here is the inherent construction of the disc/seat arrangement. Since there are two discs and flow is *over* each disc, tight shutdown is not possible, Figure 3-28.

The advantage of the three-way valve is that, as fluid flow is increased or decreased through the heat exchanger, pump head does not readily change since the fluid is diverted around the exchanger and total fluid flow is not changed.

The other three-way application, using a mixing style proportional control valve, has two installation

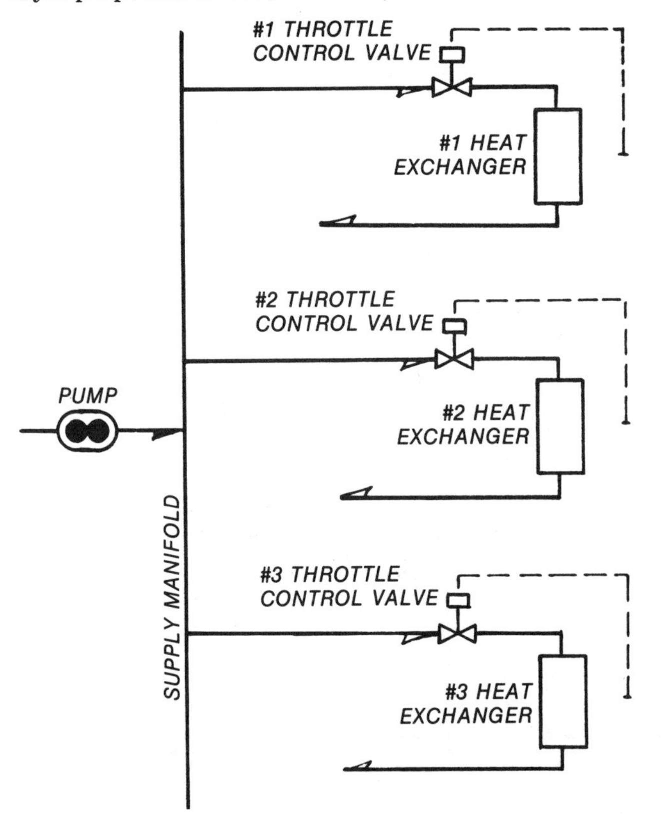

FIG. 5-34b. Parallel throttle flow circuits.

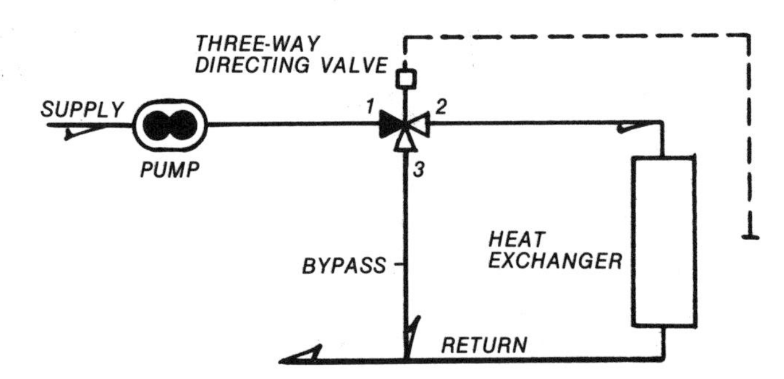

FIG. 5-35. 3-way diverting valve on heat exchanger application. Diverting valve has one inlet, two outlets.

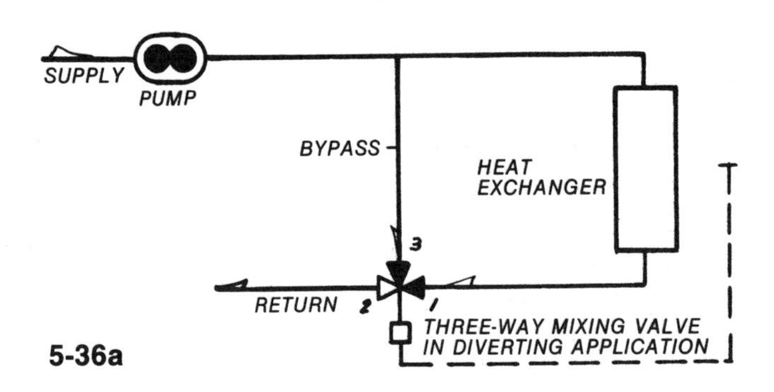

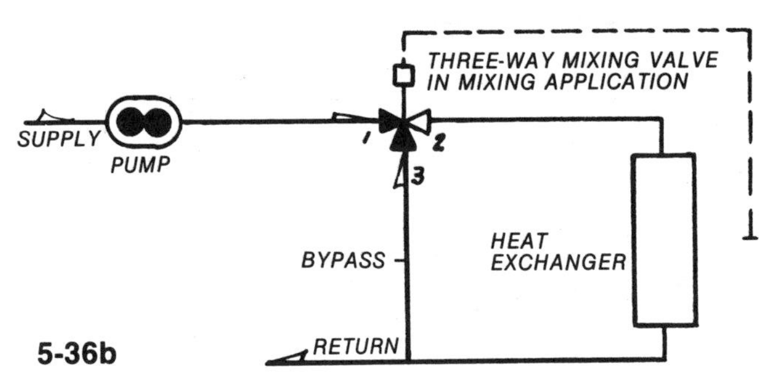

FIG. 5-36. 3-way mixing valve in both diverting (top) and mixing (bottom) applications.

possibilities:

- Diverting service
- Mixing service

Figure 5-36a shows an installation where the mixing valve is located at the heat exchanger outlet. Flow through the heat exchanger varies, as boiler water is diverted around the exchanger, but the temperature to the exchanger is held constant.

Figure 5-36b shows an application where the mixing valve is at the inlet to the heat exchanger. Here the flow through the coil is constant, but the temperature varies.

Mixing Valves in Diverting Service (variable volume control) Figure 5-37 shows a three-way valve in diverting service. Regulation is achieved by diverting boiler water around the heat exchanger. Thus, the volume through the heat exchanger is variable while the flow through the boiler is constant. However, the variable flow through the heat exchanger is at a constant temperature. The pump loop is the constant flow path. The mixing valve allows boiler water to go either through the bypass circuit or the heat exchanger. The portion passing through the heat exchanger determines the heat output.

At low loads, large temperature variations occur in the heated media, therefore, the use of this particular mixing circuit is limited. It is used primarily for special heat exchangers and coolers. If used, the more pressure drop through the control valve, compared to the entire system, the better the control of the system.

Figure 5-38 is a variation of Figure 5-37 and is used with very long piping. It is also known as an *injection circuit.* There is constant volume to both the boiler and heat exchanger and the heat output is controlled by varying the temperature of the water to the heat exchanger. The second pump, P_2, assures that the extra long runs will have steady and reasonable pressure drops. The added bypass provides constant volume to the heater circuit.

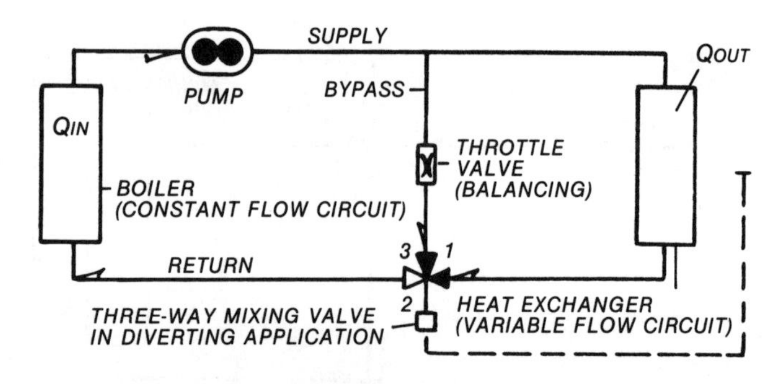

FIG. 5-37. Mixing valve in diverting application. Pump is located in constant flow portion of circuit. Throttle balancing allows bypass to have same pressure drop as heat exchanger at full load.

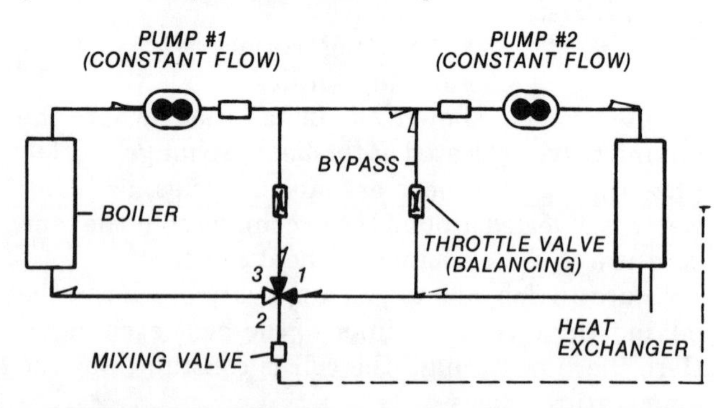

FIG. 5-38. Variation circuit with second pump to compensate for exceptionally long piping runs.

Mixing Valves in Mixing Service (constant volume control) Figure 5-39 shows a mixing valve in a mixing circuit with a circulating pump. This circuit is called *temperature control by mixing.* When the control valve is closed, the exchanger pump circulates water through the exchanger and back, through the bypass, to valve inlet #3. When the valve opens, the boiler water enters the circuit at point #1. This design is used both in air conditioning and heating because of the even temperatures over the entire heat exchanger surface. A very small temperature gradiant is experienced.

The valve mixes return heat exchanger water (cooler water) with boiler feed water (hot water) and supplies the heat exchanger with variable temperature, constant flow water. The boiler flow temperature, in turn, is constant.

Figure 5-40 shows a variation of Figure 5-39 that uses a secondary bypass circuit to moderate the impact of hot water flowing from the boiler.

Figure 5-41 shows various mixing and diverting circuits using three-way control valves in a variety of applications.

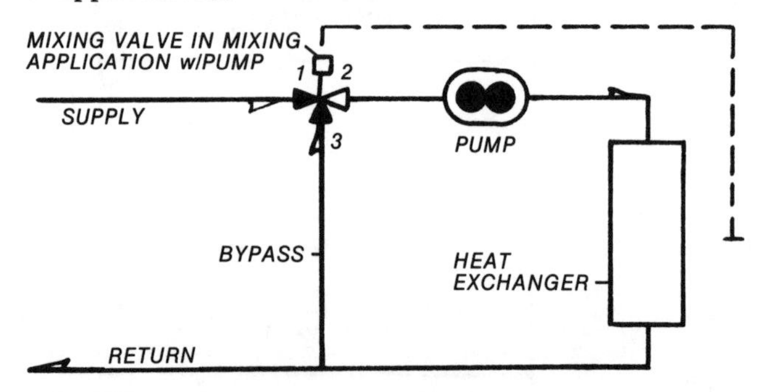

FIG. 5-39. Mixing circuit in mixing application with pump in heat exchanger circuit.

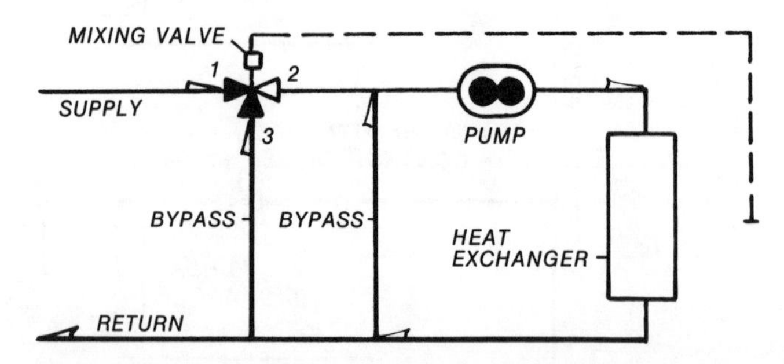

FIG. 5-40. Variation of a constant flow heat exchanger circuit with additional bypass to temper heat exchanger circuit from hot water supply.

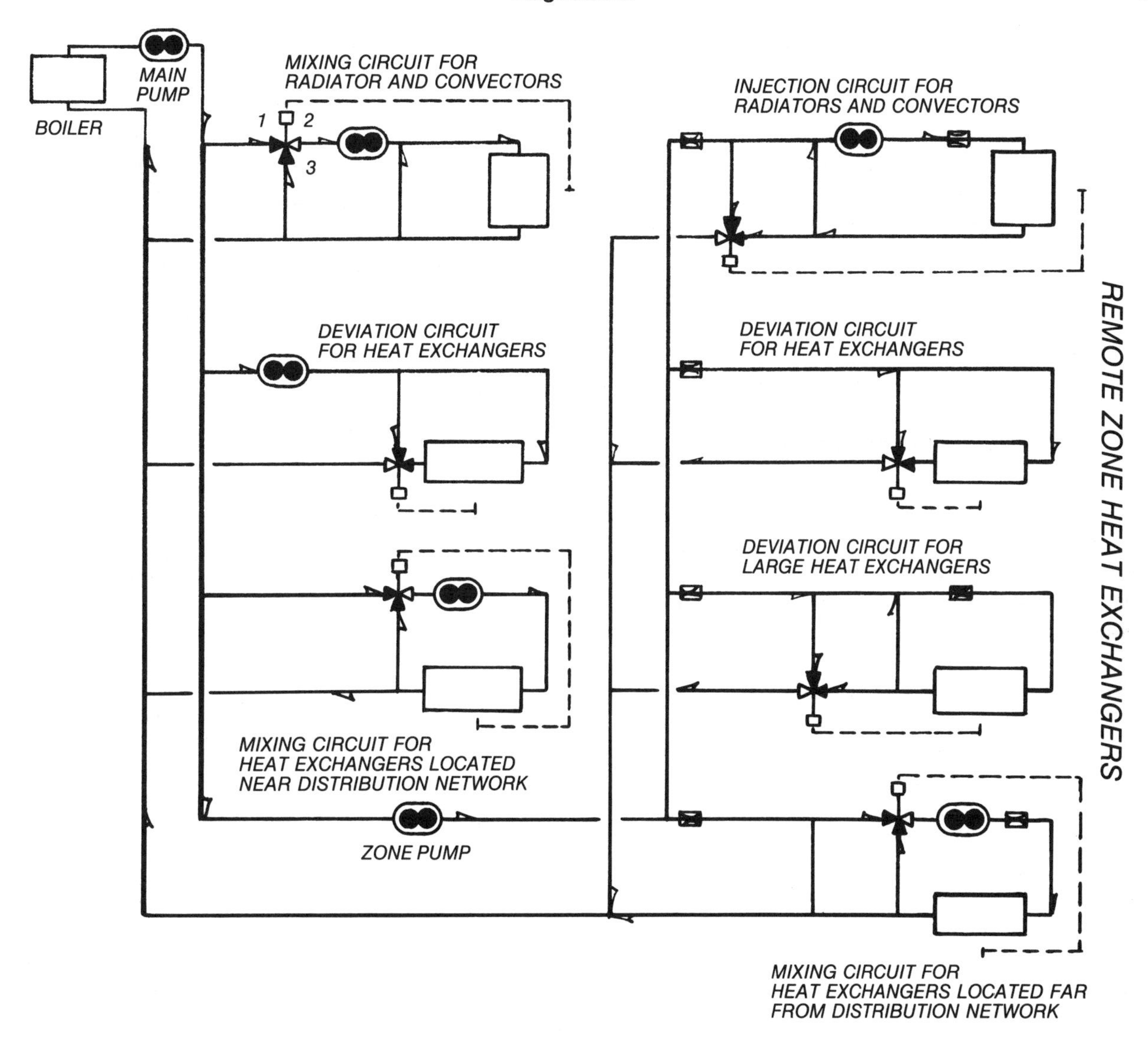

FIG. 5-41. Hydraulic circuit variations using 3-way valves and pumps to control heat exchanger performance. *Courtesy Staefa Control System Inc.*

Balancing parallel flow circuits is necessary for good control. A good design will:

- Use primary as well as secondary pumps to assure good water flow.
- Keep the difference in pressure drops between parallel circuits to a minimum
- Use balancing valves, Figure 4-30, to adjust the circuit pressure drops, so that a balance between the bypass circuits and heat exchangers can be achieved.

To properly select a three-way valve for control circuits like those just outlined, the following questions must be accurately answered. Otherwise, the valve will be too large or too small for the job.

- What is the fluid: water, air, steam or oil?
- What is the specific gravity, viscosity and specific heat of the fluid?
- What is the inlet pressure and temperature at demand load?
- What is the ambient temperature?
- What is the expected pressure drop, ΔP, across the disc (at full load and if the pump is subjected to dead head service)?
- What are the voltage requirements or specific controller inputs?
- What is the overall capacity based on maximum conditions; that is, warmest or coldest outside temperature and inside control conditions?

- Normally open or normally closed?
- C_V requirement?
- Physical data such as type ends, material of construction, weight and size?

The entire system pressure drop, from the pump outlet to system return, is made up of pipe, fittings, heat exchanger and control valve losses and is defined as:

$$\Delta H_o = \Delta P_{total} = \Delta P_{pipe} + \Delta P_{fittings} + \Delta P_{heat\ exchanger} + \Delta P_{control\ valve}$$

At maximum flow, 25 to 50% of the total system pressure drop should be across the control valve; that is:

$$\Delta P_{control\ valve} = .25\ \Delta P_{total} \text{ to } .50\ \Delta P_{total}$$

If the pressure drop across the control valve is 10% or less of the total system drop, the valve will be too large and will not control flow until almost fully closed, resulting in poor control.

Recommended pressure drops for fluids should be:

- Chilled and low temperature water systems: 25 to 50% of the total system pressure loss, H_o.
- Steam: 50% of the absolute inlet pressure without exceeding the critical pressure ratio.
- High temperature water systems: Extreme care must be taken here, since the pressure drop, if high enough, could cause downstream flashing. These systems usually are operated at 325 °F to 425 °F (80 psig to 310 psig respectively). Additional pressure, over the saturation pressure, is added by using pressurized nitrogen or air. The additional pressure may range from 15 to 45 psi over the saturation pressure.

The following example shows why this is necessary, as well as the placement of the regulator in relation to the heat exchanger, see Figure 5-42.

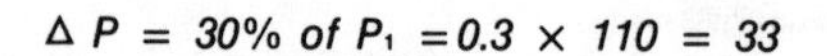

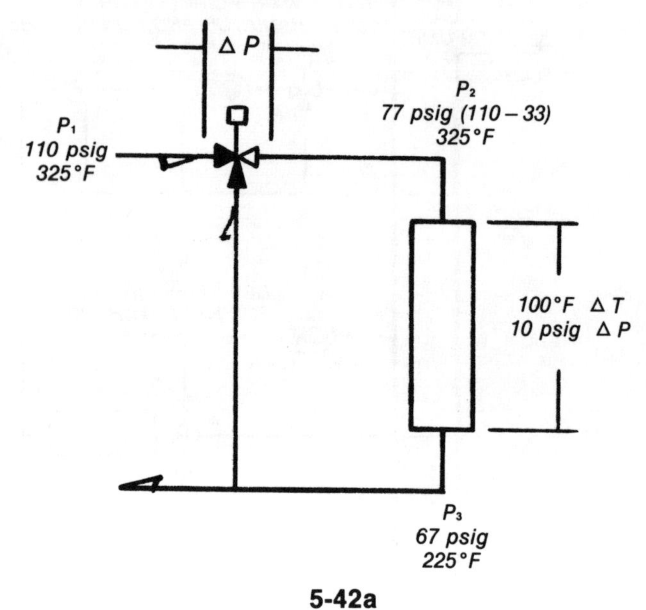

5-42a

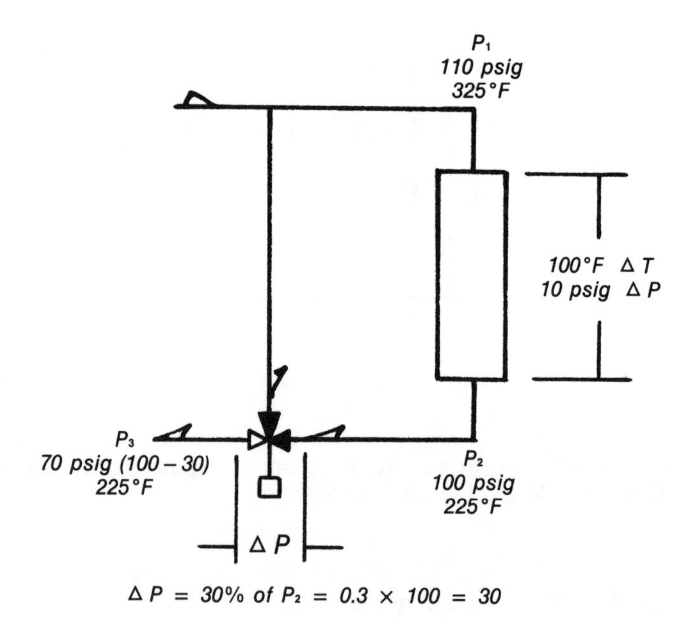

5-42b

FIG. 5-42. Effect of placing the valve at the (**a**) inlet or the outlet (**b**) of the heat exchanger in high temperature systems.

At the entrance, the valve *sees* 110 psig and a temperature of 325 °F . We will use a 30% ΔP as the pressure drop (110 x .30 = 33 psi). Note that 77 psi, downstream of the valve, is below the saturation pressure of 80 psig; therefore, flashing will occur. Either the inlet pressure will have to be increased or the pressure drop reduced.

In the second case, with the valve downstream of the heat exchanger, the entering temperature is 225 °F at 100 psig. The pressure leaving the valve is 70 psig (100 − 30). The 70 psig is well above the saturation pressure of 4 psig at 225 °F.

It can also be seen that the body sizing and material ratings become substantially less in the second example because of the lower pressure and temperatures of the fluid. The ratings were covered in Chapters 3 and 4 and should be reviewed. Table 5-26 shows ratings based on medium and high temperature water systems.

Here is an example of the selection for a water application: 5000 pounds of process water must be heated 100° in two hours using 250 °F hot water with a temperature drop through the heat exchanger of 50°. The total system pressure drop is estimated to be 35 psi and the control valve ΔP selection is 16 psi at full load.

The flow rate is determined

$$[W_p C_p \Delta T]_{process} = [W_f C_p \Delta T]_{heating\ fluid}$$

$$\frac{5000 \text{ lb} \times 1.0 \times 100°}{2 \text{ hours}} = W_f \times 1.0 \times 50\,°F$$

$$W_f = \frac{5000 \times 50}{50}$$

$$W_f = 5000 \text{ lb/hr}$$

$$W_f = 5000 \text{ lb/hr} \times \frac{1}{8.33 \text{ lb/gal}} = 600 \text{ gph}$$

$$= 10 \text{ gpm}$$

$$C_V = \frac{Q\sqrt{\varrho}}{7.9\sqrt{P_1 - P_2}} = \frac{10\sqrt{62.4}}{7.9\sqrt{16}} = 2.5$$

This can also be verified in Figure 5-43. With a ΔP of 16 psi and a flow of 10 gpm, $C_V = 2.5$. Now select a regulator with a C_V of 2.5, having a body rating capable of handling the hot water inlet temperature of 250 °F and 35 psig.

In another example, a terminal heat exchanger is heating air with saturated steam at 250 °F and at a rate of 10,000 Btu/hr. The inlet pressure is 15 psig and the outlet temperature is 150 °F. Assume a pressure drop of 30% of the 15 psig inlet pressure or 4.5 psig. The C_p of saturated steam is 0.45

$$Q = W C_p \Delta T$$

$$W = \frac{Q}{C_p \Delta T} = \frac{10{,}000 \text{ Btu/hr}}{0.45 \times 100}$$

$$W = 222 \text{ lb/hr}$$

$$P_2 = 15.0 + 14.7 - 4.5 = 25.2 \text{ psia}$$

$$C_V = \frac{W}{3\sqrt{\Delta P \times P_2}} = \frac{222}{3\sqrt{4.5 \times 25.2}}$$

$$C_V = 7.0$$

The C_V for saturated steam can be verified by using Figure 5-44. The required regulator will have a C_V of 7.0 and a body rated to handle steam at 250 °F and 15 psig.

Table 5-26. Pressure/temperature ranges of various valve types.

Table 2—Pressure-temperature ranges of various valve types.

Function	Size, in.	Class	Type	Body material	Trim	Pressure-temperature
Medium temperature water systems						
Block	2-12	125	Gate	Iron	Bronze	200 psi @ 200 F 150 psi @ 350 F
Check	2-12	125	Swing	Iron	Bronze	200 psi @ 200 F 150 psi @ 350 F
Throttle	2-10	125	Globe	Iron	Bronze	200 psi @ 200 F 150 psi @ 350 F
Block	2-12	150	Non-lube plug	Ductile iron	PTFE	250 psi @ 100 F 196 psi @ 400 F
Block	¾-10	150	Ball	Cast steel	TFE	285 psi @ 100 F 200 psi @ 350 F
High temperature water systems						
Function	Size, in.	Class	Type	Body material	Trim	Pressure-temperature
Block	2-16	300	Gate	Cast steel	Cr 13	740 psi @ 100 F 600 psi @ 500 F
Check	2-16	300	Swing	Cast steel	Cr 13	700 psi @ 100 F 600 psi @ 500 F
Throttle	2-12	300	Globe	Cast steel	Cr 13	740 psi @ 100 F 600 psi @ 500 F
Block	2-10	300	Ball	Cast steel	Reinforced TFE	740 psi @ 100 F 300 psi @ 400 F
Block	2-10	300	Non-lube plug	Cast steel	PTFE	740 psi @ 100 F 400 psi @ 400 F
Block or throttle	2-8	300	High performance butterfly	Cast steel	S.S.	740 psi @ 100 F 600 psi @ 500 F
Block	2-4	600	Through conduit	Cast steel	Stellite	1500 psi @ 100 F 1300 psi @ 500 F
Block	6-12	300	Through conduit	Cast steel	Stellite	750 psi @ 100 F 665 psi @ 500 F

Heating/Piping/Air Conditioning. Oct. 1983

Summary

While this chapter has covered a wide variety of regulators, control valves, applications and methods of selection, it should not be considered an absolute answer to regulator selection.

Each manufacturer has developed its own selection procedures, in addition to tables, charts and graphs. It is recommended that specific data be obtained from the regulator manufacturer and that it be carefully studied so that the proper selection can be based on all necessary design criteria.

This chapter not only assists in understanding the procedures and methods of selection, but may alert you to the numerous pitfalls that can be encountered in regulator selection.

NOTE: Chart is based on formula $Q = C_v \sqrt{\Delta p}$
Q = U. S. GPM, C_V = Flow Coefficient, Δp = Pressure Drop in psi.

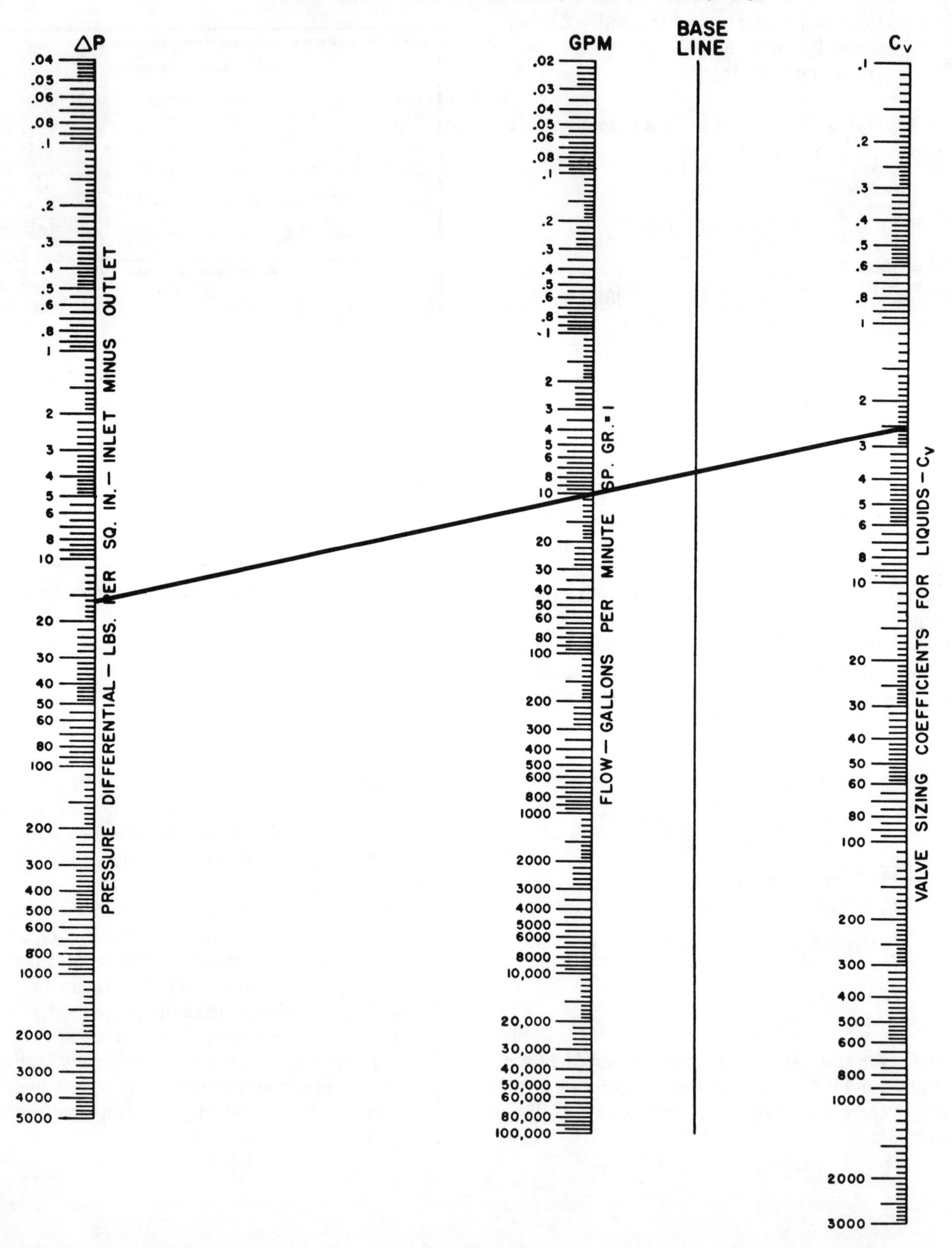

FIG. 5-43. Liquid capacity nomograph, C_V. *Courtesy Barber-Colman Co. Environmental Controls Div.*

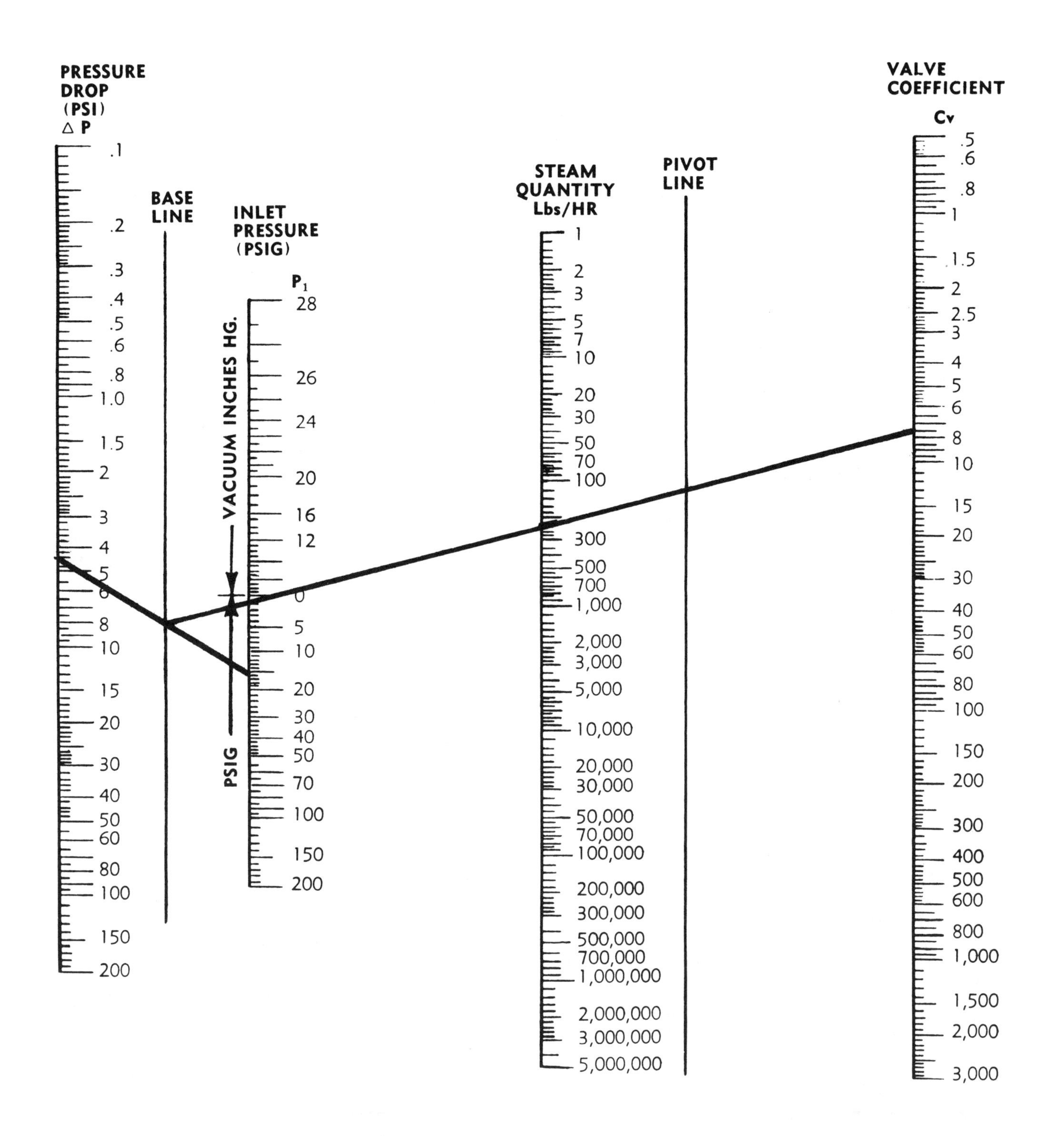

FIG. 5-44. Saturated steam nomograph, C_V. *Courtesy Barber-Colman Co. Environmental Controls Div.*

6
Applications

In this chapter we will be looking at valve applications for various refrigeration circuits, steam, hot and chilled water systems, solar systems, and general use. We will attempt to show how the labor of the previous five chapters is brought forward to final use in workable systems.

The various valves are selected, in a step-by-step method, by first denoting the type of valve and its operation in the system. The operating parameters are given and then the solution to the problem. This should give the reader a logical approach and show how a selection is made, using tables and charts supplied by valve manufacturers.

SELECT

A thermostatic expansion valve for a refrigeration system.

Operation

The thermostatic expansion valve, Figure 6-1, automatically throttles the refrigerant flow based on the temperature of the refrigerant in the suction

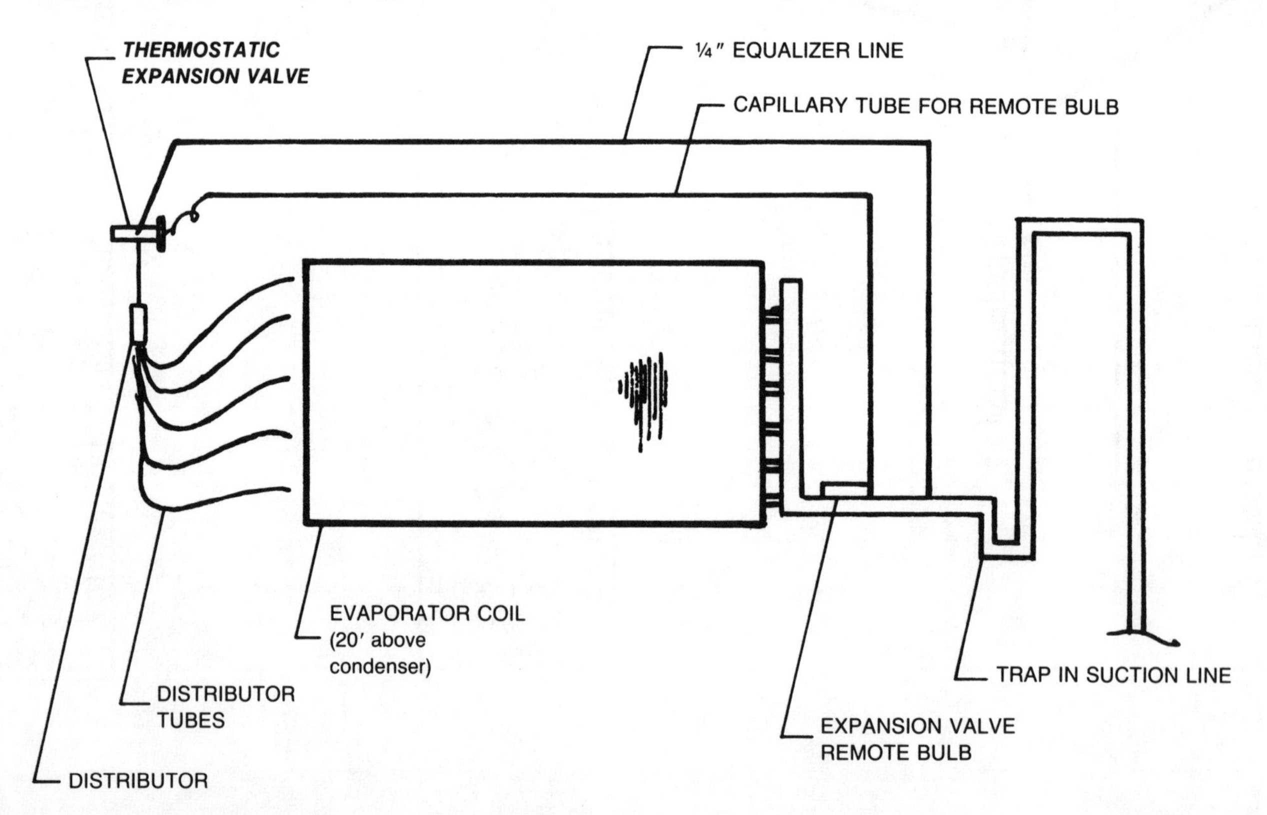

FIG. 6-1. Thermostatic expansion valve installation.

line at the remote bulb. When the load increases, the vapor temperature at the remote bulb increases and the valve opens to admit more refrigerant into the evaporator to accomodate the load. At light loading, the opposite is true: the valve throttles down.

Given

Capacity is 36,000 Btu/hr (3 tons) using R-502 at −20 °F refrigerant temperature. The evaporator is 20 feet above the air-cooled condensing unit which operates at a maximum 90 °F outside air temperature. Liquid subcooling by a suction line heat exchanger is 15 °. The condenser has a subcooling pass that allows the liquid refrigerant to leave the coil at 10 ° terminal difference, that is, 90 ° air + 10 ° = 100 °F exiting temperature. Head pressure is 248 psig.

Solution

1. Determine the pressure drop across the evaporator coil from the coil manufacturer's specifications. If the pressure drop exceeds 1.25 psi, when working with R-502 at −20 °, use an externally equalized valve per Table 6-1a.

2. Determine the pressure drop across the valve:

a. Head pressure, at the compressor, is 248 psig. The high side pressure drop is:

condenser coil	15 psi
line loss	2
valve loss in liquid line	5
drier loss	3
vertical lift (Table 6-1b)	10
pressure drop (high side)	35 psi

Valve inlet pressure is:

head pressure at condenser	248 psi
high side pressure drop	−35
	213 psi

Table 6-1a. Maximum allowable evaporator pressure drops.

REFRIGERANT	EVAPORATING TEMPERATURE Degrees F.				
	40	20	0	−20	−40
	PRESSURE DROP — psi				
12, 500	2	1.5	1	0.75	0.5
22	3	2	1.5	1.0	0.75
502	3	2.5	1.75	1.25	1.0
717 (Ammonia)	3	2	1.5	1.0	0.75

Sporlan Valve Co.

Table 6-1b. Static pressure losses due to vertical lift.

REFRIG-ERANT	VERTICAL LIFT — FEET				
	20	40	60	80	100
	Static Pressure Loss — psi				
12	11	22	33	44	55
22	10	20	30	40	50
500	10	19	29	39	49
502	10	21	31	41	52
717 (Ammonia)	5	10	15	20	25

Sporlan Valve Co.

Table 6-1c. Average distributor pressure drop at 100% loading.

REFRIGERANT	*AVERAGE PRESSURE DROP ACROSS DISTRIBUTOR
12	25 psi
22	35 psi
500	25 psi
502	35 psi
717 (Ammonia)	40 psi

Sporlan Valve Co.

Table 6-1d. Expansion valve capacities with R-502 and low evaporator temperatures.

VALVE TYPES	NOMINAL CAPACITY	EVAPORATOR TEMPERATURE DEGREES F.																	
		−10°						−20°						−40°					
		PRESSURE DROP ACROSS VALVE (Pounds Per Square Inch)																	
		100	125	150	175	200	225	125	150	175	200	225	250	125	150	175	200	225	250
G-NI-F	¼	0.25	0.28	**0.31**	0.33	0.35	0.38	0.28	0.31	0.33	0.35	0.38	0.40	0.22	0.24	**0.26**	0.28	0.30	0.32
G-S-NI-F	½	0.48	0.54	**0.59**	0.63	0.68	0.72	0.48	0.53	0.57	0.61	0.64	0.68	0.38	0.42	**0.45**	0.48	0.51	0.54
G-S-NI-F	1	0.90	1.01	**1.10**	1.19	1.27	1.35	0.89	0.98	1.06	1.13	1.20	1.26	0.67	0.73	**0.79**	0.85	0.90	0.95
G-S-H	1½	1.20	1.34	**1.47**	1.59	1.70	1.80	1.12	1.22	1.32	1.41	1.50	1.58	0.84	0.92	**0.99**	1.06	1.12	1.18
G(Ext.)-C-S	2	1.60	1.79	**1.96**	2.12	2.26	2.40	1.68	1.84	1.98	2.12	2.25	2.37	1.12	1.22	**1.32**	1.41	1.50	1.58
C-S	3	2.00	2.24	**2.45**	2.64	2.83	3.00	1.90	2.08	2.25	2.40	2.55	2.69	1.45	1.59	**1.72**	1.84	1.95	2.06
C-S-H	4	2.80	3.13	**3.43**	3.70	3.96	4.20	2.91	3.18	3.44	3.68	3.90	4.11	2.12	2.33	**2.51**	2.69	2.85	3.00
C & S(Ext.)	6	3.70	4.14	**4.53**	4.89	5.23	5.55	3.35	3.67	3.97	4.24	4.50	4.74	3.13	3.43	**3.70**	3.96	4.20	4.43
H	6½	4.35	4.86	**5.33**	5.75	6.15	6.53	4.19	4.59	4.96	5.30	5.62	5.93	3.52	3.86	**4.17**	4.45	4.73	4.98
S(Ext.)	7	4.71	5.27	**5.77**	6.23	6.66	7.06	4.27	4.67	5.05	5.40	5.73	6.04	3.98	4.36	**4.71**	5.04	5.34	5.63
P-H	9	5.00	5.59	**6.12**	6.61	7.07	7.50	5.03	5.51	5.95	6.36	6.75	7.11	3.91	4.29	**4.63**	4.95	5.25	5.53
P-H	12	8.00	8.94	**9.80**	10.6	11.3	12.0	7.83	8.57	9.26	9.90	10.5	11.1	5.59	6.12	**6.61**	7.07	7.50	7.90

Sporlan Valve Co.

Table 6-1e. Correction factors for liquid refrigerant temperature.

Refrigerant Liquid Temperature °F.	40°	50°	60°	70°	80°	90°	100°	110°	120°	130°	140°
Correction Factor	1.52	1.44	1.35	1.26	1.18	1.09	1.00	0.91	0.82	0.73	0.64

Sporlan Valve Co.

b. Determine the low side pressure drop. At −20 °F, the saturated pressure of R-502 is 15 psig, Table 5-1. At 100% loading, a distributor has a 35 psi pressure drop, Table 6-1c. Assume that the suction line losses are negligible. Therefore, the low side pressure drop is:

evaporator pressure	15 psi
distributor pressure drop	+ 35
	50 psi

c. The pressure differential across the valve is:

high side pressure (a)	213 psi
low side outlet pressure (b)	− 50
pressure drop across valve	163 psi

3. As a trial, select a nominal 3-ton valve from Table 6-1d. With a 175 psi differential and a −20° evaporator temperature, this valve will have a net capacity of 2.25 tons.

Since the condenser exit temperature is 100°, and 15° subcooling is expected from the suction-liquid heat exchanger, the temperature of the liquid at the expansion valve will be 100 − 15 = 85 °F. By interpolation between 80° and 90°, from Table 6-1e, the correction factor for 85° is 1.14. The net capacity of the valve is then 2.25 × 1.14 = 2.56 tons. The valve is slightly undersize.

The next size up, a nominal 4-ton valve, has a capacity of 3.44 tons and is adequate.

If the subcooling were 40°, the liquid temperature at the valve would be 100 − 40 = 60 °F and the correction factor would be 1.35. The capacity of the 3-ton valve would then be 2.25 × 1.35 = 3.03 tons

Table 6-1f. Subcooling needed to compensate for high side pressure drop.

REFRIGERANT	100° F. Condensing PRESSURE LOSS — psi					
	5	10	20	30	40	50
	REQUIRED SUBCOOLING — °F.					
12	3	6	12	18	25	33
22	2	4	8	11	15	19
500	3	5	10	15	21	27
502	2	3	7	10	14	18
717 (Ammonia)	2	4	7	10	14	17

Sporlan Valve Co.

Table 6-1g. Specifications, *C series* thermostatic expansion valves.

REFRIGERANT	TYPE ⑤ Internal Equalizer	TYPE External Equalizer ¼″ SAE Flare ONLY	③ NOMINAL CAPACITY Tons of Refrigeration	Thermostatic Charges Available	Standard Tubing Length - Feet	CONNECTIONS — Inches SAE Flare (Red Figures are standard and will be furnished unless otherwise specified.) INLET	OUTLET
12	CF-2	CFE-2	2	Refer to Recommended Thermostatic Charges Page 22	5	① 3/8 or 1/2	1/2 or 5/8
	CF-2½	CFE-2½	2½				
	CF-3	CFE-3	3				
	—	CFE-5	5			1/2 Only	5/8 Only
22	CV-3	CVE-3	3			① 3/8 or 1/2	1/2 or 5/8
	CV-4	CVE-4	4				
	CV-5	CVE-5	5				
	—	CVE-8	8			1/2 Only	5/8 Only
502	CR-2	CRE-2	2			① 3/8 or 1/2	1/2 or 5/8
	CR-3	CRE-3	3				
	CR-4	CRE-4	4				
	—	CRE-6	6			1/2 Only	5/8 Only

Sporlan Valve Co.

Table 6-1h. Remote bulb charges for TXV's.

REFRIGERANT	AIR CONDITIONING OR HEAT PUMP	COMMERCIAL REFRIGERATION +50°F. to −10°F.	LOW TEMPERATURE REFRIGERATION 0°F. to −40°F.
12	FCP60	FC	FZ FZP
22	VCP100 and VGA	VC	VZ VZP
500	DCP70	DC	–
502	RCP115	RC	RZ RZP

Sporlan Valve Co.

and adequate for the job. However, this choice is not made lightly, since the suction-liquid heat exchanger could overheat the suction gas. Too much superheat can cause compressor cooling problems and high discharge temperatures as shown in Figure 5-5. Stay with the nominal 4-ton valve, Type C, S, or H.

Check for flashing in the liquid line by using the 35 psi high side pressure drop and Table 6-1f. By interpolation, R-502, at 100° condensing, requires 12° subcooling. Since there is 15° of subcooling by the liquid-suction heater exchanger, and the liquid line is insulated to protect it from ambient temperatures greater than 80°, the liquid will not flash at the expansion valve inlet.

4. The expansion valve selected from Table 6-1g is the CRE-4, 3/8 × 1/2. Models S or H could also have been selected.

Per Table 6-1h, the charge in the thermostat bulb is designated RZ.

When ordering, per Table 6-1i, specify a CRE-4-RZ with 3/8 × 1/2 flare ends and a 1/4″ external equalizer.

Table 6-1i. Key to thermostatic expansion valve designations.

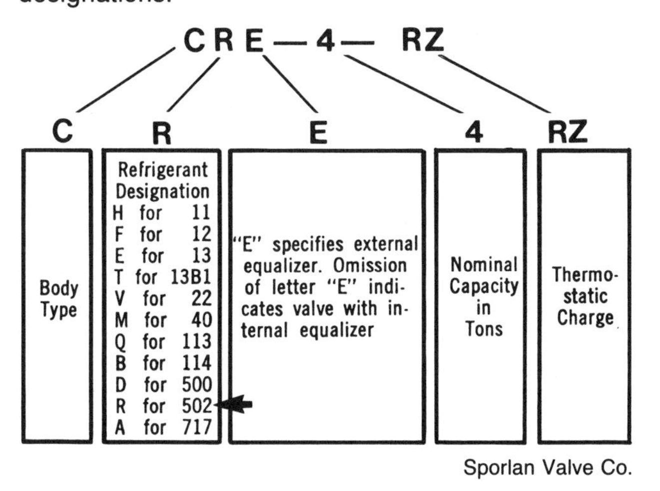

Sporlan Valve Co.

SELECT

A liquid line solenoid valve, normally closed, with manual lift stem.

Operation

The liquid line solenoid, Figure 6-2, is used for pumpdown control.

On a call for cooling, a thermostat opens the solenoid valve and allows liquid refrigerant to flow to the expansion valve downstream of the solenoid and cooling commences. When the thermostat opens, power to the solenoid is interrupted and it closes, stopping liquid flow to the low side. The refrigerant remaining in the evaporator is pumped into the condenser and the compressor shuts down when the low suction pressure switch opens.

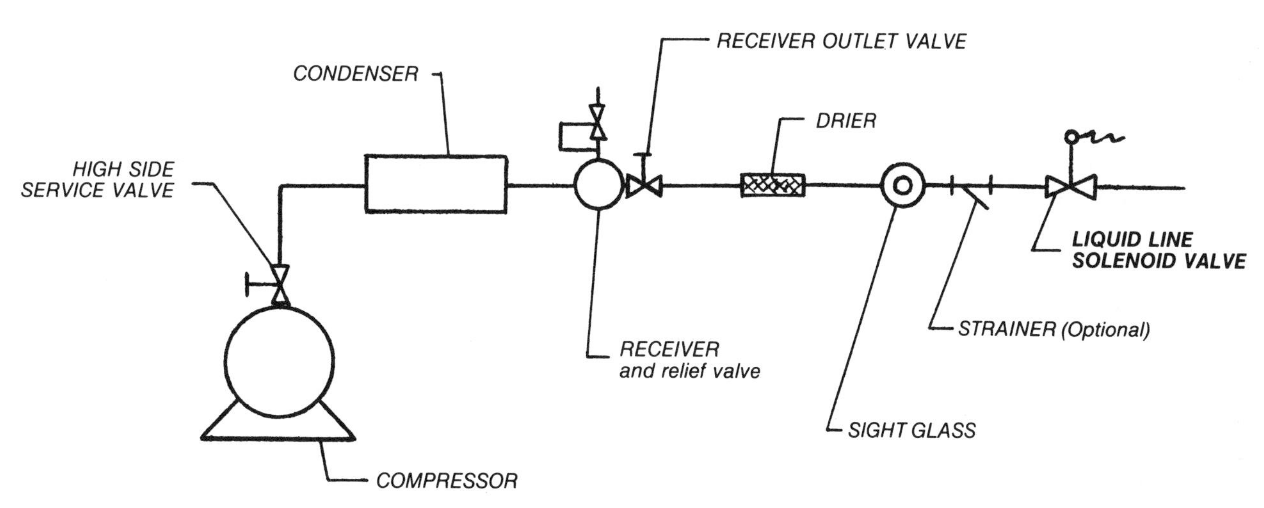

FIG. 6-2. Liquid line solenoid valve installation.

Table 6-2a. Liquid line solenoid capacities.

TYPE NUMBER			CONNECTIONS Inches	PORT SIZE Inches	TONS OF REFRIGERATION																			
WITH Manual Lift Stem	WITHOUT Manual Lift Stem	WITHOUT Manual Lift Stem			PRESSURE DROP – psi																			
NORMALLY CLOSED		NORMALLY OPEN			1				2				3				4				5			
					12	22	500	502	12	22	500	502	12	22	500	502	12	22	500	502	12	22	500	502
— — —	A3P1 A3F1 A3S1	— — —	$\frac{3}{8}$ NPTF $\frac{1}{4}$ SAE Flare $\frac{1}{4}$ or $\frac{3}{8}$ ODF Solder	.101	.7	.9	.8	.6	**1.0**	1.3	**1.2**	.8	1.2	**1.5**	1.4	**1.0**	1.4	1.9	1.6	1.2	1.6	2.1	1.8	1.3
MB6P1 MB6F1 MB6S1	B6P1 B6F1 B6S1	— — —	$\frac{3}{8}$ NPTF $\frac{3}{8}$ SAE Flare $\frac{3}{8}$ or $\frac{1}{2}$ ODF Solder	$\frac{3}{16}$	2.1	2.7	2.4	1.7	**3.0**	3.8	**3.4**	2.5	3.8	**5.0**	4.3	**3.1**	4.1	5.3	4.7	3.4	4.8	6.1	5.4	4.0
MB9P2 MB9F2 MB9S2	B9P2 B9F2 B9S2	OB9P2 OB9F2 OB9S2	$\frac{3}{8}$ NPTF $\frac{3}{8}$ SAE Flare $\frac{1}{2}$ ODF Solder	$\frac{9}{32}$	3.8	4.9	4.3	3.1	**5.0**	6.5	**5.7**	4.2	6.0	**7.5**	6.8	**5.0**	6.9	8.8	7.8	5.7	7.6	9.8	8.6	6.3
MB10F2 MB10S2	B10F2 B10S2	OB10F2 OB10S2	$\frac{1}{2}$ SAE Flare $\frac{5}{8}$ ODF Solder	$\frac{5}{16}$	5.3	6.8	6.0	4.4	**7.5**	9.0	**8.0**	6.0	8.4	**11.0**	9.5	**7.5**	9.9	12.7	11.0	8.3	11.0	14.3	12.7	9.3
MB14P2 MB14S2	B14P2 B14S2	OB14P2 OB14S2	$\frac{1}{2}$ NPTF $\frac{5}{8}$ ODF Solder	$\frac{7}{16}$	7.1	9.2	8.2	6.0	**10.0**	12.7	**11.2**	8.2	12.0	**15.0**	13.6	**10.0**	14.0	17.8	15.8	11.6	15.6	20.0	17.8	13.0

Sporlan Valve Co.

Table 6-2b. Liquid line solenoid valve dimensions and MOPD.

TYPE	STANDARD CONNECTIONS Inches	PORT SIZE Inches	MOPD psi		NOMINAL LIQUID CAPACITIES Tons of Refrigeration						STANDARD COIL RATINGS	
					REFRIGERANTS							
					12		502		22			
					Pressure Drop psi							
			AC	DC	2	3	3	4	3	4	VOLTS/CYCLES	WATTS
B10S2	5/8 ODF · 3/4 ODM	5/16	300	250	7.5	8.4	7.5	8.3	11	12.7	24/50-60 120/50-60 208-240/50-60 120-208-240/50-60	15
MB10S2												
OB10S2			275	275								
B10F2	1/2 SAE Flare		300	250								
MB10F2												
OB10F2			275	275								

Sporlan Valve Co.

Given

A 10-ton air conditioning system requires a liquid line solenoid for pumpdown control. Solenoid is to operate with a 120V AC, 60 cycle coil, R-22 refrigerant, 300 psi MOPD, 40 °F evaporator temperature and 100 °F liquid temperature.

Solution

Assume that a 3 psi pressure drop is acceptable with the valve at full load. Table 6-2a shows that a MB10S2 valve has a capacity of 11.0 tons with R-22 and a 3 psi differential. The charts are based on a 40 °F evaporator temperature and 100 °F liquid in

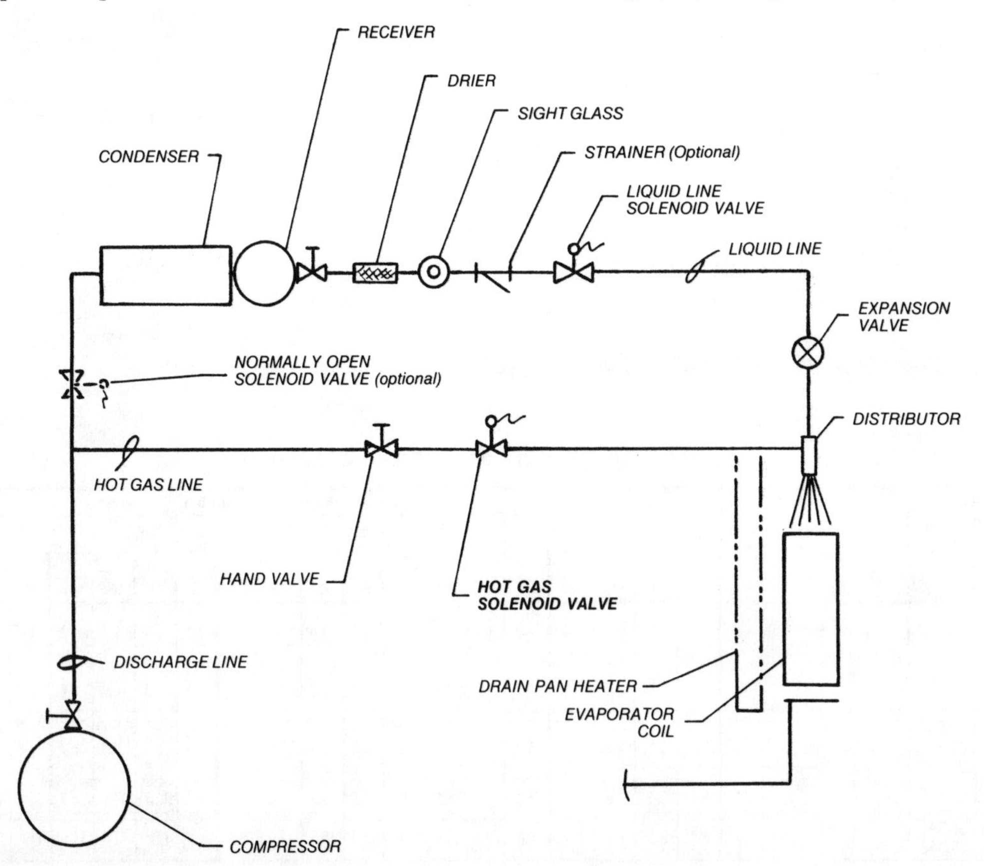

FIG. 6-3. Hot gas solenoid valve installation.

the high side. Since these are the design conditions, no further corrections are needed. Correction factors, for conditions other than those given, are a 1½% capacity reduction for every 10° below 40°F evaporator and a 5% increase for every 10°F reduction below 100°F liquid temperature.

Table 6-2b shows a MOPD of 300 for the MB10S2 valve with AC power. The order would be as follows: MB10S2 with 120V AC, 60 cycle coil, 5/8 ODF connections.

SELECT

A hot gas solenoid valve for defrost control on a refrigeration system.

Operation

The hot gas solenoid, Figure 6-3, is closed during normal refrigeration. When defrost is initiated, the liquid line solenoid closes while the hot gas solenoid opens and bypasses hot gas from the compressor and diverts it into the evaporator inlet. The evaporator becomes the condenser and the sensible and latent heat of the condensing refrigerant melts the frost on the evaporator coil. In some cases, the hot gas first passes through a pan heater coil before entering the evaporator.

Upon defrost termination, the hot gas solenoid closes, the liquid solenoid opens and refrigeration recommences. In some applications, when head pressure control valves are not used, a normally open (N.O.) solenoid valve is installed just prior to the condenser to block refrigerant flow to the condenser during defrost. At low ambients, condenser pressure is abnormally low and, during defrost, refrigerant will migrate into the condenser rather than into the evaporator, leaving insufficient hot gas to completely defrost the evaporator.

Table 6-3a. Capacities of hot gas solenoid valves.

VALVE SERIES	Discharge Gas Capacities - Tons, Pressure Drop Across Valve psi					
	2	5	10	25	50	100
Refrigerant 502						
B6	.56	.88	1.2	1.9	2.5	3.0
B9 – OB9	1.0	1.5	2.2	3.3	4.4	5.4
B10 – OB10	1.6	2.4	3.5	5.2	6.9	8.5
B14 – OB14	2.3	3.6	5.1	7.7	10.0	11.9
B19 – OB19	3.6	5.7	8.0	12.2	16.3	20.2
B25 – OB25	5.7	8.9	12.4	18.8	24.7	30.4
B33 – EB33 OB33 – MA32	9.4	14.6	20.4	31.0	40.7	50.1
MA42 – EMB42	16.8	26.3	36.6	55.7	73.1	90.0
MA50	29.6	46.3	64.5	98.2	129.0	158.8

EVAPORATOR TEMPERATURE CORRECTION FACTORS

Evap. Temp. °F.	40°	30°	20°	10°	0°	–10°	–20°	–30°	–40
Multiplier	1.00	.96	.93	.90	.87	.84	.81	.78	.75

Sporlan Valve Co.

Given

A system with a −30°F evaporator (refrigerant temperature), 100°F condensing liquid temperature, subcooled to 80°F, in a 8-ton R-502 system.

Table 6-3b. Hot gas solenoid valve dimensions and MOPD.

TYPE	CONNECTIONS Inches Sizes shown in RED will be furnished unless otherwise specified.	PORT SIZE Inches	MOPD psi AC	MOPD psi DC	STANDARD COIL RATINGS VOLTS/CYCLES	WATTS
B19P2	¾ NPT	$\frac{19}{32}$	300	250	24/50-60 120/50-60 208-240/50-60 120-208-240/50-60	15
MB19P2						
OB19P2			275	275		
B19S2	⅝ ODF - ⅞ ODM & ⅞ ODF - 1⅛ ODM		300	250		
→ MB19S2						
OB19S2			275	275		

Sporlan Valve Co.

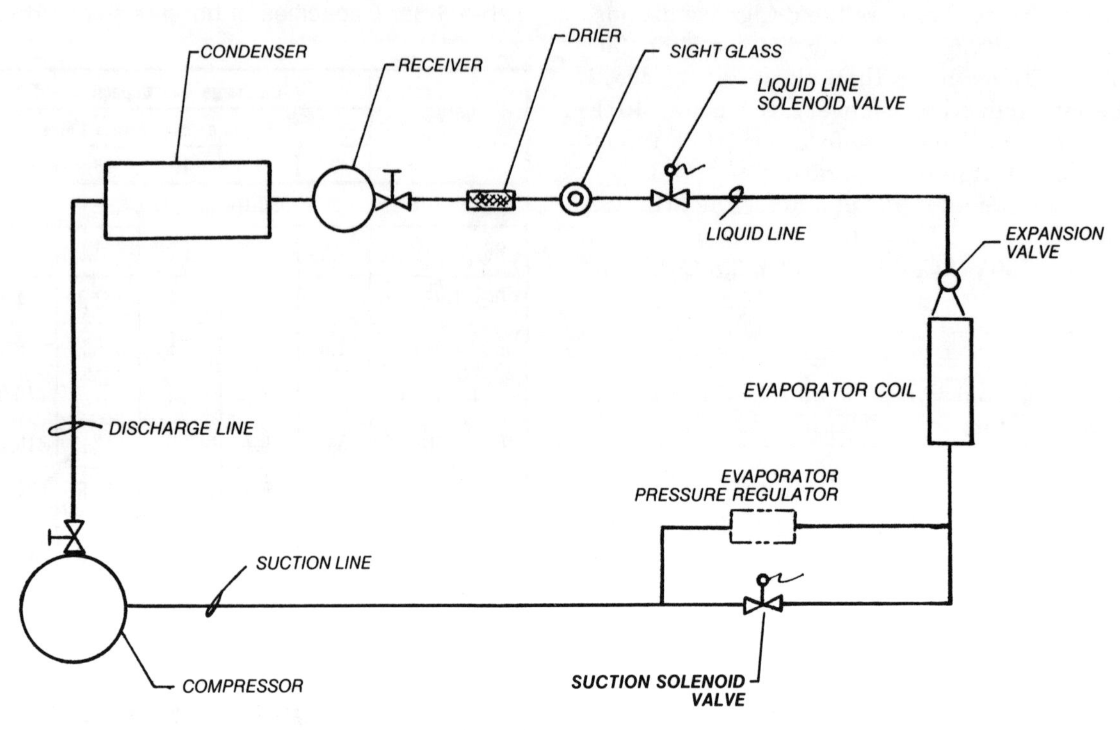

FIG. 6-4. Solenoid valve used for suction gas diversion.

Valve to be N.C. (normally closed), have a manual lift stem, and operate against a 300 psi MOPD. Valve to have a maximum of 25 psi pressure drop at full load. Coil voltage to be 24 volts AC.

Solution

With a capacity of 8 tons, Table 6-3a is used. At a 25 psi pressure drop, for hot gas application, a B19 series valve has a capacity of 12.2 tons. Table 6-3a, however, is based on a 40 °F evaporator, not a −30 °F evaporator. A correction factor must be applied. Table 6-3a shows that a −30 °F evaporator requires a .78 correction. Therefore, the net capacity is 12.2 tons × .78 = 9.51 tons.

This valve is adequate for the job and Table 6-3b shows the MOPD to be 300 with AC power.

SELECT

A solenoid valve for suction gas diversion.

Operation

This solenoid valve is used for a special dual temperture application in which the evaporator is alternately used for high humidity applications and for refrigeration, Figure 6-4. The evaporator is at 50 °F with the valve closed and at 20 °F when the N.O. valve is open to the compressor. With the valve closed, suction gas is diverted through an evaporator pressure regulator, parallel to the suction solenoid valve, and maintains the evaporator at 50 °F for humidity control.

Table 6-4a. Suction capacities of solenoid valves.

VALVE SERIES	Suction Capacity - Tons At 1 PSI Pressure Drop and Evaporating Temperatures Of:		
	40°F.	20°F.	0°F.
Refrigerant 12			
B6	.21	.17	.13
B9 – OB9	.37	.30	.24
B10 – OB10	.59	.47	.37
B14 – OB14	.88	.70	.56
B19 – OB19	1.39	1.11	.88
B25 – OB25	2.17	1.73	1.38
B33 – EB33 OB33 – DMA32	3.6	2.9	2.3
MA42 – EMB42	6.4	5.1	4.1
SMA50	11.3	9.1	7.2

Sporlan Valve Co.

TYPE	CONNECTIONS Inches Sizes shown in RED will be furnished unless otherwise specified.	PORT SIZE Inches	MOPD psi		STANDARD COIL RATINGS	
			AC	DC	VOLTS/CYCLES	WATTS
B19P2	¾ NPT	$\frac{19}{32}$	300	250	24/50-60 120/50-60 208-240/50-60 120-208-240/50-60	15
MB19P2						
OB19P2			275	275		
B19S2	⅝ ODF · ⅞ ODM & ⅞ ODF · 1⅛ ODM		300	250		
MB19S2						
OB19S2			275	275		

FIG. 6-4b. Specifications of solenoid valves used in suction line service. *Courtesy Sporlan Valve Co.*

Given

System capacity is ¾ ton at 20 °F evaporator and 100 °F liquid, with R-12 refrigerant. One psi maximum pressure drop is allowed. A 230V AC, 60 cycle coil is required.

Solution

From Table 6-4a, select a normally open OB19 solenoid valve, for R-12 at a 20 °F evaporator and a one psi pressure drop, with a capacity of 1.11 tons.

The physical dimensions are shown in Figure 6-4b.

SELECT

A N.O. and a N.C. solenoid valve for a heat reclaim system.

Operation

During normal operation, Figure 6-5, compressor discharge gas flows to the condenser. When heat reclaim is initiated by a thermostat, the normally open solenoid (N.O.) valve to the condenser closes, the normally closed solenoid valve (N.C.) is opened and hot gas flows to the heat reclaim heat exchanger. A check valve is placed in the outlet of the heat reclaim coil so that high pressure hot gas from the compressor does not backflow into the heat reclaim coil and block the flow of condensate.

Given

A 15-ton, R-22 air conditioning system operates with a 120 °F liquid temperature and a 40 °F evaporator.

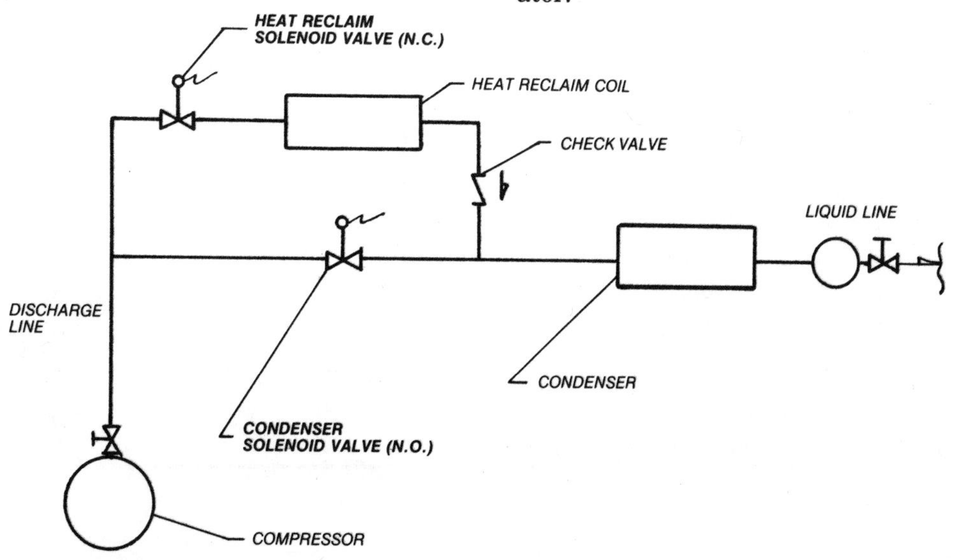

FIG. 6-5. Hot gas reclaim solenoid valve installation.

R22

Table 6-5a

Valve Type	Liquid Line Capacities & Pressure Differential, psi				Suction Line Capacities Fahrenheit degrees				Discharge Line Capacities (Tons) Pressure Differential, psi					
	1 psi	2 psi	3 psi	5 psi	40°	20°	0°	-20°	2 psi	5 psi	10 psi	25 psi	50 psi	100 psi
RB1	.69	.98	1.2	1.6	.12	.10	.08	.06	.20	.32	.45	.70	.96	1.3
RB3	3.0	4.5	5.7	7.5	.48	.39	.31	.25	.92	1.4	2.0	3.1	4.3	5.7
RB5	4.1	6.0	7.4	9.8	.60	.49	.39	.31	1.2	1.9	2.7	4.2	5.7	7.5
RB7	6.3	9.2	11.4	15.1	.91	.74	.60	.47	1.8	2.8	4.0	6.2	8.2	9.1
RB9	8.6	12	15	20	1.4	1.1	.90	.71	2.6	4.0	5.7	8.8	12	16
RB15	13	19	24	32	2.1	1.7	1.4	1.1	3.9	6.1	8.6	13	18	24
RB21	20	30	36	48	3.3	2.7	2.2	1.7	6.0	9.5	13	21	28	37
RB41	33	48	60	79	4.7	3.8	3.1	2.4	9.9	16	22	34	47	61
RA51	-	64	79	101	8.8	7.1	5.7	4.5	13	21	30	46	63	83
RA71	-	94	115	153	14	11	9.1	7.1	19	30	42	66	90	118
RA101	-	140	172	223	20	17	13	10	29	46	65	101	138	181
RA201	-	242	300	388	34	28	22	18	50	80	112	174	238	313
RA301	-	382	470	624	54	44	35	28	81	127	179	277	379	499
RA401	-	547	675	869	78	64	51	40	115	181	254	395	541	712

Parker Hannifin Corp. Jackes-Evans Controls Div.

Table 6-5b. Correction factors

Evaporator Temperature °F	Correction factor
+40	1.00
+30	.96
+20	.92
+10	.88
0	.84
−10	.80
−20	.77
−30	.74
−40	.71
Liquid Temperature °F	**Correction Factor**
+ 90	1.05
+100	1.00
+110	.95
→ +120	.90
+130	.86

Parker Hannifin Corp. Jackes-Evans Controls Div.

Solution

Table 6-5a shows that a RA-71 solenoid valve has a discharge line capacity of 19 tons with a 2 psi differential. The N.O. solenoid is placed in the condenser leg and the N.C. solenoid is placed in the heat reclaim leg. The *normal* condenser becomes oversized in the heat reclaim mode and substantial subcooling will take place.

Table 6-5b shows that the correction factor for 120 °F condensing is .90. Capacity is 19 × .90 = 17.1 tons. No correction for a 40 °F evaporator is needed.

SELECT

A 3-way hot gas defrost valve to defrost evaporators in a multiplex coil system.

Operation

In the normal refrigeration mode, Figure 6-6a, the 3-way valve passes low pressure refrigerant from the evaporator to the suction line. Upon defrost initiation, the pilot-operated, 3-way solenoid valve opens and the valve discs shift. The suction line from the 3-way valve is now blocked. Hot gas, which was previously blocked, back flows from the discharge

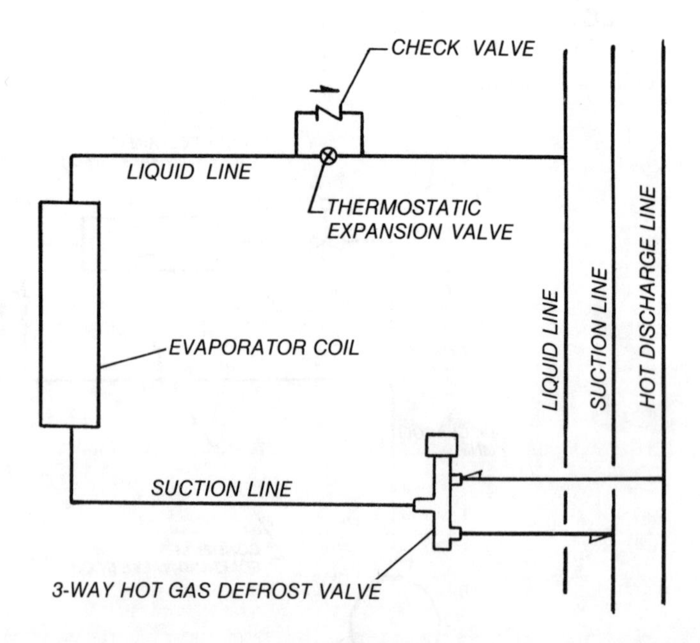

FIG. 6-6a. 3-way valve installation.

Table 6-6. Suction gas capacities of a 3-way diverting valve.

Capacities Tons at 1 psi pressure drop	Evaporator Temperatures in Fahrenheit Degrees					
	20°	10°	0°	-10°	-20°	-40°
Refrigerant 12	6.96	6.36	5.53	5.31	4.42	3.45
Refrigerant 22	10.40	9.50	8.25	7.92	6.60	5.15
Refrigerant 502	8.74	7.98	6.93	6.65	5.54	4.33

Parker Hannifin Corp. Jackes-Evans Controls Div.

line into the evaporator where it condenses while defrosting the coil. The condensed liquid leaves the evaporator through a check valve that bypasses the expansion valve and flows into the liquid line feeding the other multiplexed evaporators.

Given

Ice cream is stored in a 4-ton refrigerated case at −20 °F (−30 °F refrigerant temperature) using R-502. A 120V AC 60 cycle coil is needed.

Solution

Interpolating from Table 6-6, the capacity of the 3-way diverting valve, with R-502 refrigerant, is 4.93 tons at −30 °F. This capacity is adequate and is based on a 1 psi pressure drop in the normal refrigeration mode. Order an RS20 3-way valve with a 120V AC, 60 cycle coil. Figure 6-6b shows details of the valve.

SELECT

A crankcase pressure regulator (CPR) to protect a compressor from overloading due to abnormally high suction pressures.

Operation

The crankcase pressure regulator, located just before the compressor, Figure 6-7, will sense the suction pressure in the compressor and prevent it from exceeding a preset level. Any upstream pressures above that value, caused by high load conditions or defrosting, will be throttled. A strainer is placed at the inlet of the regulator to assure clean gas to the valve.

Given

A R-502 refrigerant system of 3 tons capacity and −30 °F temperature in the evaporator. Suction pressure is restricted to 15 psig maximum to the compressor with a maximum pressure drop of 0.5 psi through the valve. Liquid refrigerant to the expan-

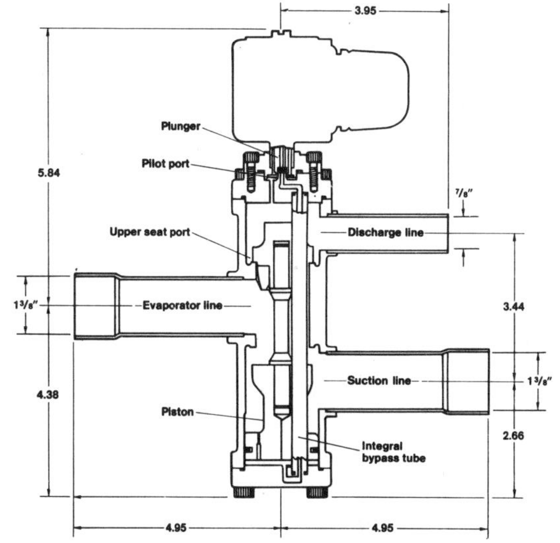

FIG. 6-6b. Interior of a 3-way diverting valve. *Parker Hannifin Corp. Jackes-Evans Controls Div.*

sion valve is subcooled from 100 ° to 90 °F.

Solution

Table 6-7 shows that a 1-5/8″ A4A crankcase pressure regulator valve has a capacity of 3.5 tons, at a 0.5 psi pressure drop, when used with R-502 at −30 °F. Subcooling from 100 ° to 90 °F adds another 7% to the capacity with R-502: 3.5 tons × 1.07 = 3.74 tons. Since only a 0.5 psi pressure drop is required, a special Low Pressure Drop (LPD) valve must be specified since standard CPR valves have a pressure drop of 2 or 10 psi.

Order CPR valve A4A, 1-5/8″, Range A, with the LPD option.

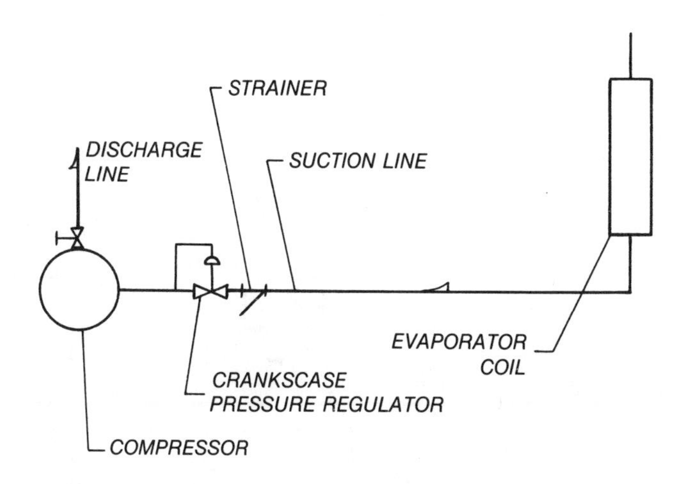

FIG. 6-7. Crankcase pressure regulator installation.

Table 6-7. Capacities of crankcase pressure regulators with R-502.

Evaporator Temp. °F & Press. psig	Regulator Pressure Drop psi	¾″ (20 mm) A4A †	1″ (25 mm) A4A	1¼″ (32 mm) A4A	1⅝″ (40 mm) A4A	2″ (50 mm) A4A	2½″ (65mm) A4A	3″ (75 mm) A4A	4″ (100 mm) A4C	5″ (125 mm) A4C	6″ (150 mm) A4C	8″ (200 mm) A4C
50°F	2	4.1	5.7	9.9	19	28	40	57	76	120	180	350
96.9	10	9.0	12	22	42	62	87	120	170	250	390	780
40°F	2	3.7	5.2	9.1	17	26	36	52	70	110	160	320
80.2	10	8.2	11	20	38	56	79	110	150	230	360	710
30°F	2	3.3	4.6	8.1	15	23	32	46	62	95	150	290
65.4	10	7.2	10	18	34	50	70	100	140	210	320	630
20°F	2	3.0	4.2	7.3	14	21	29	42	56	85	130	260
52.4	10	6.5	9.0	16	30	45	63	90	120	180	280	560
10°F	2	2.6	3.7	6.4	12	18	26	37	49	75	120	230
41.1	10	5.7	7.9	14	26	39	55	79	110	160	250	490
0°F	2	2.3	3.2	5.6	11	16	22	32	43	66	100	200
31.2	5	3.6	5.0	8.7	17	25	35	50	67	100	160	310
−10°F	2	2.0	2.8	5.0	9.5	14	20	28	38	58	89	180
22.8	5	3.2	4.4	7.7	15	22	31	44	59	90	140	270
−20°F	2	1.7	2.4	4.2	8.1	12	17	24	33	50	76	150
15.5	3	2.1	2.9	5.2	9.8	15	21	29	40	60	93	180
−30°F	0.5*	0.76	1.1	1.9	3.5	5.3	7.4	11	14	22	33	66
9.4	2	1.5	2.1	3.7	7.0	10	15	21	28	43	66	130
−40°F	0.5*	0.66	0.91	1.6	3.0	4.5	6.4	9.1	12	19	29	57
4.3	2	1.3	1.8	3.1	6.0	8.8	12	18	24	37	56	110

Parker Hannifin Corp. Jackes-Evans Controls Div.

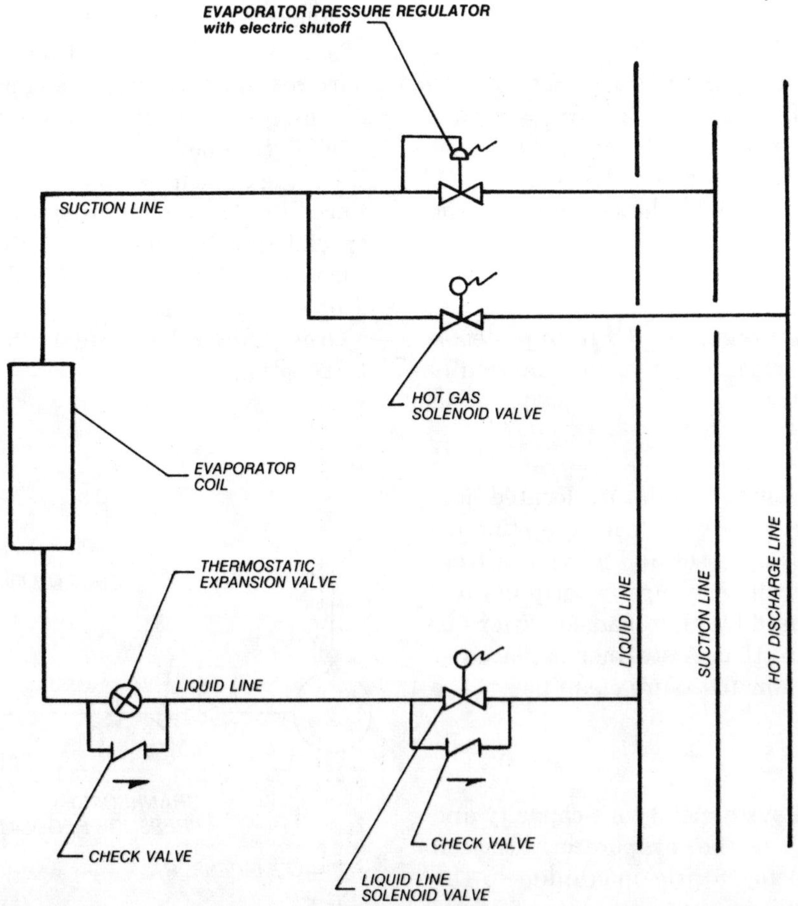

FIG. 6-8. Evaporator pressure regulator combined with gas defrost.

Table 6-8. Capacities of evaporator pressure regulators with R-502.

Valve Type	A7A									A7A1									A72					
ODS Connections Sizes in inches	5/8 7/8 1-1/8			5/8 7/8 1-1/8			7/8 1-1/8			1-1/8 1-3/8 1-5/8			1-1/8 1-3/8 1-5/8			1-3/8 1-5/8			1-5/8 2-1/8 2-5/8			2-1/8 2-5/8		
Effective Port Size Sizes in Inches	Reduced 5/8			Full 5/8			7/8			7/8			1-1/8			1-3/8			1-5/8			2-1/8		
Pressure Drop (psi)	2	5	10	2	5	10	2	5	10	2	5	10	2	5	10	2	5	10	2	5	10	2	5	10
Evaporator Temp. °F/Press. psig: 50°F 84.0	1.4	2.2	3.1	2.9	4.6	6.5	4.3	6.6	9.3	6.0	9.3	13.0	8.5	13.3	18.7	9.4	14.6	20.5	15.3	23.9	33.6	23.0	35.9	50.5
40°F 68.5	1.3	2.0	2.9	2.7	4.2	5.9	3.9	6.0	8.5	5.4	8.5	11.9	7.7	12.1	17.0	8.5	13.3	18.7	13.9	21.8	30.6	20.9	32.6	45.9
30°F 54.9	1.2	1.8	2.6	2.4	3.8	5.3	3.5	5.5	7.6	4.9	7.6	10.6	7.0	10.9	15.2	7.7	12.0	16.7	12.5	19.6	27.3	18.8	29.4	41.0
20°F 43.0	1.0	1.6	2.3	2.2	3.4	4.7	3.1	4.8	6.7	4.3	6.8	9.4	6.2	9.8	13.5	6.8	10.7	14.9	11.2	17.6	24.4	16.8	26.4	36.6
10°F 32.8	.9	1.4	1.9	1.9	3.0	4.1	2.7	4.2	5.8	3.8	6.0	8.2	5.5	8.5	11.7	6.1	9.4	12.9	9.8	15.3	21.1	14.8	23.0	31.7
0°F 24.0	.8	1.3	1.8	1.7	2.7	3.6	2.5	3.8	5.2	3.5	5.4	7.3	5.0	7.7	10.5	5.5	8.5	11.5	8.9	13.9	18.9	13.4	20.9	28.3
−10°F 16.5	.7	1.1		1.5	2.3		2.1	3.3		3.0	4.6		4.3	6.6		4.7	7.3		7.7	11.9		11.6	17.8	
−20°F 10.1	.6			1.3			1.9			2.6			3.7			4.1			6.8			10.2		
−30°F 4.9	.5			1.1			1.6			2.2			3.2			3.6			5.8			8.7		
−40°F 0.5	.4			1.0			1.4			2.0			2.8			3.1			5.1			7.7		

Parker Hannifin Corp. Jackes-Evans Controls Div.

SELECT

An evaporator pressure regulator (EPR) with electric shutoff for a refrigerated display case.

Operation

The standard evaporator pressure regulator (EPR) is used to maintain a preset temperature in one of many paralleled refrigerated cases that employ a common suction line. In this example, the case with the EPR is held at −20° while another case, which is open to the suction line, may be operating at the suction line temperature or −40°. On a rise in evaporator pressure, the EPR will open until the temperature is once again at its preset level, then close and hold until the evaporator pressure again rises.

This creates a problem during hot gas defrost. Notice, in Figure 6-8, that the EPR will open with the introduction of the hot gas and short circuit the evaporator coil that requires defrosting. Positive closure is assured by the addition of an electric shutoff to the EPR.

During refrigeration, the liquid line solenoid valve and EPR valve with electric shutoff are energized and normal refrigeration takes place. The hot gas line is blocked. Liquid refrigerant, under high pressure, flows through the liquid solenoid valve and the expansion valve to the evaporator inlet. The pressure, and therefore the temperature, in the evaporator is controlled by the EPR valve.

When defrost is initiated, the hot gas solenoid valve opens and the liquid solenoid and EPR valve are de-energized and close. Hot gas backflows through the evaporator via the suction line. The condensed refrigerant flows around the expansion valve and the liquid line solenoid, via the check valves, to the liquid line where the liquid feeds other parallel evaporator units still in their refrigeration mode. A timer usually terminates the defrost mode.

Given

A 24,000 Btu/hr (2 ton) refrigerated case in a multiple case system is operated at −20°F evaporator temperature with refrigerant R-502. The evaporator pressure regulator is to have an electric shutoff. Maximum pressure drop across the regulator at full load is 2 psi. 100°F liquid with 20°F subcooling are other operating parameters.

Solution

With R-502 at −20°F evaporator temperature, Table 6-8 shows that an A7A regulator with a 7/8″ port and 2 psi pressure drop is rated at 1.9 tons.

With 20 °F subcooling, the capacity is increased by 4% for each 10 °F subcooling or 8% total. The 1.9 tons, multiplied by 1.08, becomes a net capacity of 2.05 tons. Since 2.05 tons × 12,000 Btu/hr/ton yields 24,600 Btu/hr, this meets the 24,000 Btu/hr requirement. The model selected, therefore, is an A7AS with a 120V AC electric shutoff solenoid and 7/8″ port and line sizes.

SELECT

Control valves for condenser head pressure control and to supply hot refrigerant gas to a heat reclaim coil.

Operation

Two things are happening at once in Figure 6-9:

- Compressor head pressure is being controlled during cold weather operation
- Heat reclaim is available on demand

In warm ambients the inlet regulator will be open and hot gas will pass through the condenser, become liquid and flow into the receiver and the liquid line.

During cold winter conditions, the discharge pressure will tend to drop due to the greater efficiency of the condenser under these conditions. Below 90 ° (185 psig) the inlet pressure regulator will close and remain closed until the compressor discharge pressure is equivalent to 120 ° (280 psig). Simultaneously, the outlet sensing regulator will open and pass hot gas to the receiver to build up pressure to assure good expansion valve operation.

The two check valves prevent backflow into the condenser and heat reclaim heat exchanger.

When heat reclaim is demanded, the solenoid valve is energized by a thermostat and hot gas flows into the heat exchanger. If, as a result, the compressor discharge pressure falls below the inlet pressure regulator setting, it will close and all the discharge gas will flow to the heat reclaim heat exchanger. If the discharge pressure climbs above 280 psig (120 °), the inlet pressure regulator will open and pass the excess hot gas to the condenser.

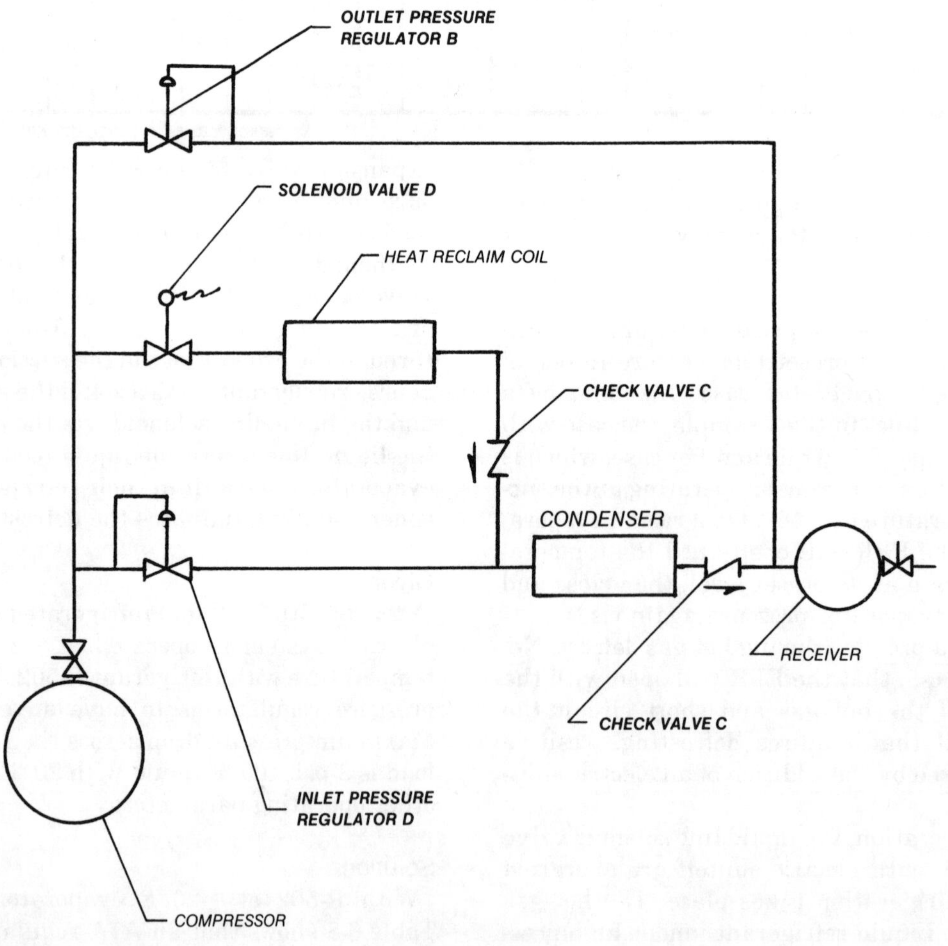

FIG. 6-9. Heat reclaim system.

Given

A 9.3 ton system, operating with R-502 and a −20 °F evaporator, is equipped with a 6.8 ton heat reclaim coil that operates between 90 ° and 120 °F. At 90°, the compressor discharge gas is bypassed from the condenser, through an outlet pressure regulator, to the receiver to maintain cold weather head pressure control.

Solution

The outlet regualtor will supply hot gas to the receiver, warming the liquid, and thereby raising the receiver pressure. Based on 9.3 tons at 90°F condensing, an A9-7/8" regulator, Table 6-9a, will handle the capacity, with spring range B, at a 10 psi pressure drop.

Table 6-9b shows a .94 correction factor for a −20 °F evaporator and R-502: 14.0 × .94 = 13.2 tons which is more than adequate. The next smaller size is only 7.4 tons and is too small.

The inlet or relief regulator that controls the flow of hot gas to the condenser coil is selected from Table 6-9c. At 90 °F condensing and 5 psi pressure drop, an A7A1-1-1/8" is rated at 10 tons with spring range D. With R-502 and a −20 °F evaporator, the correction factor is .94, Table 6-9b. Capacity is then 10.0 × .94 = 9.4 tons which is adequate for a 9.3 ton system.

Table 6-9a. Capacities of outlet pressure regulators with R-502.

Size	Type	90 °F		110 °F		130 °F	
		10 psi	20 psi	10 psi	20 psi	10 psi	20 psi
A9 OUTLET REGULATORS "B"							
5/8"†	A9	2.3	3.3	1.8	2.6	——	——
5/8"	A9	7.4	10	5.8	8.1	——	——
7/8"	A9	14	20	11	15	——	——
1 1/8"	A9	19	26	15	21	——	——

Parker Hannifin Corp. Jackes-Evans Control Div.

Table 6-9b. Pressure regulator correction factors.

Ref't	TABLE III Evaporator Temperature				
	−40 °F	−20 °F	0 °F	20 °F	40 °F
R-12	0.90	0.95	1.00	1.04	1.08
R-22	0.94	0.97	1.00	1.03	1.06
R-502	0.88	0.94	1.00	1.05	1.10

Parker Hannifin Corp. Jackes-Evans Control Div.

The check valve at the condenser outlet, selected from Table 6-9d, is a CK4A-3 size 3/4". The correction factor for this valve, with a −20 °F evaporator, is .94. Therefore, 15.0 × .94 = 14.1 tons which is more than adequate.

A solenoid valve is used to *dump* hot gas into the reclaim coil on demand. The S4A-1" valve, Table 6-9c, is rated at 10 tons with a 5 psi pressure drop and is adequate.

Table 6-9c. Capacities of inlet pressure regulators and solenoid valves with R-502.

Size	Type	90 ° to 130 °F	
		2 psi	5 psi
INLET REGULATORS "D"			
5/8"*	A7A	1.1	1.8
5/8"	A7A	2.3	3.6
7/8" & 1 1/8"	A7A	3.2	5.1
1 1/8"	A7A1	6.5	10
1 3/8" & 1 5/8"	A7A1	7.1	11
1 5/8"	A72	12	19
2 1/8" & 2 5/8"	A72	18	28
SOLENOID VALVES "D"			
3/16"	S6N	0.39	0.61
1/2"	S8F	1.8	2.8
3/4"	S4A	4.7	7.3
1"	S4A	6.5	10
1 1/4"	S4A	12	18
1 5/8"	S4A	22	34
2"	S4A	32	51
2 1/2"	S4A	45	71
3"	S4A	65	100

Parker Hannifin Corp. Jackes-Evans Control Div.

Table 6-9d. Capacities of check valves with R-502.

Size	Type	90 °F	110 °F	130 °F
CHECK VALVES "C"				
1/2"	CK4A-2	7.1	6.0	——
3/4"	CK4A-3	15	12	——
1"	CK4A-4	27	23	——
1 1/4"	CK4A-6	40	33	——
2"	CK4A-8	96	81	——
2 1/2"	CK4A-9	150	120	——
3"	CK4A-0	230	200	——
4"	CK4-16	440	370	——

Parker Hannifin Corp. Jackes-Evans Control Div.

The check valve at the heat reclaim discharge is selected on the same criteria as the solenoid valve and a ¾″ CK4A-3 at 12 tons at 110 °F is again more than adequate.

SELECT

A hot gas regulator to bypass hot gas to the evaporator inlet at the distributor.

Operation

Under light loads, evaporator temperatures fall far below their design levels. To prevent this, a hot gas regulator, Figure 6-10a, will bypass hot gas into the evaporator inlet to create a false load and maintain temperature down to 50% of capacity. At −25 °F, the saturated suction pressure of R-502 is 12.1 psig per Table 5-1. Since the regulator spring is designed to have a maximum pressure change, from fully closed to completely open, of 5 psi, the regulator should be set to begin to open at 17.1 psig (12.1 + 5 = 17.1). The regulator will be fully open and at maximum capacity at 12.1 psig or −25 °F as designed.

A hand shutoff valve is placed in front of the hot gas valve for manual pumpdown and servicing of the hot gas regulator. For automatic pumpdown, a solenoid valve is installed between the hand valve and hot gas regulator. Since the hot gas regulator operates on a reduction in suction pressure, pumpdown cannot be accomplished because a continuous bypassing would result without some way to block the hot gas line.

Given

A 50,000 Btu/hr system with an evaporator running at -25ºF an air-cooled condenser at 110ºF condensing. 50% of the capacity is to be bypassed at -25ºF.

Solution

The cycle is drawn on a P-H chart, Figure 6-10b. The compressor discharge temperature can be estimated at approximately 160 °F with a 110 °F condensing temperature. Table 6-10 shows that a reduced port A9 has a capacity of 2.18 tons × 12,000 or 26,160 Btu/hr and is adequate for this application. An external equalizer regulator will be used

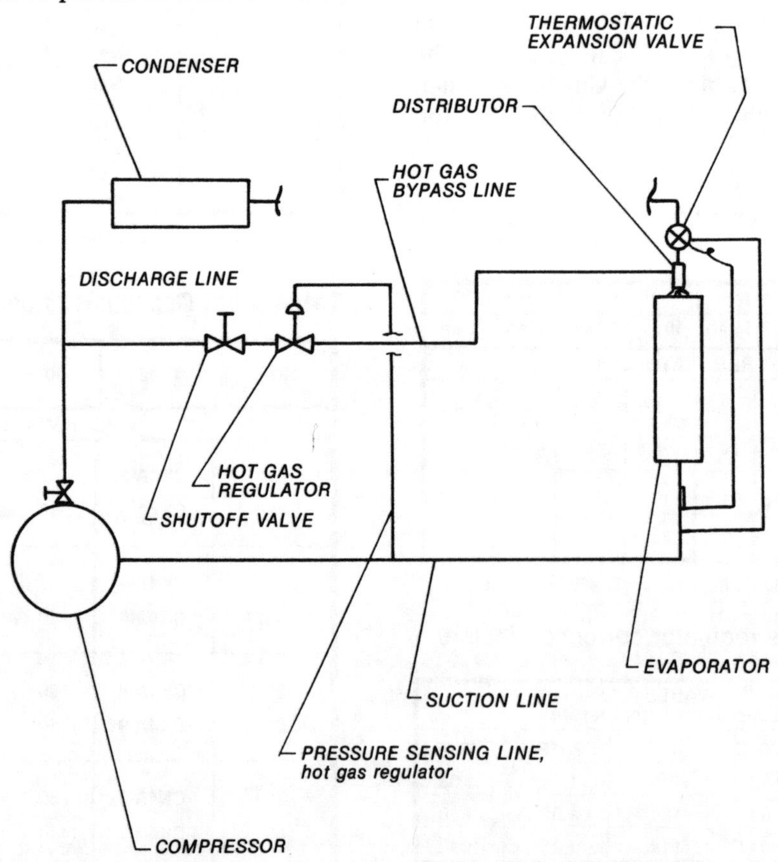

FIG. 6-10a. Hot gas bypass installation.

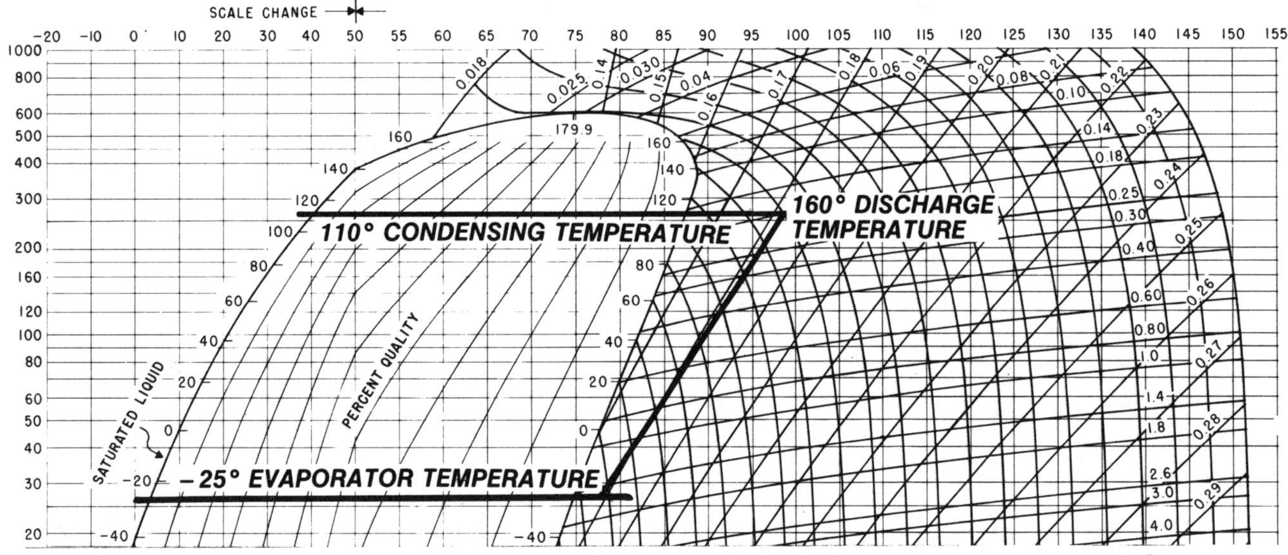

FIG. 6-10b. Estimating superheat with a R-502 P-H diagram. *Courtesy E.I. dupont de Nemours & Co.*

and designated as A9E-5/8 line size, reduced port 5/8″ size. Spring range *A*.

SELECT

A hot gas regulator and liquid injection valve to false-load a refrigeration system.

Operation

This capacity reduction system, Figure 6-11a, which may be required for humidity control, will keep the unit in operation even though there is little or no load. It is assumed that head pressure control is part of the system, because lower loads will drop condensing pressures to the point where they affect the operation of the thermostatic expansion valve.

The design keeps the controls close to the condensing unit, eliminating expensive and long hot gas lines to a distant evaporator.

Table 6-10. A9 hot gas bypass regulator capacities.

R-12

Cond. Temp. °F	*Comp. Disch. Temp. °F	Effective Port Sizes in inches: Red. 5/8	5/8	7/8	1-1/8
86°F	120	0.90	2.8	6.4	9.9
93.3	140	1.06	3.3	6.7	10.0
psig	160	1.12	3.5	7.1	11.0
	180	1.18	3.7	7.5	12.0
100°F	120	1.15	3.6	7.4	12.0
117.2	140	1.22	3.8	7.8	12.0
psig	160	1.28	4.0	8.2	13.0
	180	1.34	4.2	8.6	14.0
	200	1.40	4.4	9.0	14.0
110°F	140	1.34	4.2	8.6	13.0
136.4	160	1.44	4.5	9.1	14.0
psig	180	1.54	4.8	9.6	15.0
	200	1.60	5.0	10.0	16.0
	220	1.70	5.3	11.0	17.0
120°F	140	1.47	4.6	9.4	14.0
157.6	160	1.57	4.9	10.0	15.0
psig	180	1.66	5.2	11.0	16.0
	200	1.76	5.5	11.0	17.0
	220	1.86	5.8	12.0	18.0

R-22

Cond. Temp. °F	*Comp. Disch. Temp. °F	Effective Port Sizes in inches: Red. 5/8	5/8	7/8	1-1/8
86°F	140	1.89	5.9	12.0	18.0
158.2	160	1.95	6.1	12.0	19.0
psig	180	2.05	6.4	13.0	20.0
	200	2.14	6.7	14.0	21.0
100°F	140	2.18	6.8	14.0	22.0
195.9	160	2.27	7.1	14.0	23.0
psig	180	2.37	7.4	15.0	24.0
	200	2.46	7.7	16.0	25.0
	220	2.56	8.0	16.0	26.0
110°F	140	2.37	7.4	15.0	23.0
226.3	160	2.50	7.8	16.0	25.0
psig	180	2.62	8.2	17.0	26.0
	200	2.75	8.6	18.0	27.0
	220	2.88	9.0	19.0	29.0
120°F	140	2.60	8.1	16.0	25.0
259.9	160	2.75	8.6	17.0	27.0
psig	180	2.88	9.0	18.0	28.0
	200	3.01	9.4	19.0	29.0
	220	3.17	9.9	20.0	31.0

R-502

Cond. Temp. °F	*Comp. Disch. Temp. °F	Effective Port Sizes in inches: Red. 5/8	5/8	7/8	1-1/8
86°F	160	1.80	5.6	11.0	18.0
175.1	180	1.86	5.8	12.0	19.0
psig	200	1.98	6.2	12.0	19.0
	220	2.05	6.4	13.0	20.0
100°F	160	1.98	6.2	13.0	20.0
214.4	180	2.11	6.6	14.0	21.0
psig	200	2.24	7.0	14.0	23.0
	220	2.37	7.4	15.0	24.0
110°F	160	2.18	6.8	14.0	22.0
245.8	180	2.34	7.3	15.0	23.0
psig	200	2.46	7.7	15.0	25.0
	220	2.62	8.2	16.0	26.0
120°F	160	2.37	7.4	15.0	24.0
280.3	180	2.53	7.9	16.0	25.0
psig	200	2.66	8.3	16.0	17.0
	220	2.82	8.8	17.0	28.0

Parker Hannifin Corp. Jackes-Evans Control Div.

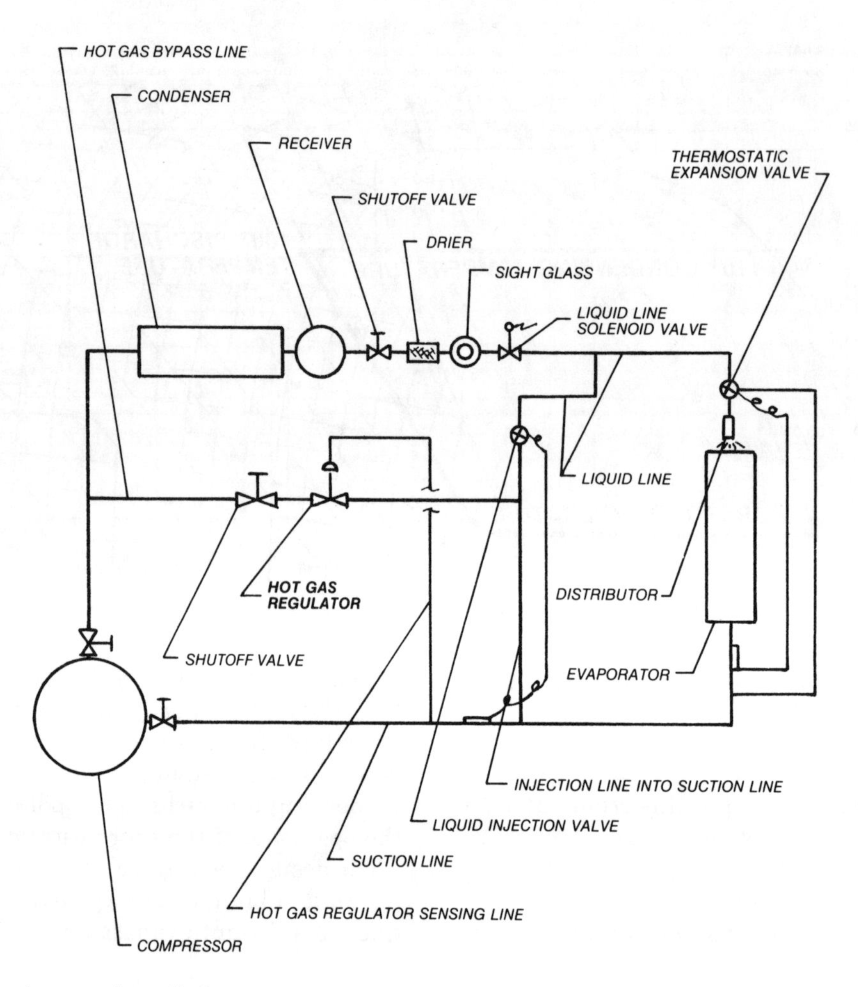

FIG. 6-11a. Hot gas bypass installation.

On a load reduction, the compressor will mechanically unload 1/3, then 2/3 of its capacity. If the load continues to drop, the accompanying drop in suction pressure will be transmitted to the regulator through the 1/4 in. sensing line. At 20° (21.0 psig with R-12, Table 5-1), the hot gas bypass regulator opens and delivers hot gas to the suction line.

The sensing bulb for the liquid injection valve is located downstream from the hot gas inlet and will measure the mixed suction gas/hot gas temperature. If this mixed temperature is above the design temperature, liquid refrigerant will be injected into the hot gas to cool it.

Most compressor manufacturers specify a maximum suction gas temperature of 65°. This is to protect the compressor from excessively high discharge temperatures and to assure adequate cooling of hermetic and semihermetic motor windings.

Location of the liquid and hot gas lines is critical since the line containing the mixed gases should be long enough to assure thorough mixing and evaporation of the liquid. Sensing bulbs, including the expansion valve remote bulb, should never be closer than 5 feet from the compressor.

As with the previous example, a hand valve is located ahead of the hot gas regulator. A solenoid valve may also be added for automatic pumpdown control. Since the liquid injection valve has it supply line downstream of the system liquid solenoid valve, an additional liquid solenoid valve is not needed.

Given

Hot gas to be injected into the suction line downstream of the evaporator just prior to the compressor suction service valve. This system is to be used on a 30-ton commercial R-12 system at a minimum evaporator temperature of 20°F and 100°F condensing temperature. The compressor can be

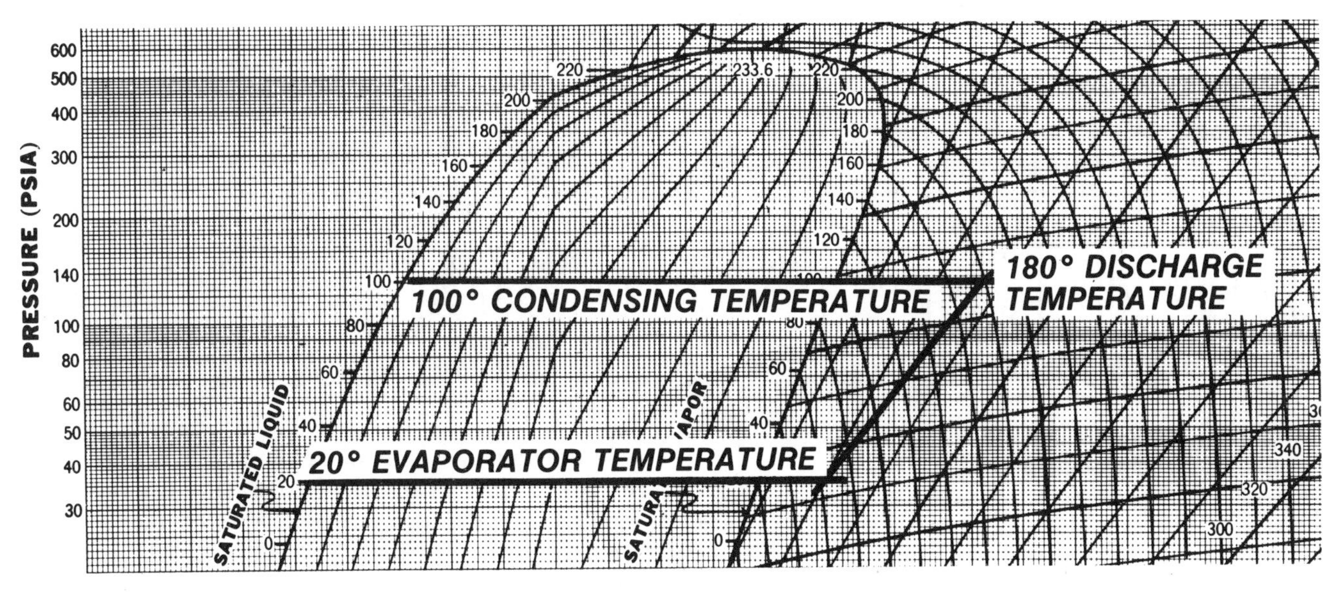

FIG. 6-11b. R-12 P-H diagram. *Courtesy E.I. dupont de Nemours & Co.*

mechanically unloaded in thirds; that is, to 20 and 10 tons respectively.

Solution

Figure 6-11b shows that with a 20 °F evaporator and 100 °F condensing temperature, a discharge temperature of 180 °F can be expected. Table 6-11a shows that, with R-12 at 100 °F condensing and

Table 6-11a. Hot gas regulator ratings with R-12.

Condensing Temperature °F.	*Compressor Discharge Temperature °F.	A9E or A9 Hot Gas Bypass Regulator Size		
		⅝	⅞	1⅛
86°F. 93.3 psig	120	2.8	6.4	9.9
	140	3.3	6.7	10.
	160	3.5	7.1	11.
	180	3.7	7.5	12.
100°F. 117.2 psig	120	3.6	7.4	12.
	140	3.8	7.8	12.
	160	4.0	8.2	13.
	180	4.2	8.6	14.
	200	4.4	9.0	14.
110°F. 136.4 psig	140	4.2	8.6	13.
	160	4.5	9.1	14.
	180	4.8	9.6	15.
	200	5.0	10.	16.
	220	5.3	11.	17.
120°F. 157.6 psig	140	4.6	9.4	14.
	160	4.9	10.	15.
	180	5.2	11.	16.
	200	5.5	11.	17.
	220	5.8	12.	18.

Parker Hannifin Corp. Jackes-Evans Controls Div.

Table 6-11b. Liquid injection valve capacities with R-12.

Condensing Temperature °F.	Compressor Discharge Temperature °F.	Suction Temperature (Saturation)			
		−40°F.	0°F.	+20°F.	+40°F.
86°F. Condensing	120	7.6	5.8	4.7	3.8
	140	8.9	7.1	6.0	5.2
	160	11.	8.8	7.7	6.8
	180	12.	10.	9.3	8.4
100°F. Condensing	120	8.7	6.5	5.1	4.0
	140	11.	8.4	7.1	6.0
	160	13.	10.	9.1	8.0
	180	16.	12.	11.	9.9
	200	17.	14.	13.	12.
110°F. Condensing	140	12.	9.4	7.8	6.5
	160	14.	12.	10.	8.7
	180	16.	14.	12.	11.
	200	19.	16.	15.	13.
	220	21.	18.	17.	16.
120°F. Condensing	140	13.	10.	8.2	6.8
	160	16.	13.	11.	9.3
	180	18.	15.	14.	12.
	200	21.	18.	16.	14.
	220	24.	20.	19.	17.

Table 6-11c. Liquid injection valve correction factors.

A9E or A9		
⅝″	⅞″	1⅛″
0.11	0.23	0.36

Parker Hannifin Corp. Jackes-Evans Controls Div.

180 °F discharge temperature, an 1-1/8″ A9E regulator has 14 tons capacity.

Table 6-11b shows that a liquid injection thermostatic expansion valve, at 100 °F condensing and 180 °F discharge, with a 20 ° evaporator, has a capacity of 11 tons. Table 6-11c shows a correction factor of .36 for a 1-1/8″ A9E. Therefore, (11 × .36) a 4-ton liquid injection valve should be used.

Table 6-11d shows the liquid injection valve capacities. For 20 °F evaporator and 100 ° condenser, the

Table 6-11d. Desuperheating thermostatic valve capacities with R-12.

VALVE TYPE NUMBERS (Including nominal capacity – tons)				STANDARD CONNECTIONS		MINIMUM ALLOWABLE EVAPORATING TEMPERATURE AT THE REDUCED LOAD CONDITION – °F.								
Internally Equalized		Externally Equalized												
SAE	ODF	SAE	ODF	G & C	S	40°			26°			20°		
REFRIGERANT 12						PRESSURE DROP ACROSS VALVE – psi								
						60	80	100	60	80	100	60	80	100
GF-¼	—	—	—	⅜" x ½" SAE	⅜" x ⅝" ODF	0.25	0.29	0.32	0.25	0.29	0.32	0.25	0.29	0.32
GF-½	SF-½	GFE-½	SFE-½			0.50	0.58	0.64	0.50	0.58	0.64	0.50	0.58	0.64
GF-1	SF-1	GFE-1	SFE-1			1.00	1.15	1.29	1.00	1.15	1.29	1.00	1.15	1.29
GF-1½	SF-1½	GFE-1½	SFE-1½			1.60	1.85	2.06	1.32	1.52	1.70	1.20	1.38	1.55
CF-2	SF-2	GFE-2	SFE-2			2.00	2.31	2.58	1.76	2.03	2.27	1.65	1.91	2.13
CF-2½	SF-2½	CFE-2½	SFE-2½		½" x ⅞" ODF	2.50	2.89	3.23	2.22	2.56	2.87	2.10	2.42	2.71
CF-3	SF-3	CFE-3	SFE-3			3.00	3.46	3.87	2.72	3.14	3.51	2.60	3.00	3.36

Sporlan Valve Co.

Table 6-11e. Thermostatic charges for desuperheating thermostatic expansion valves.

THERMOSTATIC CHARGE TYPE	REFRIGERANT			
	12	22	500	502
	MINIMUM ALLOWABLE EVAPORATING TEMPERATURE AT THE REDUCED LOAD CONDITION – °F.			
L1	——	40° thru 30°	——	40° thru 20°
L2	40° thru 30°	29° thru 0°	40° thru 20°	19° thru 0°
L3	29° thru 0°	–1° thru –25°	19° thru –10°	–1° thru –30°
L4	–1° thru–40°	–26° thru –40°	–11° thru –40°	–31° thru –40°

Sporlan Valve Co.

pressure drop across the valve is:

100°F condensing	117 psig (Table 5-1)
20° evaporator	21 psig (Table 5-1)
	96 psi difference (Use 100 psi)

A 3.36 ton capacity SF-3 style valve is used. This would be adequate to subcool the hot gas injected into the suction line and keep it below the limits for safe compressor operation.

Table 6-11e shows that the valve will require a thermostatic charge type L3 (29 °F through 0 °F). The final regulator selections would then be:

Hot gas bypass regulator:
Parker Hannifin A9E 1-1/8 in. with 1/4 in. external equalizer, R-12 refrigerant, 1-1/8 in. line size ODS.
Liquid injection device:
Sporlan Valve Co. SF-3; 1/2 × 7/8 in. ODF; internally equalized, R-12 refrigerant; L3 charge.

SELECT

Dual relief valves, that are approved by the National Board of Boiler and Pressure Vessel Inspectors, for a refrigerant receiver.

Operation

Before opening and repairing other refrigerating system components, it is customary to pump all of the refrigerant within the system into the receiver. Thus, when used as a reservoir, the receiver may be up to 90% filled with refrigerant.

Receivers must be protected from excess pressure from any cause, including fire. The Refrigeration Code requires a single pressure relief valve on all receivers with volumes of over 3 ft^3 and under 10 ft^3. There can be no other valve between the relief valve and the receiver.

Receivers with volumes of over 10 ft^3 must be protected by two parallel relief valves. These may be mounted on a three-way valve that assures that one valve will always protect the system while the other is being inspected or replaced.

Given

A receiver used in an R-12 system is 18″ in diameter by 8′ long and is ASME rated for 450 psig. The receiver is to be protected by dual pressure relief valves. Discharge piping from each relief valve to the outside is not expected to exceed 8 feet in length. Maximum operating pressure of the system is 200 psig.

Solution

As noted in chapter 4, the discharge capacity of the relief device used with R-12 is:

$$C = fDL$$
$$= 1.6 \times 1.5 \times 8$$
$$= 19.2 \text{ lb/min}$$

To prevent refrigerant loss during normal operation, the minimum valve setting will be 25% above the maximum expected operating pressure, in this case, 200 psig + 25% or 250 psig. The high pressure safety cutout on the system should be set at 90% of relief valve pressure or 225 psig.

Table 6-12a shows the available relief valve sizes

Table 6-12a. Relief valve capacities.

ASME Series	3000A	3020A	3030A	3060A
valve setting psig	3000 3001 3002 3012 3014 3015 3212 3214 3215	3020 3220 3016 3216	3030 3040 3045 3055	3060 3070 3075
capacities in pounds of air per minute				
235	9.3	17.7	36.9	88.0
→ 300	11.7	22.3 ←	46.6	111.0
350	13.6	25.9	54.0	128.7
400	15.5	29.5	61.4	146.4
425	16.4	31.2	65.1	155.3
450	17.3	33.0	68.8	164.1

Superior Valve Co.

and their capacities. The standard settings available are 235 or 300 psig. Since our nominal setting was 250 psig, we will use the next highest setting or 300 psig.

At 300 psig, the second column has a rating of 22.3 lb/min which is just above our calculated requirement of 19.2 lb/min. Per Table 6-12a, the valve

Table 6-12b. Three-way valves.

CATALOG NUMBERS	SIZE CON-NECTIONS	BODY MATERIAL	DIMENSIONS—INCHES: INLET CENTER TO				WEIGHT POUNDS
			ELBOW CENTER	CAP END	INLET FACE TO OUTLET FACES	OUTLET PORT CENTERS	
925	1/2" F.P.T.	Forged Brass	2 1/16	3 1/4	2 1/16	2 3/4	1
927	3/4" F.P.T.	Forged Brass	2 1/16	3 7/8	2 13/16	2 3/4	1
8021A	1/2" F.P.T.	Forged Steel	2 5/16	5 3/4	3 3/8	3 5/8	3 1/2
8022A	3/4" F.P.T.	Forged Steel	2 5/16	5 3/4	3 3/8	3 5/8	3
8022A-B	1" F.P.T.	Forged Steel	3 5/8	7 1/8	3 7/8	5 3/4	7
8023A	1 1/4" F.P.T.	Forged Steel	3 5/8	7 1/8	3 7/8	5 3/4	6

Henry Valve Co.

selected is a 3020A with a ½" MPT inlet and 5/8" SAE flare outlet.

The left column of Table 4-5 shows the calculated air flow in lb/min. Our calculations require 19.2 lb/min and the installation requires an 8-foot long discharge. The table shows that for 20 lb/min, a 5/8" O.D. copper tube can only be 9 feet long for a 300 psig setting. The installation requires only 8 feet of discharge pipe. Therefore, the 5/8" O.D. copper tube is of sufficient diameter.

Since the volume of the receiver is over 10 cubic feet (approximately 14 cubic feet), two relief devices are required by the ANSI/ASHRAE 15-1978 code. The two relief valves are mounted in parallel on a 3-way valve and each relief valve is to have the capability to handle the required flow.

The 3-way valve is selected per Table 6-12b and is

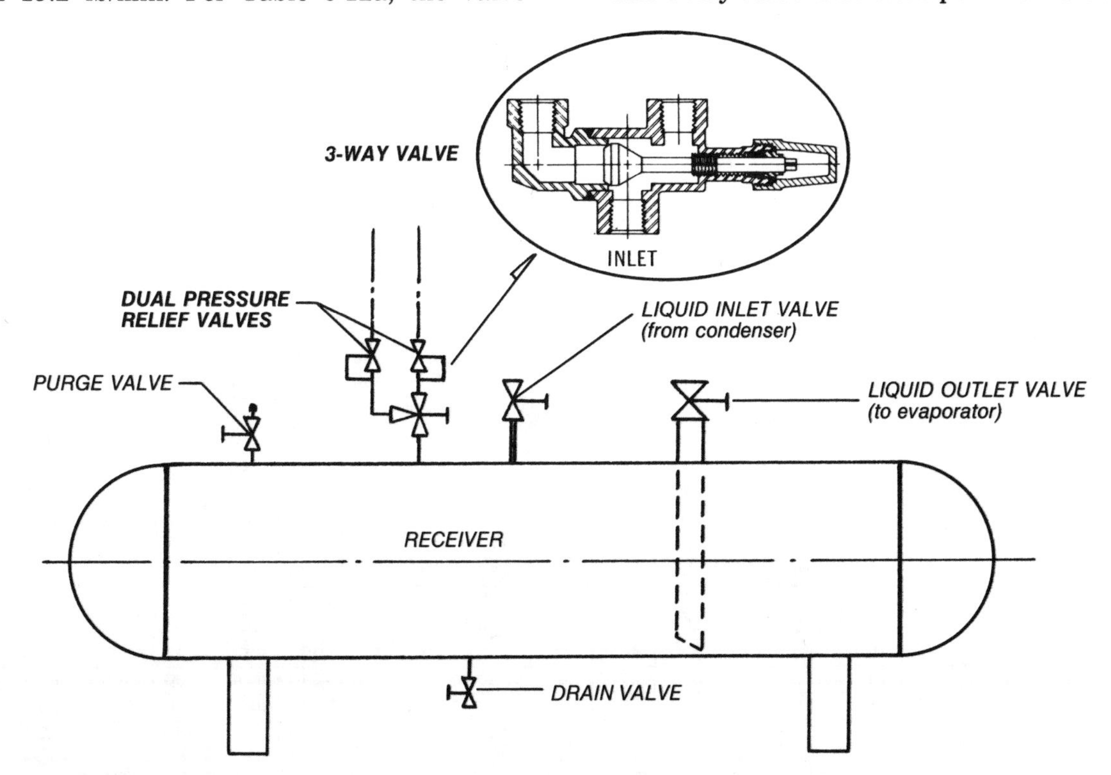

Fig. 6-12. Receiver equipped with valves.

designated No. 8021A with ½″ FPT. Two 3020A relief valves are then mounted on the 3-way valve and their discharge piping connected.

Figure 6-12 shows the receiver assembly. The liquid inlet valve is an angled type Henry 216. The outlet valve mounted on the dip tube is a Henry globe type 203. The charging or purging valve is a Henry 643 and the drain valve is a Superior 607 series.

SELECT

A self-contained and temperature-compensated steam regulator to heat water.

Given

1500 gallons of water are to be heated from 40° to 200°F using 100 psig saturated steam at the regulator inlet and 50 psig steam to the heater.

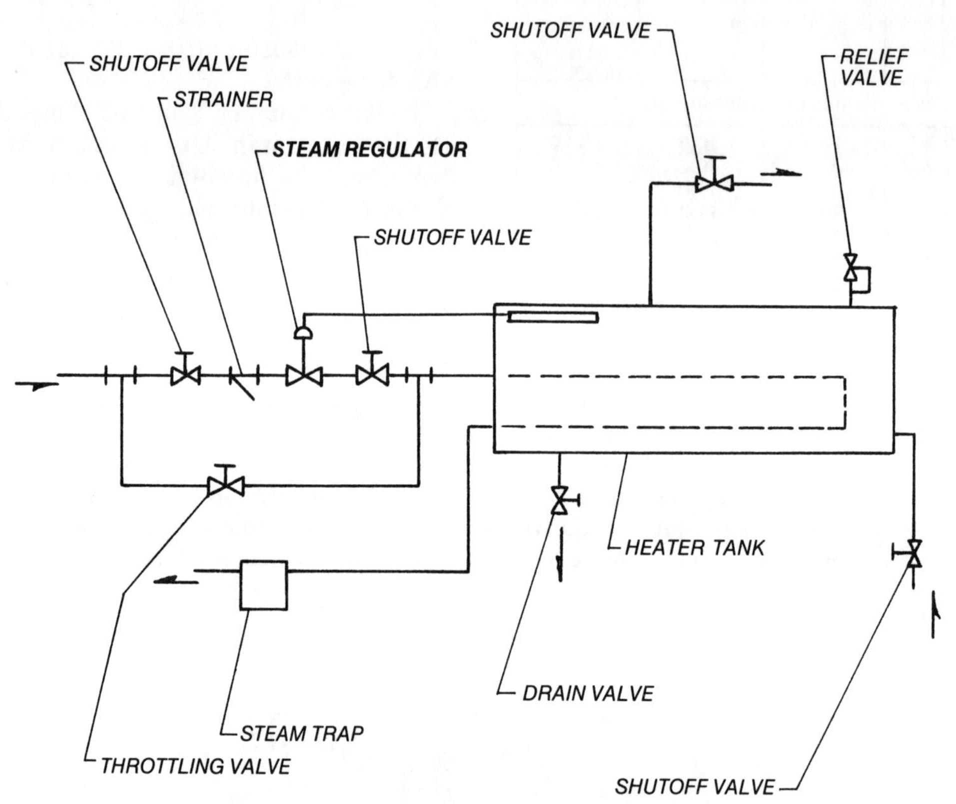

FIG. 6-13a. Steam regulator installation.

Table 6-13a. Capacities of self-operating regulators.

Temp. Rise °F.	U.S. GALLONS OF WATER HEATED PER HOUR—For fuel oil, use half the lbs. per hour of steam listed. 25	50	75	100	150	200	300	400	500	750	1000	1500	2000	3000	4000	5000	7500	10000
	POUNDS OF STEAM PER HOUR REQUIRED																	
10						17	25	33	42	63	83	120	167	250	330	420	620	830
20					25	33	50	67	83	125	167	250	330	500	670	830	1250	1670
30				25	37	50	75	100	125	190	250	370	500	750	1000	1250	1900	2500
40			25	33	50	66	100	130	170	250	330	500	660	1000	1330	1700	2500	3300
50		21	31	42	63	84	125	170	210	310	420	630	840	1250	1680	2100	3100	4200
60	12	25	37	50	75	100	150	200	250	370	500	750	1000	1500	2000	2500	3700	5000
80	16	33	50	67	100	130	200	270	330	500	670	1000	1340	2000	2700	3300	5000	6700
100	21	42	63	83	120	170	250	330	420	630	830	1250	1700	2500	3300	4200	6300	8300
120	25	50	75	100	150	200	300	400	500	750	1000	1500	2000	3000	4000	5000	7500	10000
140	29	58	88	117	175	230	350	470	580	880	1170	1750	2340	3500	4700	5800	8800	11700
160	33	66	100	133	200	270	400	530	660	1000	1330	2000	2700	4000	5300	6600	10000	13300
180	37	75	113	150	225	300	450	600	750	1125	1500	2200	3050	4500	5950	7500	11300	14950
200	42	84	126	165	250	330	500	660	840	1260	1660	2500	3400	5000	6600	8300	12600	16600

Trerice Co.

Operation

Figure 6-13a shows the regulator in the inlet steam piping with a shutoff valve and strainer before the regulator. A bypass line is shunted around the regulator to keep the system on line when the regulator must be removed for service. A shutoff valve and a relief valve are used to isolate and protect the tank during operation.

The isolation valves can be gate valves, while the bypass valve should be a throttling globe valve.

Solution

Table 6-13a shows that 2000 pounds of steam per hour are required to heat 1500 gallons of water per hour with a temperature rise of 160 °F (200 − 40).

Table 6-13b then sizes the regulator, from this particular manufacturer, to meet the heating needs. With 100 psig inlet steam, left column, and 50 psig outlet steam, second column, 2100 lb/hr of steam are delivered by a 1″, double-seated regulator for which the tables are computed. For a single-seated regulator, the capacity would have to be 2860 lb/hr: that is, 2000 ÷ .70. A 1¼″ regulator would be required to do the same job. Table 6-13c shows a Cv of 12 for the 1″ and 15 for the 1¼″ regulator.

Table 6-13b. Valve capacities in pounds per hour of saturated steam. Single seated valves have 70% of the capacities shown.

Inlet PSIG	Outlet PSIG	NOMINAL VALVE SIZE—N.P.T. ½	¾	1	1¼	1½	2	2½
10	8	75	150	225	405	565	900	1325
	5	111	221	335	600	840	1330	1990
	0	140	275	410	740	1040	1650	2475
20	15	138	280	415	775	1070	1750	2570
	10	180	365	545	1010	1410	2280	3375
	5	200	410	610	1120	1560	2520	3750
30	25	160	325	490	880	1220	1970	2950
	20	213	435	650	1175	1630	2625	3900
	10	260	530	800	1430	2000	3200	4800
40	30	240	485	755	1320	1825	2950	4425
	20	310	615	955	1660	2325	3720	5610
	15	320	640	1000	1740	2425	3880	5900
50	40	275	550	880	1420	2075	3250	5000
	30	355	710	1120	1900	2675	4200	6500
	25	375	750	1200	2050	2850	4500	6900
60	50	275	575	915	1600	2240	3525	5350
	40	365	755	1200	2100	2950	4600	7000
	30	410	850	1350	2350	3300	5200	7950
75	65	320	640	1020	1750	2410	3850	5800
	50	455	915	1490	2500	3250	5550	8400
	35	520	1040	1680	2840	3930	6300	9500
100	80	490	1000	1570	2680	3800	6000	9100
	65	600	1220	1900	3300	4680	7400	11100
	50	660	1330	2100	3620	5100	8050	12100
125	100	590	1210	1910	3250	4600	7300	10900
	80	740	1500	2370	4050	5700	9000	13550
	60	800	1620	2600	4400	6200	9800	14680

Trerice Co.

Table 6-13c.

Cv FACTORS FOR SINGLE & DOUBLE SEATED VALVES															
VALVE SIZE	1/8*	3/16*	1/4*	3/8*	1/2	3/4	1	1 1/4	1 1/2	2	2 1/2	3	4	5	6
Cv FOR SINGLE SEAT	0.17	0.35	0.7	1.4	2.8	5.6	8.4	15	21	33	49	66	125	185	327
Cv FOR DOUBLE SEAT	—	—	—	—	—	8	12	21	30	47	70	94	179	265	467

*Port Size in 1/2″ N.P.T. Valve.

Trerice Co.

Table 6-13d. Series 91000 self-operating temperature regulators.

VALVE TYPE		*SINGLE SEATED BRONZE BODY MALLEABLE IRON UNION ENDS					
VALVE SIZE		½″	¾″	1″	1¼″	1½″	2″
STAINLESS TRIM	DIRECT ACTING	91000-AES	91000-AGS	91000-AIS	91000-AKS	91000-AMS	91000-AOS
	REVERSE ACTING	91000-BES	91000-BGS	91000-BIS	91000-BKS	91000-BMS	91000-BOS
†MAX. PSI DIFF.		250	140	80	50	35	20
SHIPPING WT. LBS.		12	14	15	18	19	24

***REDUCED PORTS AVAILABLE ON ½″ REGULATORS**

VALVE TYPE		DOUBLE SEATED BRONZE BODY MALLEABLE IRON UNION ENDS					DOUBLE SEATED CAST IRON BODY FLANGED 125 LB. STANDARD				
VALVE SIZE		¾″	1″	1¼″	1½″	2″	2½″	3″	4″	5″	6″
STAINLESS TRIM	DIRECT ACTING	91000-AHS	91000-AJS	91000-ALS	91000-ANS	91000-APS	91000-CRS	91000-CTS	91000-CVS	91000-CXS	91000-CZS
	REVERSE ACTING	91000-BHS	91000-BJS	91000-BLS	91000-BNS	91000-BPS	91000-DRS	91000-DTS	91000-DVS	91000-DXS	91000-DZS
MAX. PSI DIFF.		250	250	250	250	250	125	125	40	30	20
SHIPPING WT. LBS.		14	15	18	19	24	61	83	117	180	225

†Max. PSI cannot exceed these limits. For greater PSI Diff. use double seated valves. Stainless steel trim is recommended for pressures above 50 psi.

Trerice Co.

Table 6-13e.

TEMPERATURE REGULATOR RANGES											
	NOMINAL RANGES AVAILABLE		RECOMMENDED WORKING SPAN								DUAL SCALE DIAL INDICATORS
			All Direct Action · All Double Seat · All 3 Way Valves				Reverse Action with Single Seat				
	°F	°C	°F	RANGE NO.	°C	RANGE NO.	°F	RANGE NO.	°C	RANGE NO.	°F & °C
FOR NOS. 91000 & 91400 THRU 3" VALVE SIZES & NOS. 91100 & 91200 3-WAY VALVES THRU 2" VALVE SIZE	20 to 70	−10 to 20	40 to 65	01	5 to 20	21	N.A.	N.A.	N.A.	N.A.	30 to 115 & 0 to 45
	40 to 90	5 to 30	65 to 85	02	20 to 30	22	N.A.	N.A.	N.A.	N.A.	50 to 140 & 10 to 60
	30 to 115	0 to 45	85 to 110	03	30 to 45	23	50 to 80	21	10 to 25	41	30 to 115 & 0 to 45
	50 to 140	10 to 60	110 to 135	04	45 to 60	24	80 to 105	22	25 to 45	42	50 to 140 & 10 to 60
	75 to 165	25 to 70	135 to 160	05	60 to 70	25	105 to 130	23	40 to 50	43	75 to 165 & 25 to 70
	95 to 190	35 to 85	160 to 185	06	70 to 85	26	130 to 155	24	50 to 65	44	95 to 190 & 35 to 85
	115 to 210	45 to 95	185 to 205	07	85 to 95	27	155 to 175	25	65 to 80	45	115 to 210 & 45 to 95
	140 to 240	60 to 115	205 to 235	08	95 to 115	28	175 to 205	26	80 to 95	46	140 to 240 & 60 to 115
	155 to 250	65 to 120	210 to 245	09	100 to 120	29	200 to 215	27	95 to 100	47	155 to 250 & 65 to 120
	200 to 280	95 to 135	245 to 275	10	120 to 135	30	215 to 245	28	100 to 120	48	200 to 280 & 95 to 135
	225 to 315	100 to 155	275 to 310	11	135 to 155	31	245 to 280	29	120 to 140	49	225 to 315 & 110 to 155
	255 to 370	125 to 185	305 to 365	12	155 to 185	32	275 to 335	30	135 to 165	51	255 to 370 & 125 to 185
	295 to 420	145 to 215	365 to 415	13	185 to 215	33	335 to 385	31	165 to 195	51	295 to 420 & 145 to 215
	310 to 440	155 to 225	415 to 435	14	215 to 225	34	385 to 405	32	195 to 205	52	310 to 440 & 155 to 225

Trerice Co.

Table 6-13f. Bulbs available on series 91000 temperature regulators.

TYPE	MATERIAL	FITTING NUMBERS	
		8 FEET	20 FEET
Union Hub Bushing	Brass	**H01** (formerly 50-5S1)	**H01**
	316 Stainless Steel	**H04** (formerly 50-5S6)	**H04**
Well	Brass	**W01** (formerly 53-5S2)	**W31**
	Steel	**W02** (formerly 53-5S3)	**W32**
	316 Stainless Steel	**W04** (formerly 53-5S6)	**W34**

Trerice Co.

A self-operating, non-indicating 91000 series temperature regulator is selected from Table 6-13d. The choice is between a 1″, double-seated, 91000-AJS regulator and the 1¼″, single-seated 91000-AKS model. Both are direct acting (decreased steam flow as bulb temperature increases) and have stainless steel trim.

Steam pressure into the regulator is 100 psi and out is 50 psi, a differential of 100 − 50 or 50 psi. Table 6-13d shows that the maximum differential is 250 psi for the 1″ and 50 psi for the 1¼″ regulator.

Table 6-13e recommends a working range of 185 °F to 205 °F (low range) with 200 °F water.

Figure 6-13b shows the physical dimensions for the 1″ and 1¼″ 91000 series regulator and Table 6-13f shows the bulb selection.

The designation and ordering information is as follows:

- 1″ double-seated, non-indicating regulator
- Union ends, malleable iron
- Bronze body
- Control water heating to 200 °F
- Stainless steel trim, Model 91000-AJS
- Direct acting
- 100 psig steam inlet, 50 psig outlet pressure
- Bulb No.W01 with 8-foot long capillary tube with armor braid over capillary.
- Brass socket

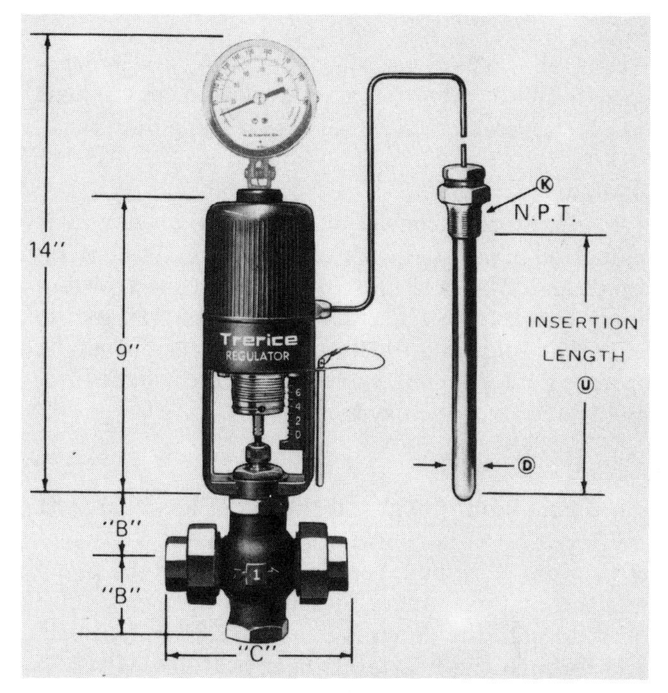

SCHEDULE OF DIMENSIONS		
Valve size - inches	Dimension "B" (max)	Dimension "C"
1	2¼	5 15/16
1¼	2⅝	7 3/16

FIG. 6-13b. Dimensions of series 91000 regulators. *Courtesy Trerice Co.*

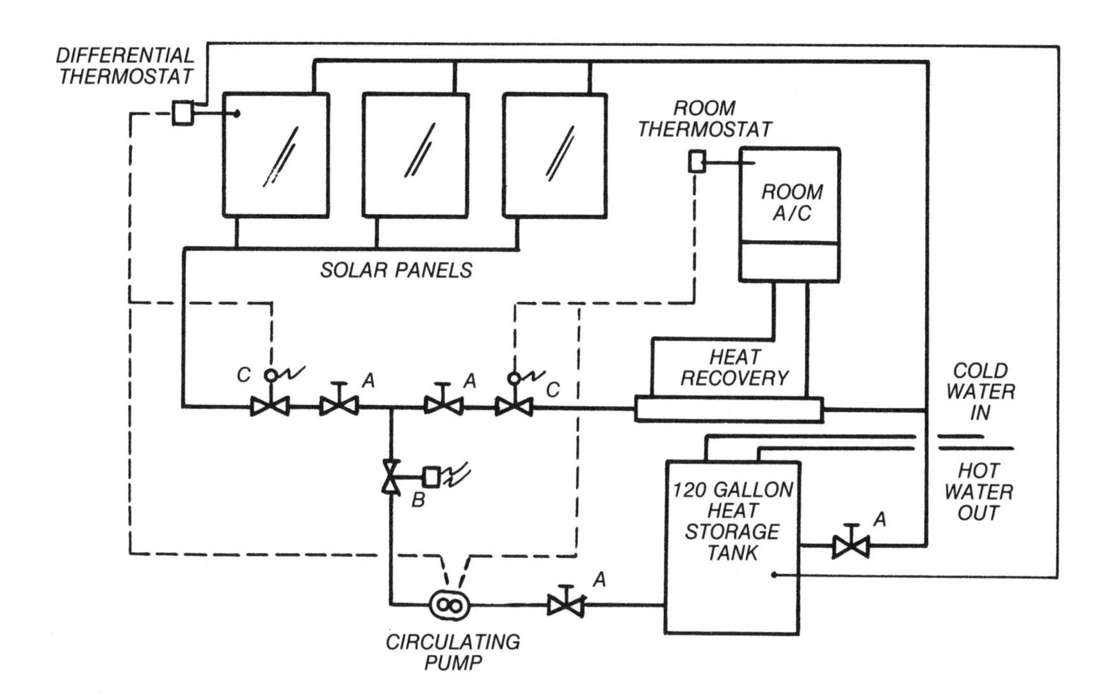

FIG. 6-14a. Schematic of a solar/heat recovery system.

SELECT

Hand shut-off valves, a solenoid valve, and a motor operated ball valve for a solar system and a heat recovery system to heat water in a 120 gallon tank.

Operation

A solar system, consisting of three solar panels, is connected in parallel with a heat recovery system on an air conditioner, Figure 6-14a. The closed system heats water to 180 °F and stores it in a 120 gallon tank. The solar and the heat recovery system can be operated independently or together. Individual thermostats control each system.

Given

A 6 gpm pump capable of 20 feet of head (20/2.31 = 8.66 psi) circulates water to the system. Sixty feet of nominal ¾″ type L copper pipe connect the panel and the heat exchanger. At rated flow, the pressure drop through the panel is 8 feet (8/2.31 = 3.46 psi), and through the heat exchanger is 10 feet (10/2.31 = 4.33 psi). There are 8 long-radius, 90° elbows.

Solution

Using Figures 4-21, 4-22, and 4-23, the pressure drop through the ¾″ pipe and 8 elbows, at 6 gpm, must be determined. Since the heat exchanger has the higher resistance, it will be used as the controlling pressure drop circuit (worse case) in selecting valve sizes. Table 4-23 is for schedule 40 pipe, but a reasonable estimate can be made since the inside diameter of ¾″ schedule 40 pipe is .822″ and that of type L copper is .785″.

From Figure 4-21, the L/D ratio of a long radius elbow is 20 and its equivalent length, from Figure 4-21, is 1.3 feet. The equivalent length of 8 elbows is 10.4 ft. which, when added to 60 ft of pipe, gives a total of 70.4 ft.

At 6 gpm, per Table 4-23, the pressure drop is 3.84 psi per 100 feet of pipe. The system pressure drop is then $3.84 \times 70.4/100$ or 2.70 psi. Since the heat exchanger has a pressure drop of 4.33 psi, the total pressure drop is 7.03 psi. This leaves 1.63 psi (8.66 − 7.03 = 1.63) to be shared by the three shut-off valves, the motorized ball valve, and the solenoid valve.

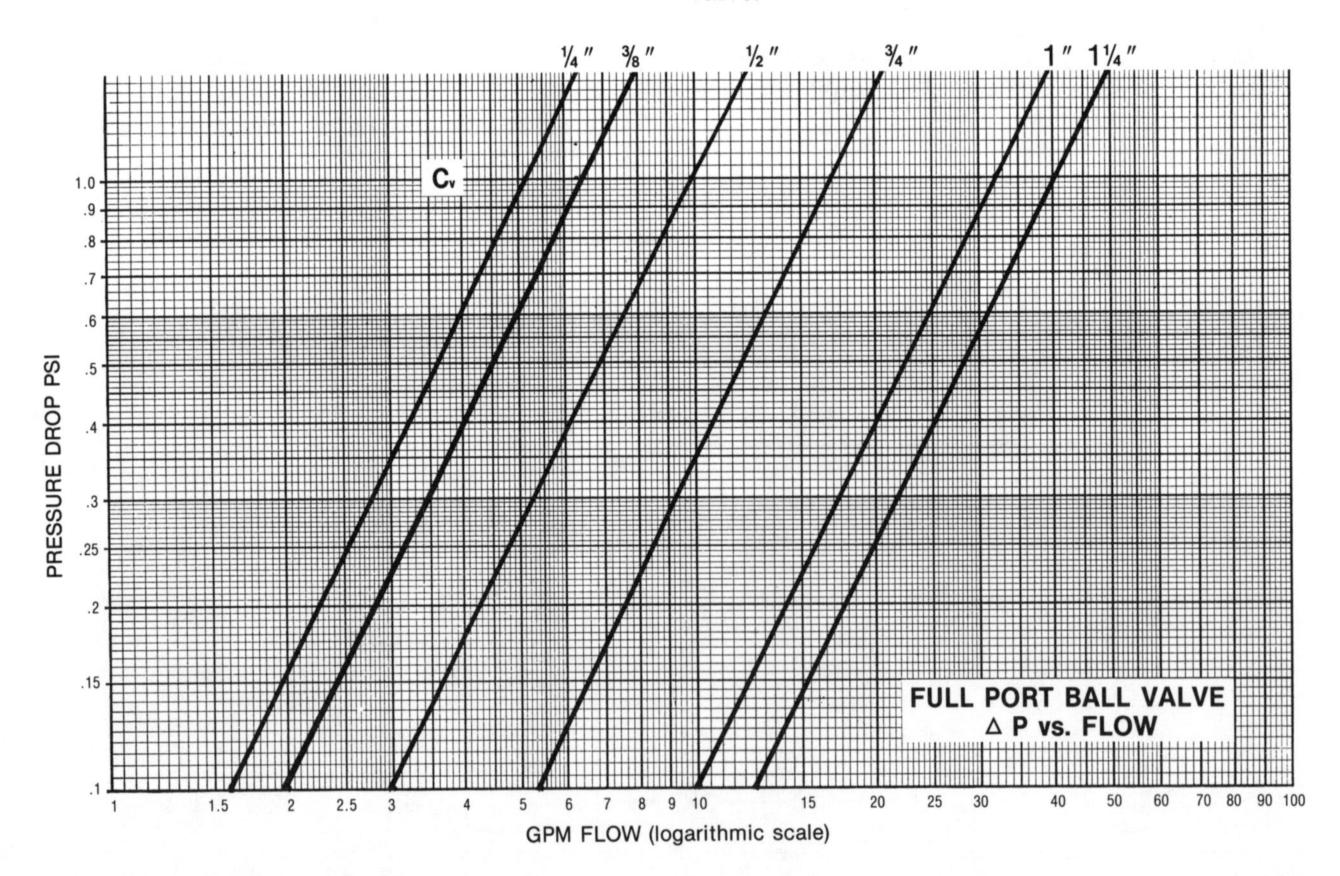

FIG. 6-14b. Pressure drop vs. flow for ball valves.

Figure 6-14b shows a pressure drop vs. flow curves for one manufacturer's ball valves. The ball type is selected for the shut-off valves because of its straight through flow and low pressure drop characteristics. Remembering that Cv is the gpm flow at one psi pressure drop, a ¾″ valve has a Cv of 18, and a ½″ valve has a Cv of 10. At 6 gpm, the ¾″ valve has a pressure drop of .13 psi and is adequate. The three shut-off and the one motorized ball valve have a combined pressure drop of .52 psi (4 × .13), leaving 1.11 psi for the solenoid valve (1.63 − .52). If a pump with a slightly higher head were available, ½″ ball valves could be used to slightly reduce the cost.

Using Table 4-14, note that, by interpolation, the pressure drop through the ¾″ solenoid valve, at 6 gpm, is 1.08 psi and through a 1″ valve it is .38 psi. A 1″ valve is recommended.

Specify the following valves for the heat exchanger circuit:

A 3 - ¾″ bronze ball valves, full ported, with FPT ends, Teflon seats, rated at a minimum of 150 lb.
B 1 - ¾″ bronze, motorized ball valve with a 120 vac motor, 9 second open-to-close operation, and a NEMA 1 enclosure.
C 2 - 1″ normally-closed industrial solenoid valves with FPT ends, 1″ port size, 120 vac coil, negligible pressure drop, rated for 180° maximum water temperature, and a NEMA 4 enclosure.

Select

Air pressure regulators for three individual pieces of equipment.

Operation

Figure 6-15a shows a compressed air line serving a laboratory that requires ultra-clean, dry air for instrumentation, to pressurize a sterile vat, and to drive an agitator motor near an explosive environment.

Given

The laboratory requires 15 scfm of compressed air, at 25 psig ± .2 psi at a −40 °F dew point, for the instruments. The vat needs 5 scfm, at 15 psig, with a

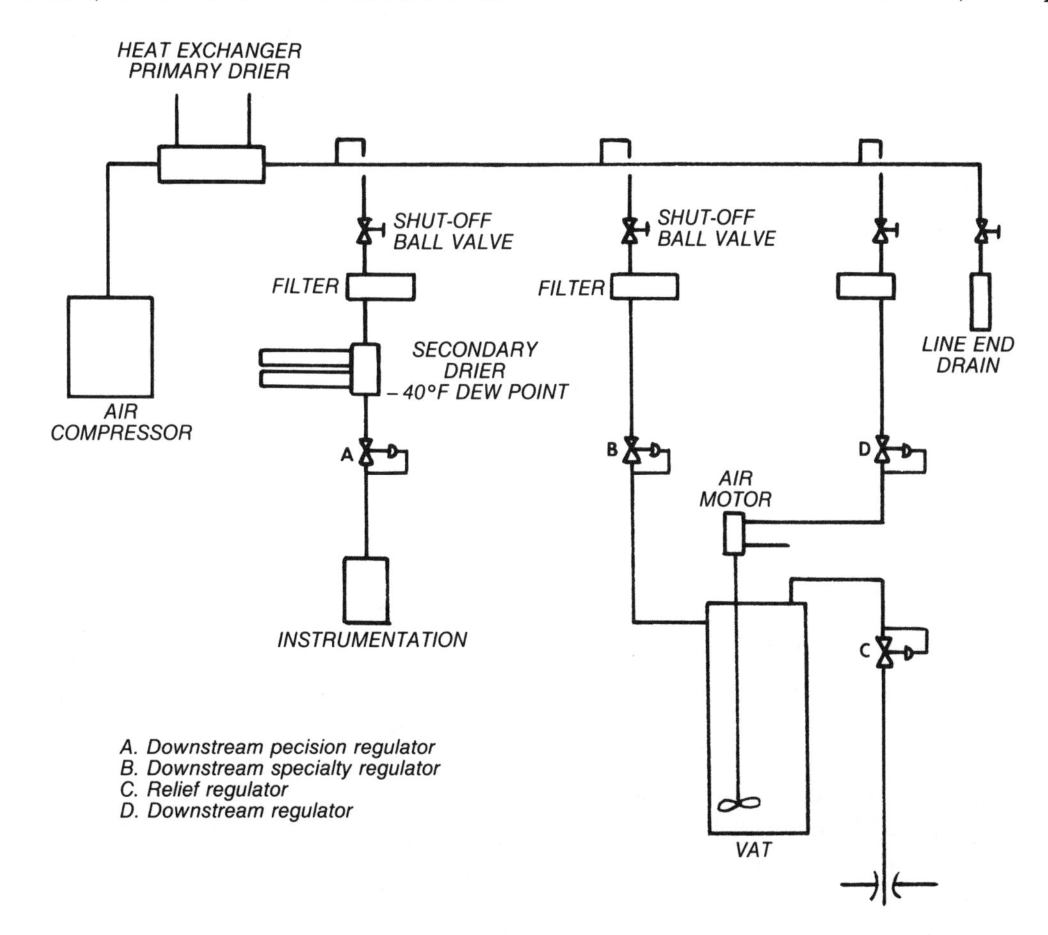

FIG. 6-15a. Schematic of a compressed air supply system.

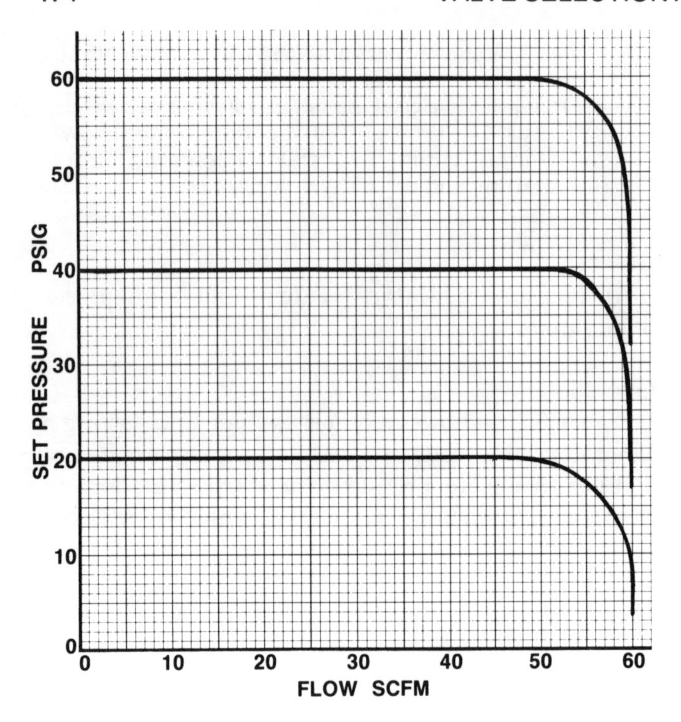

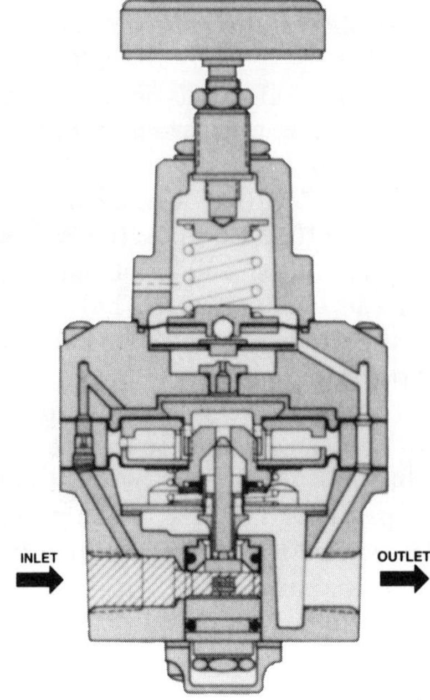

FIG. 6-15c. Precision regulator. *Courtesy Fairchild Industrial Products Co.*

MODEL 81		FLOW S.C.F.M. DROOP % OF SET PRESSURE					
RANGE PSIG	SUPPLY PRESSURE / SET POINT	1/2%	1%	2%	5%	10%	MAX. FLOW
	100 / 3	23	23.5	24	24.5	26	31
	140 / 3	*	*	24	25	26.5	31.5
	100 / 5	24	24.5	26	30.5	33	35
	140 / 5	24	24.5	26	30.5	33.5	35.5
	100 / 10	25	26	30	37	41	44
	140 / 10	25	26	30	37	41	46
100	100 / 20	41	43	45	50	54	60
	→ 140 / 20	40	42	46	52	58	65
	100 / 40	53	54	55	55.6	56	61
	140 / 40	64	69	76	78	79	80

FIG. 6-15b. Flow vs. pressure in a high precision pneumatic regulator.

relief regulator rated at 7 scfm at 15 psig and temperatures to 250 °F. The supply regulator must have a relieving feature. The air-driven agitator motor requires 30 scfm at 70 psig. The main air line is at 150 psig and capable of supplying 75 scfm.

Solution

A precision air regulator (A) will be selected for the laboratory instruments. Special purpose regulators (B, C) will be selected for the vat and a general purpose regulator (D) will be selected for the air motor.

Figure 5-18 showed the droop characteristics of regulators. In this case, the precision and special purpose regulators will be selected for straight line performance. The general purpose regulator can have considerable droop since high accuracy is not required.

Figure 6-15b shows the curves for a high precision, pneumatic regulator, Figure 6-15c, designed as a multi-stage, pilot-operated power stage in a force-balanced system. The slightest change in the system's pressure causes the pilot nozzle to throttle over a very narrow band. This throttling of the pilot stage is followed by instant response by the main or power stage and immediate correction of any deviation in downstream pressure.

If the system pressure rises above the set point, the pilot nozzle immediately reduces pressure to the power stage, causing exhaust valving to open and relieve the excess pressure. If the downsteam pressure falls below the set point, the pilot stage increases pressure to the power stage, causing the main supply to open and provide additional pressure.

This regulator has a ½% variance from the set pressure. At 25 psig, the variance is .13 psi, meeting the requirment of ± .2 psi.

The ordering specifications for this regulator are as follows:

Line pressure	150 psig
Regulated pressure	25 psig ± .2 psi
Flow	15 scfm
Inlet/outlet port	¼″ FPT
Diaphragm	Nylon
Body	Die cast aluminum
Gage	0–50 psig at outlet
Mounting	Panel
Handle	Tamper proof

For the vat, a stainless steel, downstream-sensing regulator (B) is used to control inlet air and a stainless steel, upstream-sensing, relief regulator (C) is selected for exhaust air control. This set of specialty regulators can handle 20 scfm with 1.5 psi droop. Here are the ordering specifications for the two regulators:

Line pressure	150 psig
Regulated pressure	15 psig
Flow	5 scfm
Temperature	250 °F
Body	316 ss
Diaphragm	Teflon (reinforced)
Trim	316 ss
Inlet/outlet port	¼″ FPT
Gage	0–30 psig
Mounting	Panel
Handle	Tamper proof

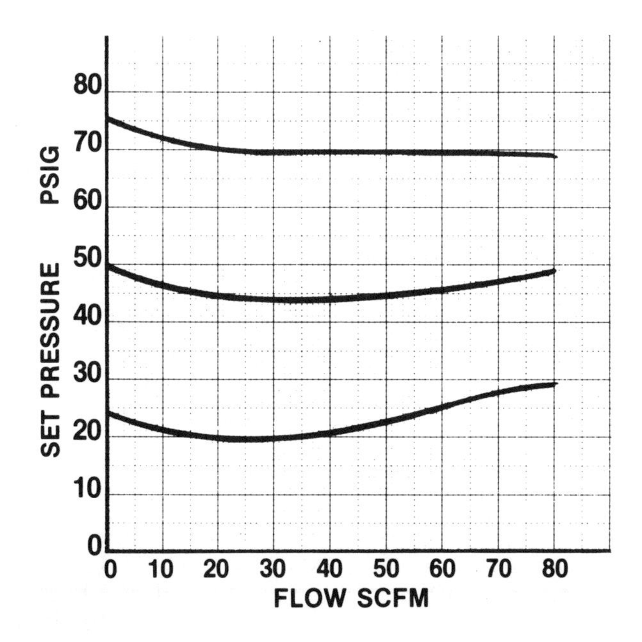

FIG. 6-15d. Set pressure vs. flow in the air motor regulator.

Inlet pressure	15 psig
Outlet pressure	Atmospheric
Flow	7 scfm
Temperature	250 °F
Body	316 ss
Diaphragm	Teflon (reinforced)
Trim	316 ss
Inlet/outlet port	¼″ FPT
Gage	0–30 psig
Mounting	Panel
Handle	Tamper proof

Figure 6-15d show curves for 1/4″ and 3/8″ regulators capable of delivering 30 cfm, at 70 psig, for the air motor. A delivery rate of 30 cfm is within the straight line portion of the curve, so droop will not hamper performance. Here are the ordering specifications for regulator D:

Inlet pressure	150 psig
Outlet pressure	70 psig
Flow	30 scfm
Body	Zinc die cast
Diaphragm	Buna N
Trim	Brass
Inlet/outlet port	3/8″ FPT
Gage	0–100 psig (low side)
Mounting	In-line
Handle	Standard tee

SELECT

Two 3-way mixing valves, with actuators, to control hot water flowing to the parallel heating terminals shown in Figure 6-16. Manual flow controllers are to be sized to balance the pressure drops in the bypass and return lines.

Operation

When maximum heat output is required, the 3-way valve allows a full flow of hot water through the heating terminal. As the temperature approaches the thermostat set point, the 3-way valve diverts a portion of the hot water into a bypass line, thus reducing water flow and the heat output of the coil.

Balancing valves in the bypass line match the pressure drop through the terminal so that there are no changes in pump head when water is diverted around the heating terminal.

The balancing valve in each return line is in series with the heating coil and is used to fine tune the water flow to each of the parallel circuits.

Given

Water, at 200 °F and 25 psig is pumped to parallel

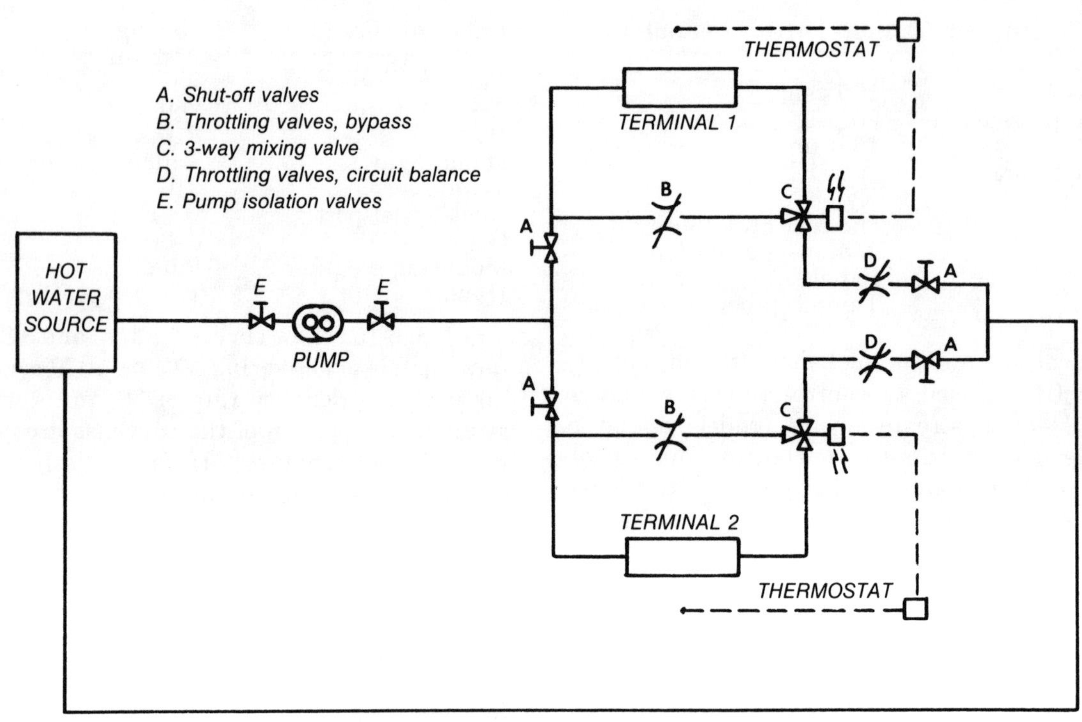

FIG. 6-16. Hot water system with heating terminals and 3-way valves.

heating terminals. Terminal 1 requires 100 gpm at 8 feet of head. Terminal 2 requires 30 gpm at 5 feet of head. 120 volt AC power is available for the actuators. Ambient conditions are dry and 95 °F maximum.

Solution

As discussed in Chapter 4, the control valve should cause the major pressure drop in the system to achieve linear output.

The highest pressure drop is through terminal 1. Therefore, it will be sized first. The valve pressure drop will be initially sized for 3 times that of the heating terminal, that is 3 × 8 feet of head = 24 feet. Knowing the flow and the pressure drop (24 ft ÷ 2.31 ft/psi = 10.39 psi), C_V may be calculated using Table 4-14.

$$C_V = \frac{Q\sqrt{\varrho}}{7.9\sqrt{P_1 - P_2}} = \frac{100\sqrt{60.1}^*}{7.9\sqrt{10.39}} = 30.4$$

*Density of water at 200 °F

Table 6-14a shows a 1½" mixing valve with a C_V of 33. At 100 gpm, per Table 6-14b, the valve has a head of 9.3 psi or 9.3 × 2.31 = 21.5 feet of head. Therefore, the total pressure drop is:

Terminal 1	8.0 feet at 100 gpm
Control valve C	21.5 feet at 100 gpm
Shut-off valves A, E	2.0 feet at 100 gpm
Balance valve D (wide open)	2.0 feet at 100 gpm
Piping and fittings	4.0 feet at 100 gpm
	37.5 feet at 100 gpm

The second terminal is now sized to match the pressure drop of terminal 1 so that the two circuits

Table 6-14a. C_V ratings of mixing valves.

P Code	Valve Size		
-2	1/2" or 5/8"	2	2
-3			
-4		4	4
-6	3/4"		6.8
-8	1"		12
-9	1-1/4"		16
-10	1-1/2"		33
-11	2"		55
-12	2-1/2"		
-13	3"		
-14	4"		
-15	5"		
-16	6"		

Barber-Colman Co.

have essentially the same pressure drop.

Terminal 2	5.0 feet at 30 gpm
Shut-off valves A, E	2.0 feet at 30 gpm
Balance valve D (wide open)	1.5 feet at 30 gpm
Piping and fittings	3.0 feet at 30 gpm
	11.5 feet at 30 gpm

The pressure drop through the control valve for terminal 2 should be 26.0 feet (the terminal 1 pressure drop less the terminal 2 pressure drop equals 37.5 − 11.5 = 26.0) and 26.0 divided by 2.31 = 11.3 psi. The Cv is then:

$$Cv = \frac{Q\sqrt{\varrho}}{7.9\sqrt{P_1 - P_2}} = \frac{30\sqrt{60.1}^{*}}{7.9\sqrt{11.3}} = 8.8$$

Table 6-14b. Water capacities in gpm.

ΔP / Cv*	2	3	4	5	10	15	20	25	30	35
	Differential Pressure (PSI)									
.4	.57	.69	.80	.89	1.26	1.55	1.79	2.0	2.2	2.4
.95	1.3	1.7	1.9	2.12	3.0	3.7	4.3	4.8	5.2	5.6
1.3	1.8	2.2	2.6	2.9	4.1	5.0	5.8	6.5	7.1	7.7
1.4	2.0	2.4	2.8	3.1	4.4	5.4	6.3	7.0	7.7	8.3
1.7	2.4	2.9	3.4	3.8	5.4	6.6	7.6	8.5	9.3	10.1
2	2.8	3.5	4.0	4.5	6.3	7.8	8.9	10	11	12
2.2	3.1	3.8	4.4	4.9	7.0	8.5	9.8	11	12	13
2.4	3.4	4.2	4.8	5.4	7.6	9.3	10.7	12	13	14
2.5	3.5	4.3	5.0	5.6	7.9	10	11	13	14	15
3.3	4.7	5.7	6.6	7.4	10.4	13	15	17	18	20
3.6	5.1	6.2	7.2	8.1	11.4	14	16	18	20	21
3.8	5.4	6.6	7.6	8.5	12.0	15	17	19	21	22
4	5.7	6.9	8.0	8.9	12.7	15	18	20	22	24
5	7.1	8.7	10	11	15	19	22	25	27	30
5.5	7.9	9.5	11	12	17	21	25	28	30	33
6	8.5	10.4	12	13	19	23	27	30	33	36
6.2	8.8	10.7	12	14	20	24	28	31	34	37
6.8	9.6	11.8	14	15	22	26	30	34	37	40
7.4	10.5	12.8	15	17	23	29	33	37	41	44
7.5	10.6	13.0	15	17	24	29	34	38	41	44
8	11.3	13.9	16	18	.25	31	36	40	44	47
8.2	11.6	14.2	16	18	26	32	37	41	45	49
8.5	12.0	14.7	17	19	27	33	38	43	47	50
9	12.7	15.6	18	20	28	35	40	45	49	53
10.5	15	18	21	23	33	41	47	53	58	62
11	16	19	22	25	35	43	49	55	60	65
► 12	17	21	24	27	38	46	54	60	66	71
15	21	26	30	34	47	58	67	75	82	89
16	23	28	32	36	51	62	72	80	88	95
17.4	25	30.1	35	39	55	67	78	87	95	104
25	35	43	50	56	79	97	112	125	137	148
30	42	52	60	67	95	116	134	150	164	177
► 33	47	57	66	74	104	128	148	165	181	195
35.8	51	62	72	80	113	139	160	179	196	212
40	57	69	80	89	126	155	179	200	219	237
42	59	73	84	94	133	163	188	210	230	248

Barber-Colman Co.

Consulting Table 6-14a, the closest choice is a 1″ valve with a Cv of 12. Turning to Table 6-14b, a valve with a Cv of 12 has a capacity of between 27 and 38 gpm. Interpolating, at 30 gpm the pressure drop will be 6.36 psi (14.7 ft). The total pressure drop through the terminal 2 system is then:

Terminal 2	5.0 feet at 30 gpm
Shut-off valves A, E	2.0 feet at 30 gpm
Balance valve D (wide open)	1.5 feet at 30 gpm
Control valve	14.7 feet at 30 gpm
Piping and fittings	3.0 feet at 30 gpm
	26.2 feet at 30 gpm

The difference between the 37.5 ft pressure drop for terminal 1 and the 26.2 ft pressure drop for terminal 2 is 11.3 ft. This difference will be eliminated by adjusting balance valve D. As a result, the pressure drop through terminal 2 will equal that through terminal 1.

Terminal 2	5.0 feet at 30 gpm
Shut-off valve A	2.0 feet at 30 gpm
Balance valve D (throttled)	12.8 feet at 100 gpm
Control valve	14.7 feet at 30 gpm
Piping and fittings	3.0 feet at 30 gpm
	37.5 feet at 30 gpm

Table 6-14c. Catalog description of mixing valves.

FITTING		Flare	Screw
Size		5/8″ O.D.	1/2″-2″
Part Number		VB-9312-0-4-P	VB-9313-0-4-P
Flow Type		Mixing	Mixing
Material	Body	Bronze	Bronze
	Seat	Bronze	Bronze
	Stem	Stn. Steel	Stn. Steel
	Plug	Brass	Brass
	Packing	Spring loaded Teflon "V" Rings	Spring loaded Teflon "V" Rings
	Disc	None	None
Pressure (psig)	Static	250	250
	Recommended Differential *	35	35
Fluid Temperature °F(°C)	Min.	40° (4°)	40° (4°)
	Max.	281° (138°)	281° (138°)

Barber-Colman Co.

Notice that pressure drop through the 1″ control valve is only 14.7 ft while the pressure drop through the total system is 37.5 ft. This is 39% of the total and less than the 50% required for good control. The next smaller valve, Table 6-14a, has a C_V of 6.8. The pressure drop through the ¾″ valve,Table 6-14b is 20 psi or 46.2 ft of head. There are three alternatives:

1. Since 46.2 exceeds the entire head loss of zone 1, use balance valve D to increase the zone 1 pressure drop to equal that created by the ¾″ valve.
2. Ask the valve manufacturer if a 1″ with a restricted port and a 8.7 C_V is available.
3. Use the standard 1″ valve, knowing that control is being sacrificed because it is slightly too large.

The specifications for the 3-way mixing valves used for zone control, from Tables 6-14a and 6-14c, are as follows:

Terminal 1

$C_V = 33$
Size: 1½″ FPT (screw)
Mixing service
Bronze body
Bronze seat
Stainless steel stem
Spring-loaded Teflon V-rings
250 psig rating at 281 °F
35 psig differential

Terminal 2

$C_V = 12$
Size: 1″ FPT (screw)
Mixing service
Bronze body
Bronze seat
Stainless steel stem
Spring-loaded Teflon V-rings
250 psig rating at 281 °F
35 psig differential

Both the 1″ and the 1½″ proportional-controlled actuators will be selected from Tables 6-14d, 6-14e, and 6-14f. A VB-9313-0-4-P 3-way valve has already been selected. If the actuator is to be factory assembled to the valve, as suggested, a VP-8174-6XX-2-P actuator, rated at 281 °F, is selected from Table 6-14d. The Table lists the actuator in the 300MP series, with a SU (stem up) capability of 150 psig close-off pressure for the 1″ valve and 35 psig for the 1½″ valve, when using a the VB-9313-0-4-P mixing valve.

The details are listed in Table 6-14e. A series 300 actuator has a CCW stem up rotation and is rated for 120v. The actual part number is MP-471-600. Table 6-14f shows that this actuator has a 90 second, 180 ° stroke CCW, and a 50 lb-in torque capability.

Table 6-14d. 3-way valve close-off pressure ratings.

			Series 200		Series 300		Series 350	
			MP-5XXX Spring Return		MP— Spring Return		MS-8XXXX Spring Return	
Port ⟶			SU*	SD*	SU*	SD*	SU*	SD*
Part Number	**P Code**	**Size**						
VB-804-0-2-P	-12	2-1/2″			15	15	15	15
	-13	3″			10	10	10	10
	-14	4″			6	6	6	6
	-15	5″						
	-16	6″						
VB-817-0-2-P	-12	2-1/2″			125	125	125	125
	-13	3″			125	125	125	125
	-14	4″						
	-15	5″						
	-16	6″						
VB-9312-0-4-P	-2-4	5/8″	110	100				
VB-9313-0-4-P	-2-4	1/2″	110	100	250	250	250	250
	-6	3/4″	55	55	220	200	220	200
	-8	1″	35	35	150	140	150	140
	-9	1-1/4″	22	22	100	95	100	95
	-10	1-1/2″			35	33	35	33
	-11	2″			35	33	35	33
VB-9323-0-4-P	-4	1/2″	250	250	250	250	250	250
	-6	3/4″	250	250	250	250	250	250
	-8	1″	250	250	250	250	250	250
	-9	1-1/4″	250	250	250	250	250	250
	-10	1-1/2″			250	250	250	250
	-11	2″			250	250	250	250
VB-9332-0-4-P	-2-3-4	5/8″	42**	45**				

TO SELECT AN ASSEMBLED BODY & LINKAGE

		Series 200	Series 300	Series 350
Linkage	1/2″—1-1/4″	AV-600	AV-391	AV-430
	1-1/2″—2″		AV-392	AV-430
	2-1/2″—4″		AV-329	AV-430
	5″—6″			
Body & Linkage (Factory Assembled)		VB-9312-200-4-P	NA	NA
		VB-9313-200-4-P		
		VB-9323-200-4-P		
		VB-9332-200-4-P		
Body, Linkage & Actuator (Factory Assembled) (For specific Actuator Assembly Codes, see following pages)		VS-9312-2XX-4-P	VP-8044-6XX-2-P	VS-8043-35X-2-P
		VS-9313-2XX-4-P	VP-8174-6XX-2-P	VS-8173-35X-2-P
		VS-9323-2XX-4-P	VS-9313-3XX-4-P	VS-9313-35X-4-P
		VS-9332-2XX-4-P	VS-9323-3XX-4-P	VS-9323-35X-4-P
Maximum Ambient		140° (60°)	136° (57°)	140° (60°)
Resultant Fluid		200° (93°)	260° (186°)	281° (138°)
VB-804-0-2-P VB-817-0-2-P	Max. Fluid		300° (149°)	300° (199°)
	Result. Ambient		100° (38°)	134° (56°)
VB-9312-0-4-P VB-9313-0-4-P VB-9323-0-4-P VB-9332-0-4-P	Max. Fluid	281° (138°)	281° (138°)	281° (138°)
	Result. Ambient	115° (46°)	125° (52°)	140° (60°)

*SU = Stem Up, SD = Stem Down.
**35 psi in neutral position (both ports closed).

Barber-Colman Co.

Table 6-14e. List of actuator part numbers.

Series	Valve Linked	Voltage	Hz	Aux. Switch	Actuator Code	Actuator P.N.	VA	Watts	Wiring Fig. #
200	Stem up	24	50/60	—	201	MP-5213	18	10	1
				SPDT	202	MP-5213-500			
		120		—	211	MP-5210			
				SPDT	212	MP-5210-500			
		240		—	221	MP-5211			
				SPDT	222	MP-5211-500			
300	CCW Stem up	120	60	SPDT	311	MP-461-600	50	28	2
					312	MP-471-600			
350	Stem up	24	50/60	—	351	MS-83013	36	11	1
				—	352	MS-81113			
		120		—	353	MS-83010	37		
				—	354	MS-84110			
				SPDT	355	MS-83010-500			
				SPDT	356	MS-84110-500			
		240		—	357	MS-83011	39		
				SPDT	358	MS-83011-500			
400	Stem up	120	60	SPDT	414	MP-481-600	50	28	2

Barber-Colman Co.

The balance valves are selected using pressure drop curves found in Figures 6-16b, c, and d. The required gpm and the pressure drop are known and the shut-off valve setting are show at the top of the graphs. Reading the curves:

For the terminal 1 bypass, with 100 gpm at an 8′ head — 2½″ valve set at 7.7.

For the terminal 1 return line, with 100 gpm at a 2′ head — 3″ valve set at 8.0.

For the terminal 2 bypass, with 30 gpm at a 5′ head — 2″ valve set at 2.7.

For the terminal 2 return line, with 30 gpm at a 13.9′head — 2″ valve set at 1.9.

Table 6-14f. Actuator specifications.

Auxiliary Switch Electrical Rating	120 Volts	240 Volts
Running Current	5.8 Amps	2.9 Amps
Locked Rotor	34.8 Amps	17.4 Amps
Non-Inductive	12 Amps	6 Amps

1 6 5
Factory Settings
CCW CW

Actuator Number	Electronic Drive Part Number	Power Supplies In Drive	Power Supply Required for Drive	Input (28 Watts		Torque Lb-In.	Nominal Damper* Area Sq. Ft.		Timing Sec.	Stroke	Spring Return
				Volts 60 Hz	Amps		Parallel	Opposed			
MP-461-600	CP-8301-120	20 Vdc 50 mA	None	120	.5	50	28	36	90	180°	Clockwise
MP-471-600				120	.5	50	28	36	90	180°	Counterclockwise
MP-481-600				120	.5	220	122	157	130	180°	—
MP-2110-601				120	.5	50	28	36	25	180°	—
MP-461-621	CP-8301-101	None	20 Vdc 35 mA	120	.5	50	28	36	90	180°	Clockwise
MP-471-621				120	.5	50	28	36	90	180°	Counterclockwise
MP-481-621				120	.5	220	122	157	130	180°	—
MP-2110-621				120	.5	50	28	36	25	180°	—

*Damper ratings are nominal and based on standard (not low leakage) dampers at 1″ W.C. pressure and 2,000 FPM velocity.

Barber-Colman Co.

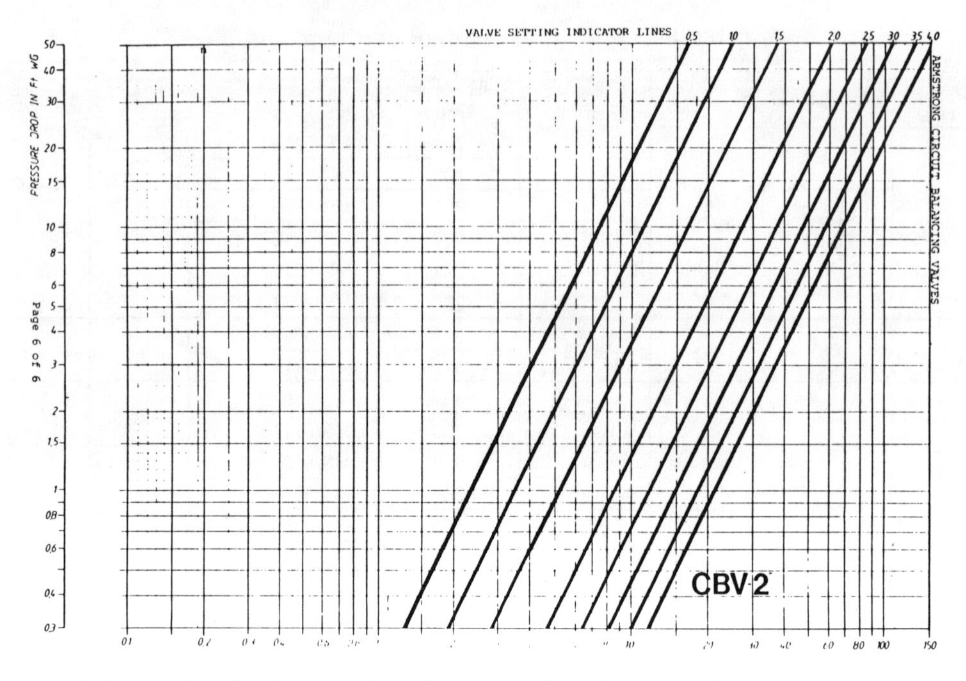

FIG. 6-16b. Settings for 2″ balancing valve. *Courtesy Armstrong Pumps Inc.*

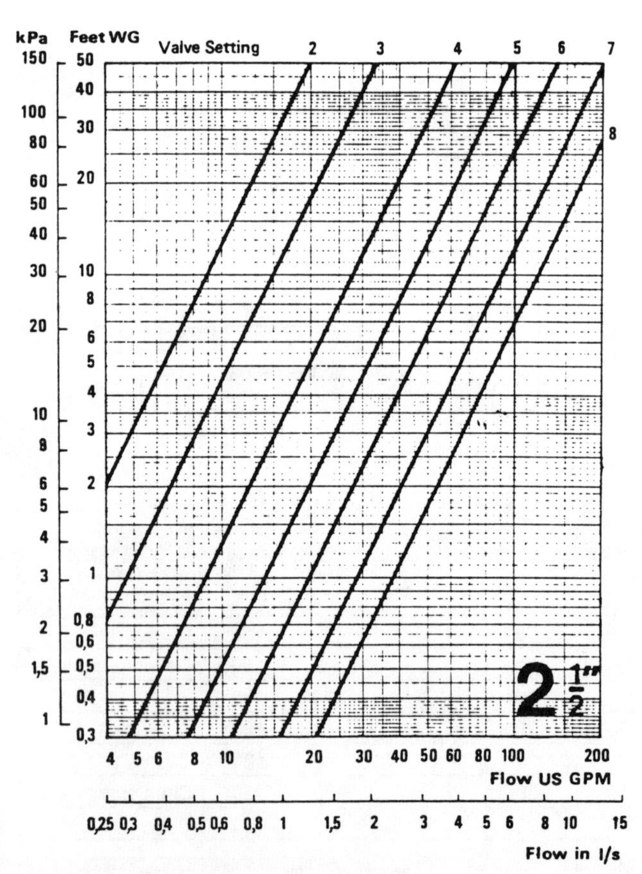

FIG. 6-16c. Settings for 2½″ balancing valve. *Courtesy Armstrong Pumps Inc.*

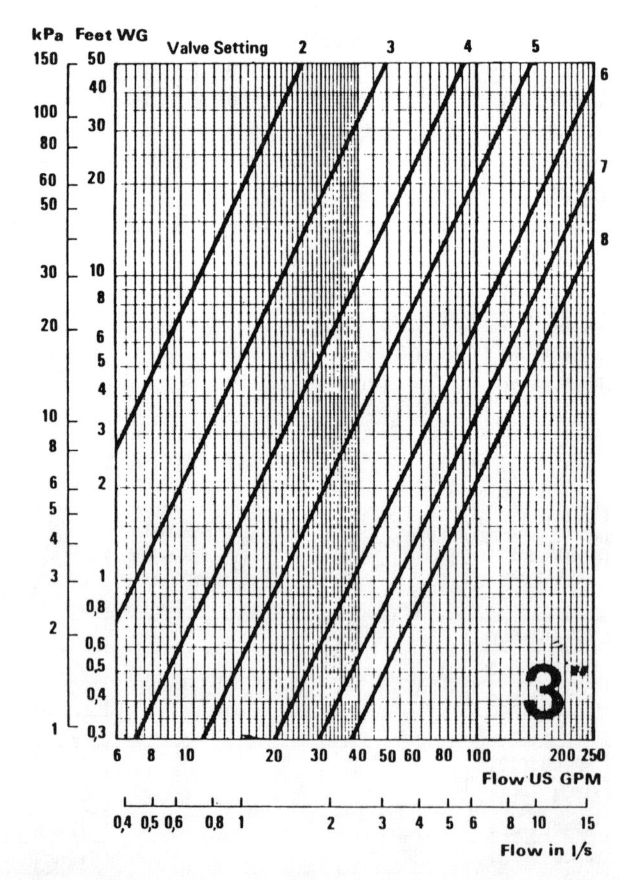

FIG. 6-16d. Settings for 3″ balancing valve. *Courtesy Armstrong Pumps Inc.*

7
Valve Installation and Service

Regulators and valves, if properly selected, will give good reliable service. However, if improperly installed or maintained, reliability will be severely reduced. Proper location, orientation, accessibility, installation, piping, and service are a must for valves to give long trouble-free service.

The cost of repairing improperly installed valves can be extremely high in terms of manhours, downtime, failed industrial processes and human discomfort.

Here are some tips and suggestions on the proper installation and troubleshooting of valves.

Valves and Piping Systems

Refrigeration valves are most commonly mounted by soldering, or with pipe threads or flanges.

Soldering

Many valves are soldered to copper tubing which comes in a number of wall sizes and with either hard or soft temper. Wall sizes are identified as follows:

- Type K, heavy wall, hard and soft temper
- Type L, medium wall, hard and soft temper
- Type M, light wall, hard temper only
- Type DWV, light wall, hard temper only

Type L is most commonly used for refrigeration service and the soft temper variety is available in 50 to 100 feet long coils. The hard temper type is usually supplied as 20-foot-long rigid tubes.

Only cleaned and sealed copper tubing should be used for refrigerant service. Manufacturers go to extra expense to see that the tubing is purged with nitrogen to remove oxygen and then sealed tightly to exclude dirt and debris during shipping and storage. When using this tubing, the end caps should be replaced after a length is cut. Many job sites are anything but clean and, in some cases, sand and dust are flying everywhere. It is easy to contaminate the tubing and, eventually, the system.

Always use clean tubing and check the inside before use, especially if an unsealed tube is being used.

Preparing the copper tubing is a major factor in reliable soldering. It consists of the following:

- ☐ Cutting the tube
- ☐ End preparation
- ☐ Cleaning
- ☐ Fluxing
- ☐ Joining
- ☐ Heating
- ☐ Soldering
- ☐ Clean-up
- ☐ Leak test

Tube cutters are commonly used to cut the copper tubing. However, the cutter must not be overtightened in an attempt to make the cut with only one or two turns around the tube. Too much pressure will produce a *rolled-in* edge, Figure 7-1, which, if deep enough, can cause fluid restriction and unnecessarily high pressure drops, reducing overall system capacity.

Never use a hack saw since chips can be left in the tubing that may eventually plug valve passages or block a moving part.

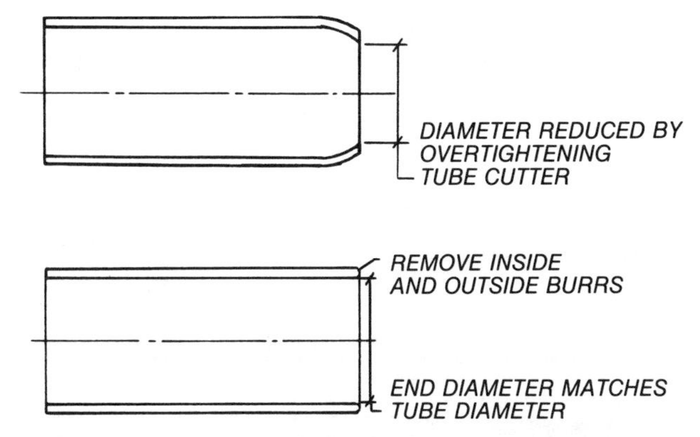

FIG. 7-1. Improper (above) and proper (below) methods of preparing tube ends for brazing.

The cut edge should be the same diameter as the tube. The inside and outside burrs are removed and the tube ends are cleaned *bright* with emery paper to remove copper oxides, dirt and grease. After cleaning, do not touch the tube with the fingers, hands, or greasy and dirty cloths. Oils in the skin can inhibit the flow of solders, causing a leaky or weak solder joint.

Flux should be added evenly, but sparingly, and only to the outside of the tube. Applying too much flux can introduce flux into the system where it can mix with the piped fluids and eventually cause corrosion, system problems and possibly failure. This is a good example of *a little more* being detrimental.

The joint should have several thousandths of an inch clearance. A larger gap will prevent good capillary action, producing a weak joint, and could even allow solder to enter the system and plug the valves.

Too tight a joint will also inhibit capillary action and a weak, leak-prone joint, without good solder penetration, will result.

Heating the joint is one of the most critical procedures. A large difference in mass is probably the worst condition since more heat must be applied to the heavier mass of the valve than to the lower mass of the tube end.

The type of flame, solder melting point and point of heat application are all critical to proper joining and reliability.

Table 7-1 shows a manufacturers' list of solders and brazing materials with recommended fluxes. The chart recommends the filler metal based on the materials to be joined such as copper-to-copper, copper-to-brass, copper-to-steel and stainless steel. These are the materials most commonly found in the general heating and refrigeration industry.

The difference between solders and brazing filler metals is the temperatures at which they melt. Solders melt at temperatures under 840 °F, brazing materials over 840 °F. The filler metals do not melt at one single temperature, but over a temperature range. They start to melt at a temperature, called the *solidus*, and are completely molten at a slightly higher temperature, called the *liquidus* temperature. At the lower range, the filler metal is not as fluid as it is at the higher range. For example, *Stay-Brite 8* starts to melt at 430 °F and is completely liquid at 535 °F. The *fluidity rating* shows, by comparison, how fast the filler metal flows within the melting range. The smaller the temperature difference between the solidus and liquidus, the greater the fluidity rating.

The soldering technique depends upon the components in question and the solder used. Copper tube, wrought copper, light-weight forged brass or cast bronze fittings can all be soldered.

Table 7-1. Solders and brazing materials with recommended fluxes.

METALS TO BE JOINED	FILLER METALS		MELTING RANGE		FLUIDITY RATING	FLUXES	TORCHES & FLAMES
	Solders	Brazing Filler Metals	Solidus	Liquidus			
Copper or Brass to Copper or Brass	Stay-Brite™		430°F	430°F	10	Stay Clean™ Soldering Flux	POL Torch Acetylene-Air
	Stay-Brite 8		430°F	535°F	8		
		Stay-Silv™ 0	1310°F	1475°F	5	No flux required for copper to copper joints with the phosphorus bearing filler metals. For brass and other alloys of copper use Stay Silv (white) Brazing Flux.	Oxy-Acetylene (neutral flame)
		Phoson +®	1190°F	1205°F	10		
		Dynaflow™	1190°F	1465°F	3		
		Stay-Silv 15	1190°F	1480°F	3		
Copper or Brass to Steel or Stainless	Stay-Brite		430°F	430°F	10	Stay Clean Soldering Flux	POL Torch Acetylene-Air
	Stay-Brite 8		430°F	535°F	8		
		Safety-Silv™ 1200	1145°F	1200°F	8	Stay Silv (White Brazing Flux)	Oxy-Acetylene (slightly reducing flame)
		Safety-Silv 1350	1150°F	1340°F	5		
		Safety-Silv 1370	1250°F	1370°F	6.5		
		Stay-Silv 45	1125°F	1145°F	9		
		Stay-Silv 35	1125°F	1295°F	7.5		
Steel or Stainless to Steel or Stainless	Stay-Brite		430°F	430°F	10	Stay Clean Soldering Flux	POL Torch Acetylene-Air
	Stay-Brite 8		430°F	535°F	8		

J.W. HarrisCo., Inc.

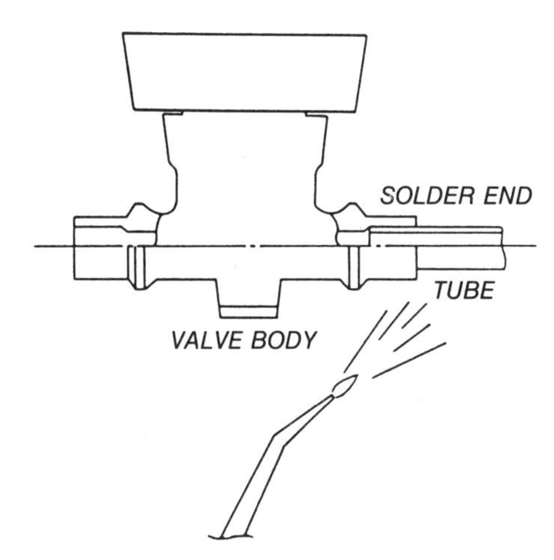

FIG. 7-2. Direct flame away from valve body to avoid overheating internal valve parts.

Figure 7-2 shows the flame being directed back and forth between the tube and fitting. The copper tube readily conducts heat to the tube end that rests within the fitting. The fitting is then heated by direct flame. At the proper temperature, the filler metal is applied and begins to flow. The flame is then directed to the fitting at the back of the joint. Heat *pulls* the filler metal into the joint by capillary action.

Prolonged heating, or too hot a temperature, is detrimental since liquid solder can flow into the system. On the other hand, insufficient heat will leave a *cold* joint that will be structurally weak and probably leak.

Most valves are either cast bronze or forged brass. However, cast iron and steel valves with flanged ends are often used on larger systems. Some flanges have extended copper ends while others have copper-plated steel ends.

Usually, the manufacturer will offer this caution: "Valve body must be wrapped with a wet cloth while soldering to avoid damage to synthetic internal parts. While the valve's steel-body construction will help avoid overheating of internal parts due to conduction, convection of heat up through the tubing and valve body, while making bottom connections, can damage the internal parts. If possible, bench-assemble to connecting tubing with stub tubes in a horizontal plane."

"The torch tip should be sized for the job and large enough to avoid prolonged heating. Overheating can also be minimized by directing the flame away from the valve body."

The above is taken from Sporlan Valve Installation Instruction sheets that are packed with each valve. It is important to read these sheets before installing any valve and to follow the manufacturer's recommendations. In some cases, the valve is small enough to be disassembled. However, if disassembled, the internal parts should be set on a clean cloth.

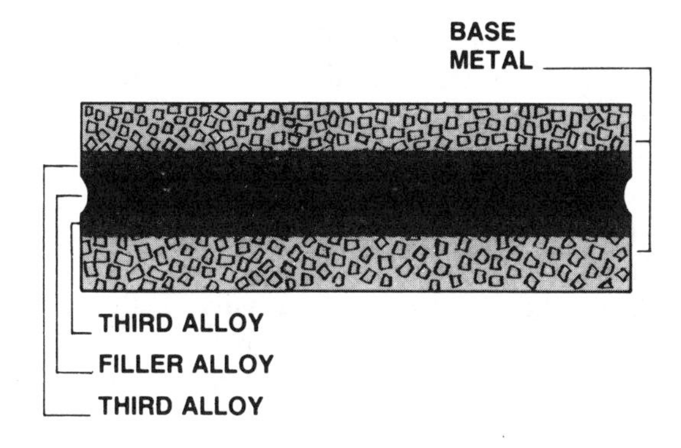

FIG. 7-3. Filler metal must metallurgically bond with the base metals, forming a third alloy. *Courtesy J. W. Harris Co., Inc.*

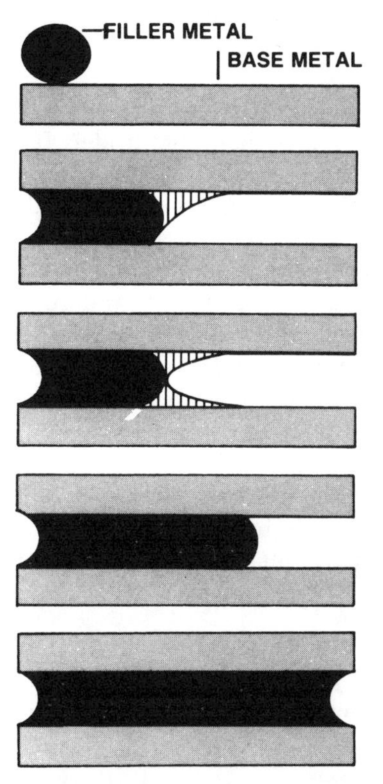

FIG. 7-4. The natural shape of molten filler material, due to surface tension, is a round drop. However, in a joint, the filler metal has a greater attraction for the base metals than for itself. Therefore, the filler first *wets* the base metal then advances to reform its original curvature in a process called *capillary attraction*. *Courtesy J.W. Harris Co., Inc.*

If laid on or in a pile of chips, dust or dirt, these close tolerance parts can not be expected to work properly when reassembled.

When shipped, some valves are assembled only finger-tight so that they are easily disassembled for soldering.

In some cases, the instructions will give body temperature limits of 250 to 300 °F. J.W. Harris Co. has outlined some of the procedures and characteristics of brazing in Figures 7-3 and 7-4.

The following rules should be followed in most cases:

- ☐ Keep the joint and valve parts clean
- ☐ Do not overflux or overheat
- ☐ Direct the flame away from valve body
- ☐ Wrap the valve in a wet cloth
- ☐ If disassembly is in order, keep the parts clean and make sure the body is clean and free of flux, solder and moisture prior to reassembly.
- ☐ Clean the finished joint to remove excess flux, prevent corrosion and create a professional-appearing final product.

Threading

Threaded pipe, used with screw-type valve ends, is usually in 2″ pipe sizes and under. It is used most often in fluid systems. Pipe threads can be unreliable if poorly installed and should be sparingly used in refrigerant systems.

Table 7-2 shows the recommended pipe thread length, based on American Standard threads. Pipe threading beyond these values can cause valve damage by allowing the pipe end to run too far into the valve. This causes distortion of internal seats and passages in the valve, Figure 7-5.

To thread and install pipe:

- ☐ Deburr the internal diameter so that restrictions are eliminated.
- ☐ Thread the pipe to the length shown in Table 7-2.

Table 7-2. American Standard pipe threads

Pipe size in.	"A" Dimension in.
1/8	1/4
1/4	3/8
3/8	3/8
1/2	1/2
3/4	9/16
1	11/16
1¼	11/16
1½	11/16
2	3/4

"A"

FIG. 7-5. Pipe threads that are cut too long will distort the gate valve seat as seen at left. *Courtesy Jenkins Bros.*

- ☐ Throughly clean out chips and oil after threading.
- ☐ Apply thread sealant to the male thread only and make sure that the sealant is compatible with the fluid in question. Special sealants, such as *Leak-Loc®*, are used for refrigerant service.
- ☐ Do not overapply the sealant. Excess sealant can enter the system and plug valve parts.
- ☐ Use the proper tools to assemble the pipe; use a pipe wrench on steel pipe, Figure 7-6, and a smooth jawed wrench on copper, brass or valve components where the metal will not stand the stress of a pipe wrench.
- ☐ Do not put the valve or fitting in a vise and overtighten. There is a good possiblity that the valve

FIG. 7-6. Proper tools are a pipe wrench on the pipe and a smooth jaw wrench on the valve body. *Courtesy Jenkins Bros.*

FIG. 7-7. Flange joint should be aligned so that the pipe center lines are concentric. Note pipe hangers used to support valve. *Courtesy Jenkins Bros.*

will be distorted and lose its *roundness*. Permanent damage results in leakage.

Flanges

Flanged joints, found on larger equipment and many ammonia systems, are joined to the piping and then bolted to the valve body. Many flanges have tail pieces that are soldered to the pipe and, in nonammonia systems, copper is used. The bolted flange fits into a gasketed recess. The most important and necessary facts to remember about bolted flanges are:

- ☐ The flanges must be aligned and concentric, Figure 7-7.
- ☐ Piping must have adequate suppport, especially with larger valves, because of the weight of the valve.
- ☐ Alternately tighten the bolts that are diagonally across from one another. This allows a more even pullup on each flange and a better *squaring* of the flange against the gasket.
- ☐ Put a thread lubricant, such as *Never-Seez®*, on each bolt prior to threading.

Flare Fittings

Flare fittings are used on smaller copper lines and are usually restricted to soft copper tubing not over ¾" in diameter. A flaring tool is used to flare the end and the flare conforms to the 45° angle machined in the flaring block.

The flare matches the valve end design and is sandwiched between it and a flare nut, which is run onto the tube before flaring. Before flaring:

- ☐ Remove any burrs from the inside diameter of the tube end.
- ☐ Begin with the proper length of tube protruding from the flaring block so that the final flare will exactly match the mating surfaces.

FIG. 7-8. Installation of Victaulic swing check valve. *Courtesy Victaulic Co.*

FIG. 7-9. Primore Rotolock assembly. *Courtesy Primore Sales, Inc.*

Overflaring will probably produce cracks in the flare, while underflaring would not allow sufficient flare-to-nut surface area for good sealing and pressure-retention, causing leakage in both cases.

Special connections

Patented connections are available, such as the grooved ends used with Victaulic's *Grooved Piping*

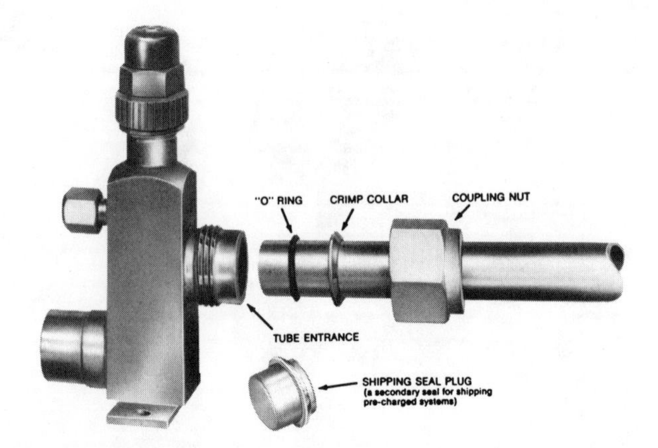

FIG. 7-10. Primore Mec-lock assembly. *Courtesy Primore Sales, Inc.*

Method, Figure 7-8, or Primore's Rotalock®, Figure 7-9, and Mec-Lock®, Figure 7-10, systems.

Whenever special fittings are used, read the instructions, use the recommended torque values, and *do not take short cuts*!

Valve and piping supports

Valves and their associated piping must be adequately supported and secured. Some valves come with their own fastening brackets so that they will not put stress on the adjacent piping system. If piping runs are long, it is recommended that supports be installed at specified intervals. ASHRAE gives the recommended spacing for hangers in Table 7-3.

Cleanliness, orientation, location

Reading the manufacturer's instructions, that are packed with the valve, is a must. The first thing many mechanics do, after opening the valve's box, is to throw away the instruction sheet. This is a no-no!

Cleanliness is imperative. A clean, dry system must be maintained. This is especially true with regulators, where close tolerances are needed for good operation. Many regulators have small bleed ports and passages to divert flow during pilot operation. Most operational problems can be traced to the presence of dirt, scale from solder fluxes and moisture, in other words, to poor housekeeping.

Orientation is also critical to some valves. Many globe valves, solenoids, and regulators have arrows indicating flow direction. Installed backwards, the valve will not function. Be careful to check for valve markings prior to installation.

Location is important. Many valves and regulators will not work properly if located in the wrong place. Shutoff valves should isolate driers, solenoids or

Table 7-3. Maximum spacing and rod size for hangers for steel and copper pipe

Pipe size	Maximum span for hangers	Minimum rod diameter
1/2″	5′	3/8″
1″	7′	3/8″
1½″	9′	3/8″
2″	10′	3/8″
3″	12′	1/2″

regulators that might have to be removed or replaced for or during service. Again, read the instructions.

Refrigeration Valves

Refrigeration valves that require the most critical installation are expansion valves, solenoids and some regulators. These will be discussed in order of importance. First, the expansion valve:

The expansion valve should be located as close to the evaporator as possible. It may be mounted in any position. However, in many cases where a distributor is used, the distributor is mounted directly to the valve outlet and both outlet and distributor are mounted so that flow is vertically downward.

For expansion valves with thermostatic remote bulb (power element) and external equalizer, the bulb and equalizer line must be properly located and fastened to the suction line at the evaporator exit.

Good thermal contact, on a clean straight area of suction pipe, is needed along the entire length of the

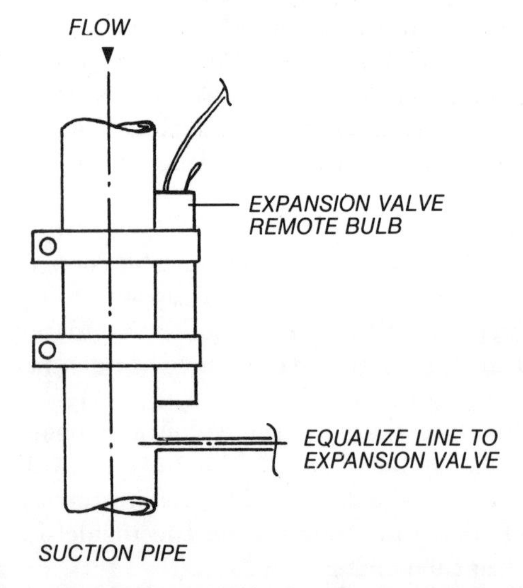

FIG. 7-11. Proper location and mounting of the remote bulb on a vertical suction pipe.

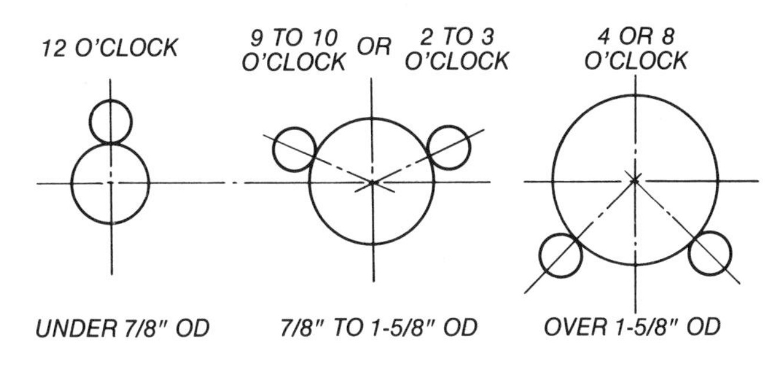

FIG. 7-12. End views showing location of remote bulbs on horizontal suction lines of various diameters.

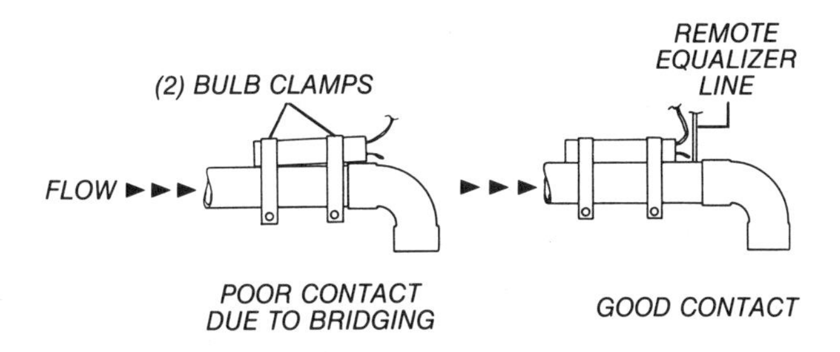

FIG. 7-13. Location of a remote bulb on a horizontal run prior to a trap.

remote bulb. A horizontal suction line is preferred. If vertical installation cannot be avoided, the bulb should be mounted with the capillary tube at the top, Figure 7-11. Never mount the bulb on the evaporator suction header since the bulb can never sense all the circuits.

With suction lines greater than 7/8" O.D., the bulb should be midway on the side of a horizontal run. With lines under 7/8" O.D., the bulb may be located at any point on the tube except the very bottom. If the bulb is attached to the bottom of the suction line, the layer of oil flowing from the evaporator will cause erratic temperatures and result in poor valve performance, Figure 7-12.

If a trap is located at the evaporator outlet, after a short horizontal run, locate the bulb on the horizontal run, not the trap. Do not have any portion of the bulb resting against an elbow or fitting. *Bridging* results and little or no tube contact is made. A recommended remote bulb installation is shown in Figure 7-13.

An external equalizer line must be located downstream of the remote bulb. If any other valves are used, the remote bulb and equalizer must be located on the evaporator side of these valves.

Usually, a manufacturer has thoroughly tested all its systems for proper performance. If field replacement is necessary, care should be taken to see that all components are replaced, following the manufacturer's instructions.

Another problem with thermostatic expansion valves is proper superheat. Superheat is the number of degrees by which the refrigerant vapor exceeds the saturated temperature at the operating pressure. The presence of superheat assures that all the liquid refrigerant has been vaporized before it reaches the compressor.

Normal superheat depends on operating temperature ranges of the system. Superheats can be in the range of:

High temperature applications	10 to 14 °F
Medium temperature applications	8 to 10 °F
Low temperature applications	2 to 4 °F

Superheat may vary with fluctuations in the load on the evaporator. If there are minor load fluctuations, the thermostatic expansion valve will adjust to maintain a constant superheat with limited cycling. At higher loads, the valve opens and meters more refrigerant into the evaporator, to handle the load and reduce the superheat. At lower loads, it will reduce refrigerant flow to meet the lower demand and still maintain the correct superheat.

Most equipment manufacturers will not install an extra port in the evaporator suction line so that true suction pressure can be measured. However, if care is taken, a reading at the compressor suction valve can approximate the evaporator exit pressure.

For high temperature systems, a 2 to 5 psi pressure drop can be expected through the suction line and components. For low temperature applications, a 0.5 to 1.5 psi pressure drop can be expected. The estimated suction line pressure drop is added to the compressor suction reading to estimate the evaporator exit pressure.

To measure superheat, determine the temperature of the refrigerant at the remote bulb and the pressure of the refrigerant at the evaporator exit.

Figure 7-14, a R-22 P-H chart, plots the refrigerant as it leaves the evaporator and enters the suction piping and compressor. The saturation curve represents 100% gaseous refrigerant and no superheat, point *A*. If superheat is present, the evaporator exit temperature is at point *B*. If a suction/liquid heat exchanger is used, it will add additional superheat, point *C*, before the refrigerant vapor enters the compressor suction valve. Note the pressure differential, ΔP, created by the suction line and regulators.

Table 5-1 shows pressures in psig and related tem-

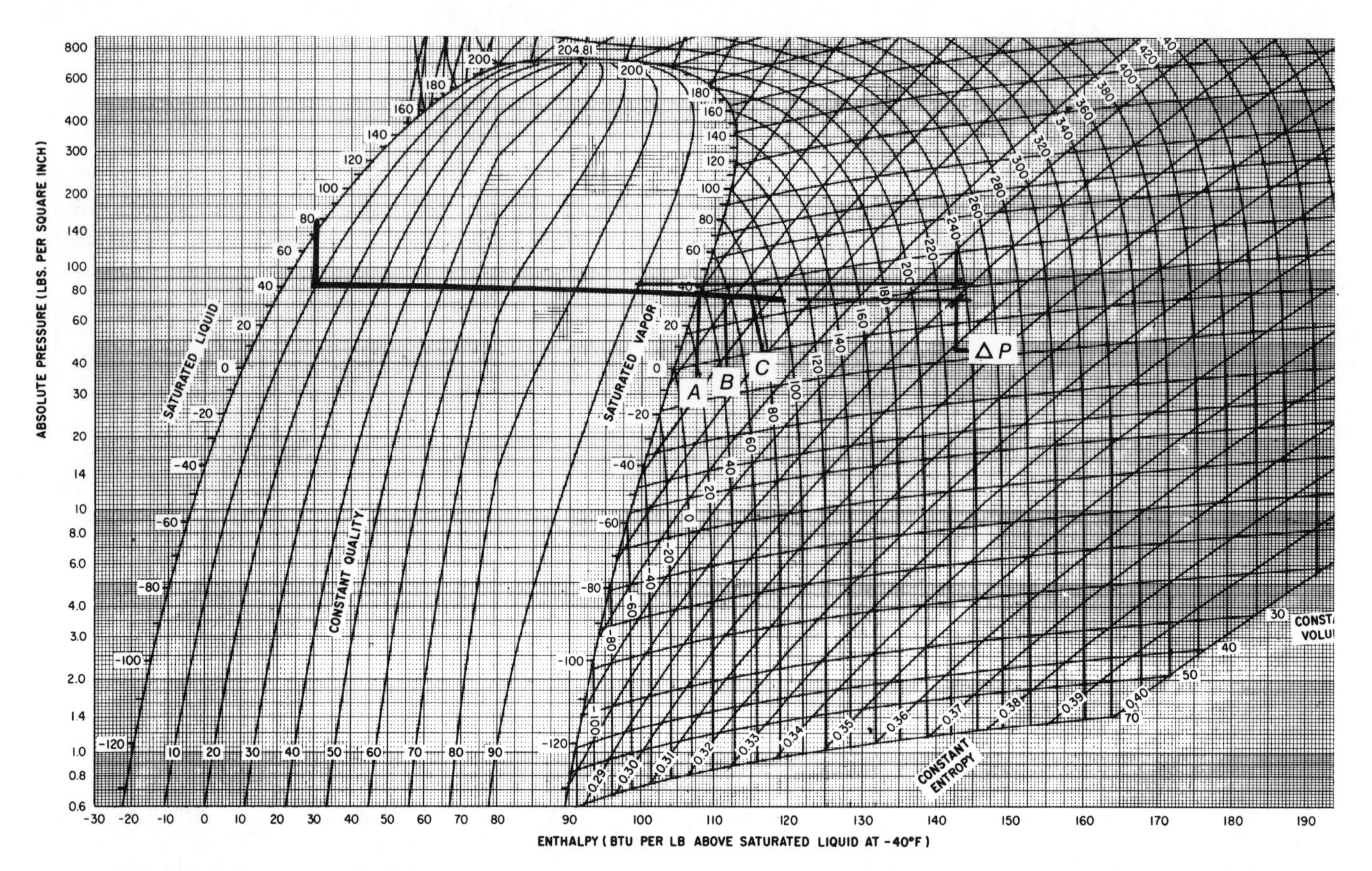

FIG. 7-14. Quality of the refrigerant vapor in the suction line: (A) 100% saturated, no superheat (B) normal superheat (C) additional superheat from a suction/liquid heat exchanger.

peratures for *saturated conditions only.* At 40 °F, the saturated pressure is 68 psig or, adjusting for absolute pressure, 68 + 14.7 = 82.7 psia. Remember, the P-H chart shows absolute pressures (psia) while Table 5-1 shows gage pressures (psig).

If a well-insulated temperature probe measures 52 °F at the remote bulb, the superheat is 52 °F − 40 °F = 12 °F, normal for this system.

If the evaporator exit pressure is read at the compressor suction valve, and not corrected for line losses, the calculated superheat will be incorrect. The greater the suction pressure drop, the more erroneous the answer. For example, point *C* of Figure 7-14 is at 78 psia or 78 − 14.7 = 63.3 psig and has a saturated temperature, Table 5-1, of 36.5 °F. If no allowance is made for line drop, the superheat is assumed to be 52 ° − 36.5 °F = 15.5 °F, not 12 °F. The correct evaporator exit pressure is the compressor inlet pressure of 63 psig plus the estimated suction losses of, say, 4.0 psi. At 67 psig, the saturated temperature is 39 °F. The superheat is then 52 °F − 39 °F = 13 °F, a more accurate value.

If the superheat is not in the desired range, check to see that the remote bulb is properly installed. Figure 7-15 shows the proper suction piping, equalizer line and remote bulb location for evaporators above and below the suction main.

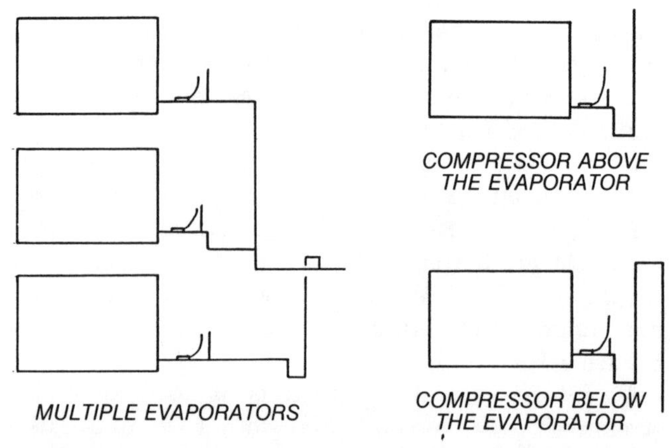

FIG. 7-15. Correct suction piping, equalizer line and remote bulb locations with various evaporator configurations.

Table 7-4. Troubleshooting expansion valves

PROBLEM	PROBABLE CAUSE	POSSIBLE SOLUTION
High superheat	Blocked refrigerant flow due to moisture and oil frozen at the valve seat.	Dry system, add or change drier.
	Dirt or foreign material.	Clean system, add strainer/filter.
	Insufficient refrigerant.	Find leak, correct, recharge system.
	Lack of subcooled liquid at valve entrance.	Liquid line pressure drop is excessive. Too much heat to liquid line. Plugged drier or liquid line components blocked.
	Equalizer line in wrong location.	Read manufacturer's installation instructions, revise installation.
	Remote bulb system has lost charge or bulb has wrong charge for the application.	Replace bulb.
	Undersized valve.	Replace with proper size valve.
	High superheat setting.	Measure superheat and readjust.
	Inlet pressure is too low.	Over condensing, check head pressure. Loss of charge. Leak.
Low superheat	Foreign material in valve.	Clean system, add strainer.
	Valve too large for application.	Use correct valve size.
	Incorrect bulb installation.	Read manufacturer's installation sheet, adjust bulb location.
	Low superheat setting	Readjust setting.
Hunting	Valve wrong size for the application.	Check capacity, replace valve.
	Poor remote bulb location.	Read manufacturer's installation instructions, correct.
	Ice, oil, or scale in valve.	Clean and correct. Replace or add strainer.
No regulation	Plugged equalizer line.	Open line.
	Valve body damaged.	Check for hammer blows, replace if necessary.
	Foreign material.	Check valve strainer, clean system.

Superheat swings of more than 3 °F are called *hunting* and can be caused by a number of factors. Table 7-4 shows a trouble shooting chart for various expansion valve problems and possible solutions.

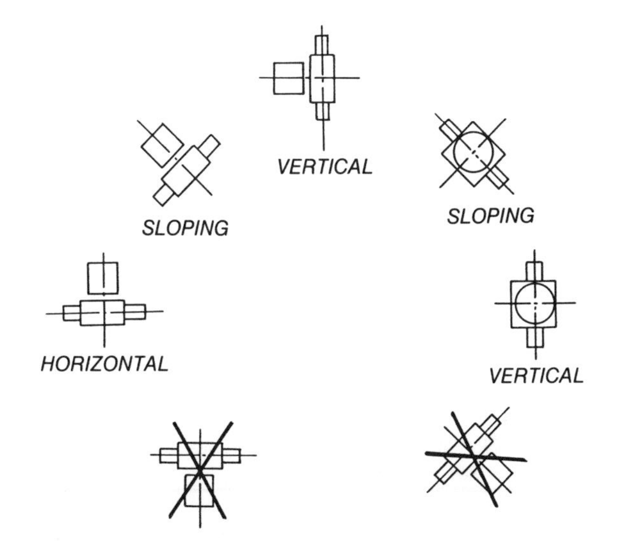

FIG. 7-16. Correct solenoid valve positions. Lowest two positions should never be used.

Solenoid valves

Most solenoid valve manufacturers will stamp an arrow or mark on the *inlet* of their valves to show the direction of flow.

If a solenoid is to be soldered into the system, it may be necessary to disassemble it prior to installation. Cleanliness is extremely important. The coil should be located above the horizontal center line, never below the valve body, Figure 7-16. It is recommended that a filter or strainer preceed the valve.

Soldering and threading rules apply.

The solenoid should be located in a cool place where heat buildup is minimum. *Hot spot* locations will reduce coil life.

Table 7-5 shows a guide for trouble shooting some of the problems encountered.

Regulators

Regulators are available in numerous configurations, sizes and styles, including both upstream and downstream types.

Regulator capacity is never matched to line size, it is a function of port size. In most cases, the regulator modulates over a range of differential pressures of

Table 7-5. Troubleshooting solenoid valves

PROBLEM	PROBABLE CAUSE	POSSIBLE SOLUTION
Won't open NC type	Coil burnout or open electrical circuit.	Check voltage vs. nameplate voltage. Check leads for open or broken wire. Check coil continuity. Check temperature of location.
	Low voltage.	Check voltage.
	Foreign material.	Clean system, install strainer.
	Parts binding.	Check for burrs, polish or replace parts.
	Excessive fluid pressure, exceeding MOP rating.	Reduce pressure or install valve with proper MOP rating.
	Low pressure drop.	Replace with valve having recommended pressure drop for system capacity.
	Cracked diaphragm.	Replace diaphragm.
	Plugged pilot orifice.	Clean pilot port.
Won't close NC type	Coil not de-energized.	Check electrical controls.
	Manual opening stem screwed in.	Turn stem out.
	Foreign material.	Clean parts, add strainer.
	Plugged pilot port.	Clean pilot port, add strainer.
	Damaged disc or seat.	Replace parts.
Leakage	Foreign material.	Clean and check diaphragm and seat.
Coil burnout	Wrong voltage.	Compare nameplate voltage and actual voltage.
	Over voltage.	Voltage should be ±10% of rating.
	Incomplete magnetic circuit.	Missing parts, check manufacturer's parts list.
	Mechanical interference.	Check for burrs or parts nor operating smoothly.
	Wrong coil for application.	Check fluid and ambient temperature vs. coil insulation rating.

Table 7-6. Troubleshooting regulators

PROBLEM	PROBABLE CAUSE	POSSIBLE SOLUTION
Won't open	Foreign material.	Clean system, clean and check strainer.
	Adjusting stem is in too far.	Back off stem, check degree of regulation with pressure gauges.
	Wrong spring range.	Match spring range value to application.
	Pilot passages plugged.	Check and clean passages.
	Pilot solenoid not opening (if used).	If regulator uses satellite pilot solenoid, check to see if it is opening.
	External equalizer line plugged.	Check for pinched, plugged or unconnected equalizer line.
	Electronic controls bad (if used).	Follow manufacturer's recommendations and check for power, voltage levels and polarity.
Won't close	Diaphragm failure.	Replace diaphragm.
	Power element lost charge (if used).	Replace element.
	Adjusting stem turned out too far.	Screw stem in, check degree of regulation with pressure gauges.
With regulators in head pressure service, also check the following:		Dirty air-cooled condenser.
		Air blocked off or fan motor not working.
		Too much refrigerant.
		Non condensibles (air) in the condenser.

from 6 to 10 pounds.

With hermetic regulators that cannot be disassembled, care must be taken not to overheat the regulator during installation or the bellows will leak. In this event, the regulator must be replaced. Dirty conditions during assembly will make the regulator inoperative and foreign material such as flux, solder and dirt cannot be allowed into the system.

Regulators that can be disassembled, usually on larger tonnage systems, have inlet strainer screens to protect the internal parts. However, cleanliness is still important during assembly. R-11 is circulated through the piping of large tonnage systems to remove construction or foreign material.

Factory repair and rebuilding services are available for large tonnage regulators.

Table 7-6 is a regulator troubleshooting guide.

Heat pump reversing valves

Reversing valves are used in heat pumps to reverse the fluid flow; that is, to switch the roles of the condenser and evaporator. Consequently, it is a key element in heat pump performance.

Assume that the reversing valve solenoid is energized during cooling and de-energized in the heating mode, although some are designed just the opposite. Fluid flow directions can be traced in Figure 7-17.

Two moving parts are necessary for successful operation: the pilot solenoid valve and the valve slide mechanism or piston.

Because of the close tolerances of the moving parts, any restriction to movement can *hang-up* the reversing valve and cause the system to malfunction.

Reversing valve problems can be classified as follows:

- ☐ Pilot solenoid valve problems
- ☐ Slide valve hangup
- ☐ Leakage in the slide piston

Electrical problems can cause the solenoid valve to malfunction. The voltage to the coil should be within $\pm 10\%$ of its rating and wire connections and coil continuity should be checked. The coil should energize during the cooling mode and take one to two seconds to operate with a clear, clean *swooshing* sound.

Charging the heat pump with refrigerant is critical since undercharging or overcharging will create problems with both the system and the reversing valve. Heat pump manufacturers are very explicit as to the amount of charge since undercharging is one of the most frequent problems.

In both the heating and cooling mode, the compres-

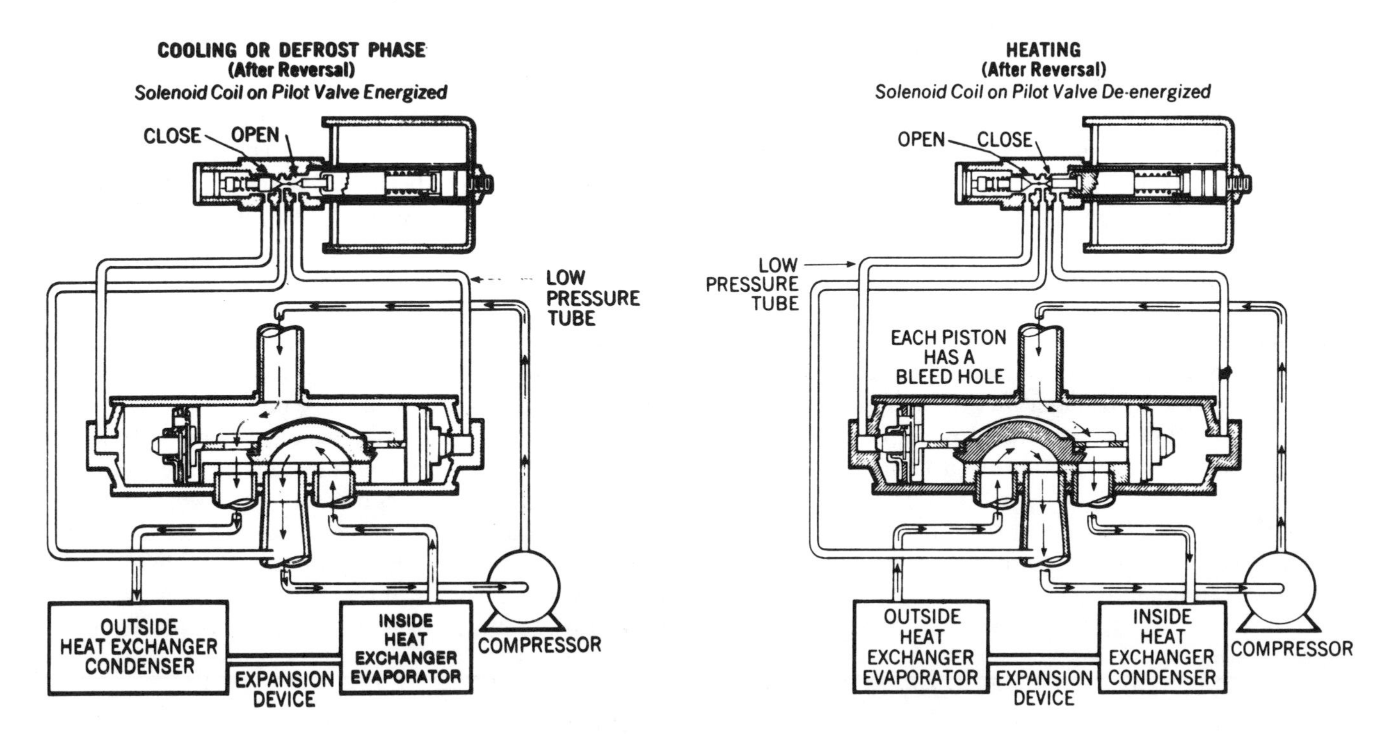

FIG. 7-17. Refrigerant flow through a heat pump reversing valve while (left) cooling or during defrost and (right) while heating. *Courtesy Ranco Controls Div.*

Table 7-7. Troubleshooting reversing valves

PROBLEM	PROBABLE CAUSE	POSSIBLE SOLUTION
Won't shift	Pilot solenoid not operating.	Check coil leads for cleanliness and tightness. Check for voltage at coil and coil continuity.
	Low refrigerant.	Check for leaks, add refrigerant and recheck cycling.
	Clogged pilot solenoid passages.	Try to energize by raising head pressure to increase pressure differential and clear tube passages. If unsucessful, remove valve and clean with R-11.
Hangs up	Low refrigerant.	Check for leaks, add refrigerant and recheck cycling.
	Body damage.	Replace valve.

sor discharge and suction line on the valve will maintain the same temperatures. Only the pipes going to the condenser and evaporator will change temperature between heating and cooling since these two lines *flip-flop* in function.

Pressure gages on the compressor service valves and temperature probes on the copper lines to and from the valve can verify the shifting of the valve.

Replacing any component with multiple lines certainly is not easy. Cleanliness is most important. If the lines are not to be unsoldered, but cut, a tubing cutter should be used, not a *hack saw*! Chips from a saw will almost always cause the new valve to fail. The slightest chip will either hangup the slide or block passages in the pilot solenoid.

Any compressor burnout, where the refrigeration system has been contaminated, will affect valve life and performance. All burnout residue must be removed prior to restarting of the system. A fine strainer in the compressor discharge line aids in keeping foreign material from the valve.

Table 7-7 shows a trouble-shooting chart for reversing valve problems.

Refrigerant safety

Every few years an article tells about someone being asphyxiated by refrigerant gases. If a few simple rules were followed, these deaths could have been avoided. Awareness of potential problems and of the hazards is critical.

A number of steps are necessary to remember when working with refrigerants such as R-12, R-22, R-502 and ammonia.

Room volume

ASHRAE Standard 15-1978, *Safety Code for Mechanical Refrigeration*, specifies the maximum amount of refrigerant that can be safely allowed per 1000 cubic feet of enclosed space:

R-12: 31 pounds per 1000 cubic feet
R-22: 22 pounds per 1000 cubic feet
R-502: 30 pounds per 1000 cubic feet

Most refrigerant mechanics know or can determine the amount of refrigerant in a system. With air conditioning systems, the volume of fresh, *makeup* air is adequate and a problem rarely arises. However, in enclosed, confined equipment rooms or manufactured cooler or freezer enclosures, volumes are small and asphyxiation is a definite possibility.

It is advisable to carry a leak detector and activate it during service. If the detector starts to sound, it is time to leave the enclosed space. Light-headedness is also a sign that oxygen is depleted and one should leave the area immediately.

Rupture in refrigerant systems

Another problem that arises is that of hydrostatic expansion of refrigerants being contained between two shutoff valves. If the refrigerant is isolated, or adjacent shutoff valves are closed, liquid refrigerant can be trapped. Ambient heat or opening the line or joint for valve repair or removal (removal of pressure) causes rapid expansion of the liquid. The expansion of the liquid exerts enormous forces on the tubing, piping, valves and joints. If the bursting pressure is exceeded, a rupture occurs. Under reduced pressure, the liquid flashes to a vapor, with a tremendous increase in volume, and severe personal and physical damage can result. A *pump-down* sequence should be initiated prior to any service work and the serviceman should have a complete knowledge of where the liquid is contained, at what pressures and temperatures, and which valves are closed.

Leak checking

A number of leak detectors are available, including electronic, halide torch and bubble solutions.

The electronic detector is extremely accurate and can pick up traces of refrigerant measuring only sevral ppm or ounces per year. If there is a large leak, or the leak has occurred in an enclosed space, the electronic detector is too sensitive to really pinpoint the leak.

A halide detector is less sensitive and sometimes more useful if the leak is in a confined space. The flame of a halide detector breaks down the halogen refrigerant and creates a poisonous gas (phosgene). Care must be taken not to inhale these fumes.

Leaks are indicated by traces of oil escaping with the refrigerant or an actual *hissing* sound may be heard, if other noises are minimal.

Bubble solutions are also good when an area of leak is suspected. However, soap solutions can contaminate the joint. The joint will have to be unsoldered, cleaned and reassembled.

Ammonia Systems

Personnel operating and servicing ammonia systems are usually better aware of and better prepared to repair leaks. Because of the strong fumes and usually large refrigerant volumes encountered, personnel have available portable fans, ammonia gas masks and portable air supplies

Common sense must be used when checking for leaks. Figure 7-18 is a good example of a manufacturer taking the extra precaution to notify and warn personnel of the hazards of refrigerant leaks.

Solder (filler metals)

Looking at Table 7-1, under the brazing filler metals column, the skull and crossbones symbol next to Stay-Silv 45 and Stay-Silv 35 indicate that

WARNING

DANGER EXISTS IF LARGE AMOUNTS OF REFRIGERANT LEAK INTO THE FISHHOLD. REFRIGERANT REPLACES AIR, AND COULD CAUSE SUFFOCATION. LARGE AMOUNTS CAN ACCUMULATE IF A LEAK IS ALLOWED TO CONTINUE AND MORE REFRIGERANT IS ADDED TO THE REFRIGERATING UNIT TO STAY IN OPERATION.

TO AVOID POSSIBLE INJURY:
1. DO NOT OPERATE REFRIGERATING EQUIPMENT KNOWN TO HAVE A LEAK. REPAIRS SHOULD BE MADE AS SOON AS POSSIBLE AFTER A LEAK IS DETECTED
2. LEAKS CAN BE DETECTED BY:
   A) LOW COMPRESSOR SUCTION AND DISCHARGE PRESSURES
   B) VAPOR BUBBLES IN THE LIQUID LINE SIGHT GLASS
   C) POOR COOLING
3) DO NOT ENTER THE FISHHOLD WHEN THE UNIT IS SUSPECTED OF HAVING A LEAK, EXCEPT WHEN WEARING ARTIFICIAL BREATHING APPARATUS

FIG. 7-18. Safety sticker. *Courtesy Thermo King Corp.*

these metals, because they contain cadmium, give off poisonous fumes when heated. Proper ventilation is required when using them and, if ventilation is not possible, the mechanic should use a respirator.

When unfamiliar with a product, read the manufacturer's label and specifications to check for any poisonous materials.

Safety and relief valves

Safety valves are wired and sealed to show that the valve has been set and rated per National Board Standards. When servicing equipment with safety valves, make sure that the seal is still in place and someone has not altered the valve setting.

Relief and safety valves must be reconditioned in authorized repair shops. The National Board has set up a *VR* program to assure that the repair facility has authorized personnel, trained to repair safety valves, and that they document these programs.

Replacement components must meet the original manufacturer's specifications and any remachining must not alter the pressure rating or the capacity of the valve. Overhauled valves must be tested, properly stamped, and sealed.

Nonrefrigerant valves

Nonrefrigerant valves can be found in hot and chilled water systems and cooling tower systems. One of the problems associated with these systems is corrosion.

Water Treatment

One way to minimize corrosion and keep valves in good working order is to treat the water. Biological and corrosive effects were discussed in Chapter 4.

Once a hot and chilled water system is treated, a monthly check should be sufficient to keep the chemicals at adequate levels. Makeup water is rarely added except in case of servicing and repairs.

Any treatment must protect against scale, pitting and corrosion. In many cases, several different metals are used and galvanic corrosion protection is also required. Treatment must:

- ☐ Adjust the hardness level of the water
- ☐ Maintain the pH within a safe working range
- ☐ Control dissolved oxygen
- ☐ Not attack valve or pump packing, gasket or seal materials.

Above all, the system should be designed to quickly eliminate entrained air to keep oxygen levels at a minimum.

Open systems, such as cooling towers, are by far

the worst to care for since, every time the water circulates, new forms of contamination are introduced from the air which is laden with molds, bacteria, dust, acids and oxygen.

Chemicals should be added to these systems to treat the hardness of the makeup water, algae, bacteria, mold and fungus. There are many suppliers of water treatment chemicals, test kits, and automatic injection equipment who can tailor the treatment to individual systems.

Valve servicing

Numerous problems can arise, over a period of *online* service, that will eventually require replacement or maintenance of nonrefrigerant valves, including replacement of packing, seats, discs, stems and liners.

Inexpensive valves are usually discarded and replaced with another of equal or better quality. It is more economical to rebuild expensive and larger-size valves.

Packings

Packings are contained in a chamber just below the packing gland. The walls of this chamber, as well as the surface of the stem, must have highly polished surfaces for the stem to turn or slide without galling. When removing the packing, care should be exercised to make sure that:

- ☐ The finish of the chamber and stem are not scratched, gouged or dented.
- ☐ Erosion or corrosion has not damaged these surfaces.
- ☐ The type and number of packing rings can be identified, and that the packing material is compatible with the flowing fluid. See the Appendix for material compatibility.

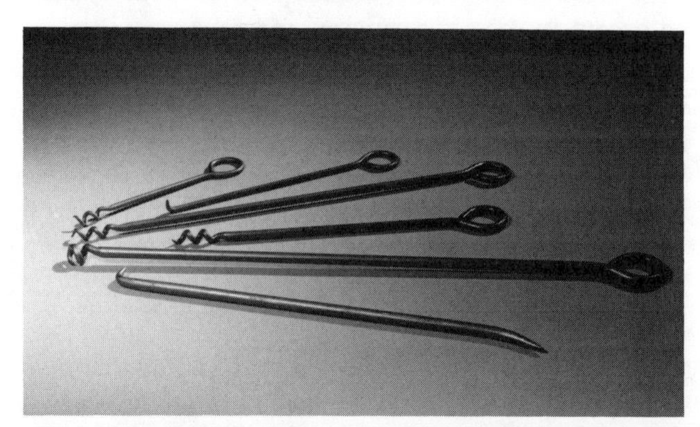

FIG. 7-19. Stiff packing extractor set. *Courtesy A.W. Chestern Co.*

Packing extractor tools are shown in Figure 7-19 and 7-20. If used improperly, they can damage internal surfaces so care and common sense must prevail.

New packing rings may be purchased presized or the rings can be cut and formed from bulk packing stock. Figure 7-21 shows a cutter used to make new packing rings.

When replacing the packing:

- ☐ Check the chamber walls for signs of grooves, pits or damage to the finish.
- ☐ Clean the parts thoroughly or, if necessary, replace.
- ☐ See that the ring is concentric within its chamber.
- ☐ Stagger the ring joints so that they do not align.
- ☐ Do not touch the parts with bare hands since salt from the skin can initiate corrosion

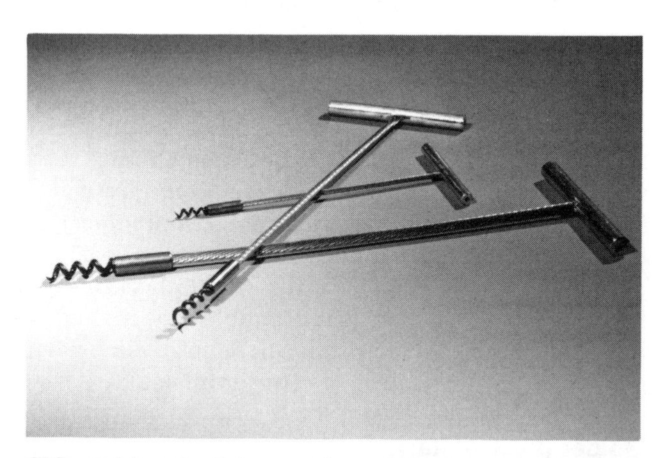

FIG. 7-20. Flexible packing extractor set. *Courtesy A.W. Chestern Co.*

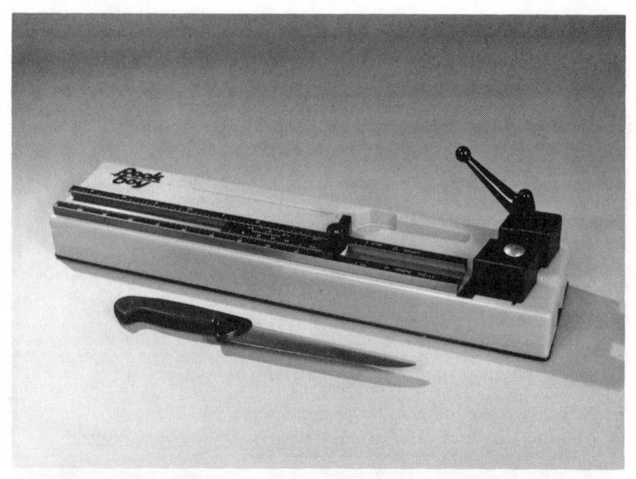

FIG. 7-21. Ring packing cutter. *Courtesy A.W. Chestern Co.*

- ☐ Evenly preload the packing by tightening the bolts on the gland in a balanced sequence.
- ☐ At startup, wait until all the fluid comes up to pressure and temperatures stabilize. Then, make slight and even adjustments to the packing gland to stop any leaks.

Discs and Seats

Globe and gate valves must have the mating disc and seat in good condition for tight sealing. Depending upon the quality, the valve may have a replaceable disc or both a replaceable disc and seat.

Figure 7-22 shows a valve with a replaceable disc and a seat that is integral with the valve body. Once the bonnet is unscrewed or unbolted, the upper portion of the valve and disc can be removed. Disc replacement is easy and, if the seat is in good condition, the valve can be quickly restored to service.

In some cases, the integral seat is reground, using grinding compound and the disc, Figure 7-23. The grinding compound is abrasive and the valve must

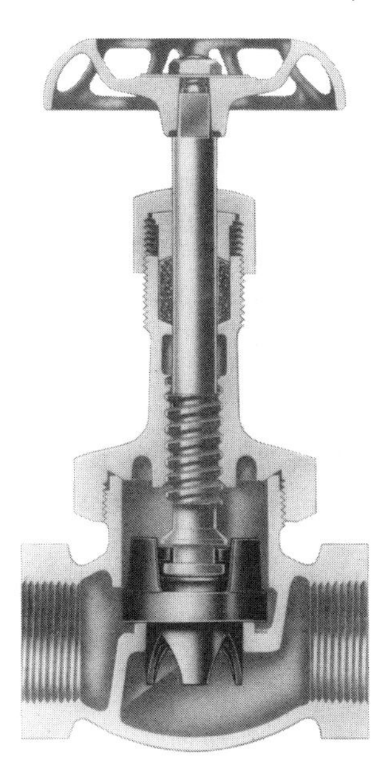

FIG. 7-22. Replaceable disc. Seat is integral with valve body. *Courtesy Jenkins Bros.*

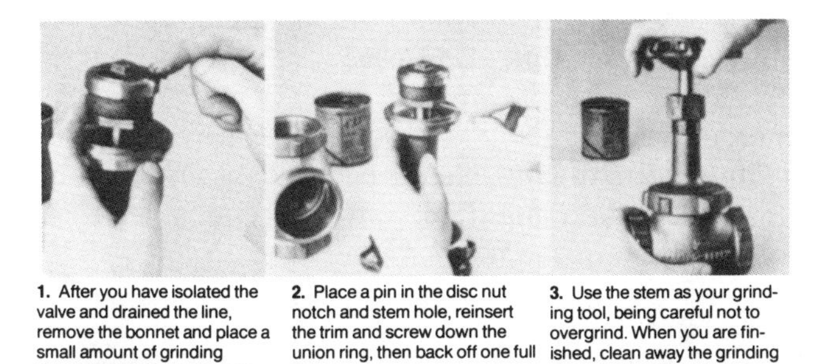

1. After you have isolated the valve and drained the line, remove the bonnet and place a small amount of grinding compound on the metal disc.
2. Place a pin in the disc nut notch and stem hole, reinsert the trim and screw down the union ring, then back off one full turn.
3. Use the stem as your grinding tool, being careful not to overgrind. When you are finished, clean away the grinding compound, remove the pin, lubricate the threads and reassemble the valve.

FIG. 7-23. Procedure for regrinding disc and seat in a globe valve. *Courtesy Jenkins Bros.*

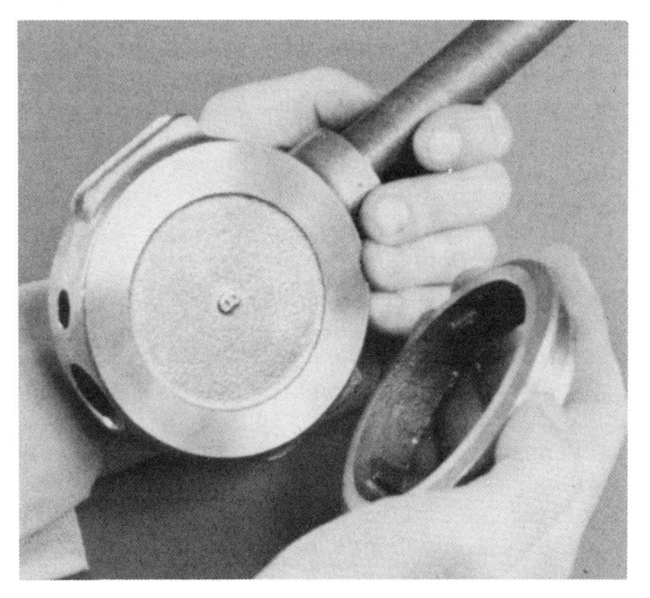

FIG. 7-24. Examining opened gate valve disc and seat. *Courtesy Jenkins Bros.*

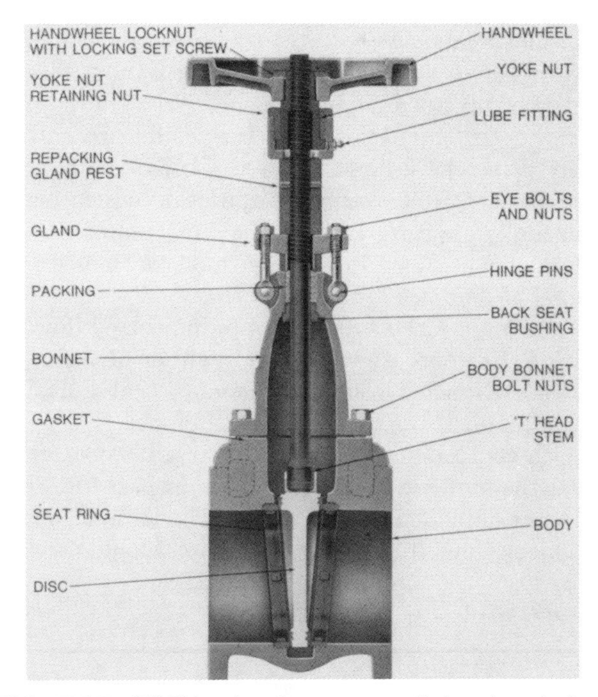

FIG. 7-25. OS&Y valve has grease fitting just below the hand wheel. *Courtesy Lunkenheimer, Div. of Conval Corp.*

be thoroughly cleaned so that any grinding residue is not washed downstream to contaminate the other valves.

Valves with a replaceable seat and disc can have both replaced at the same time. Figure 7-24 shows a disc and seat being examined prior to replacement.

Lubrication

Valves with exposed stems and screws should be oiled or greased on a regular basis. Valves with exposed screws, such as the OS&Y valve, have grease fittings, Figure 7-25. Regular maintenance should call for replenishing the grease in these fittings.

When the valve is reassembled or when initially installed, the bonnet threads and gland screws should receive a film of lube to ease disassembly at a future date. Never-Seez® is a good example of such a lube.

Cost of Leaks

The cost of leaks from neglected valves can be substantial. Table 7-8 shows the dollar cost of leaks in air, steam and water systems, based on the diameter of the leak. Not only are leaks costly, in terms of lost energy, but severe erosion of the valve parts will take place in short order, adding considerably to maintenance and repair costs.

Valve Installation

Most valves should be installed with their bonnet and stem in an upright position. With the bonnet upright, the cavities in the bonnet do not collect fluid. With the bonnet below the body, the bonnet acts as a trap for sediment and scale which could eventually migrate into the packing material and cause stem and packing damage along with degradation of the surface finishes.

Some globe valves have arrows indicating flow direction. However, flow may go in either direction, if desired. Knowledge of the particular valve design and manufacturer can make a difference.

With continuous flow or throttling, it is better to have the flow up under the disc. In case the disc becomes detached from the stem, the loose disc will not block flow. If flow were opposite, the plug would drop onto the seat and stop flow entirely.

Table 7-8. Cost of leaking air, steam, and water, based on 1976 energy prices.

SIZE OF LEAK	AIR		STEAM		WATER	
Diameter Inches	Number of cubic feet per month at 75 psi pressure	Total cost of waste per month at 18c per 1000 cubic feet	Pounds wasted per month at 160 psi pressure	Total cost of waste per month at $1.27 per 1000 lb.	Gallons wasted per month at 60 psi pressure	Total cost of waste per month at 31c per 1000 gallons
1/2″	13,468,000	$2,424.00	1,219,280	$1,548.49	1,524,100	$472.47
3/8″	7,558,500	1,360.53	684,290	869.05	855,360	265.16
1/4″	3,366,990	606.05	304,820	387.12	381,020	118.12
1/8″	824,570	148.42	74,650	94.81	93,310	28.93
1/16″	213,000	38.34	19,280	24.49	24,110	7.47
1/32″	52,910	9.52	4,790	6.08	5,990	1.86

Lunkenheimer, Div. of Conval Corp.

High temperature fluids under the disc, when closed, may cause the stem to cool and contract just enough to allow a continuous seepage of fluid and eventual erosion of disc and/or seat.

Experience has proved that flow over the top of the disc results in better service and longer life. Pressure above the disc usually produces tighter sealing since the fluid pressure augments the closing force bearing on the disc.

Valves should be located so that the bonnet and stem are not sticking out into traffic. Levers and hand wheels should be readily accessible and valve identification tags can assist personnel in making faster decisions in case a valve needs to be quickly opened or closed.

Summary

With a good installation, the valve is easily accessible, is identified, and is located in a safe area away from traffic.

Repairs should not be initiated without a thorough knowledge of the system fluids including their pressures, temperatures and corrosive and/or toxic effects. Special breathing or protective gear may be needed for safety.

In making repairs, cleanliness, the proper replacement parts and qualified, and if necessary certified, mechanics are the major requirements.

MATERIAL COMPATIBILITIES

Ratings are based on chemical resistance only. Extreme temperatures high or low, abrasives, air, pressure or mechanical stress or vibrations are not taken into account. Check with manufacturer upon application. Data is for general information only.

	ELASTOMERS					PLASTICS					METALS				
	Viton	Buna N	Neoprene	Ethylene Propylene	Silicon	Teflon	CPVC 210°F	Poly-Propylene	PVC 140°F	PVDF 280°F	Carbon Steel, 75°F	Ductile Iron, 75°F	Monel 75°F	Alum. BRZ, 75°F	316SS
Dry Air	EU	EU	EU	EU	EU	EU	EU	EU	EU	EU	EU	E	EU	EU	EU
Calcium Chloride Brine	EU	EU	EU	EU	EU	EU	EU	EU	EU	EU	E	E	EU	E	E
Diesel Fuels	EU	E	*	LU	*	EU	*	*	*	EU	EU	EU	EU	EU	EU
Dow Therm	EU	LU	LU	*	*	EU	*	*	*	*	EU	EU	EU	EU	EU
Ethylene Glycol (antifreeze)	EU	EU	EU	*	*	EU	EU	EU	EU	EU	EU	EU	EU	EU	EU
Freon 11	LU	LU	G	LU	*	EU	EU	*	G	*	EU	EU	EU	EU	EU
Freon 12	EU	LU	E	E	*	EU	*	G	LU	LU	EU	EU	EU	EU	EU
Freon 22	LU	LU	E	EU	*	EU	*	*	LU	*	EU	EU	EU	EU	EU
Freon 113	E	LU	G	LU	*	EU	*	*	EU	*	EU	EU	EU	EU	EU
Freon 114	EU	LU	G	*	*	EU	*	*	*	*	EU	EU	EU	EU	EU
Gasoline, Automotive	EU	G	LU	LU	*	EU	*	LU	G	LU	EU	EU	EU	EU	EU
Hydraulic Oil (petroleum)	LU	EU	*	LU	*	*	*	*	*	*	EU	EU	*	EU	EU
Propylene Glycol	EU	EU	*	*	*	EU	*	*	*	*	EU	EU	*	EU	EU
Sea Water	EU	EU	EU	EU	*	EU	EU	EU	EU	EU	G	G	EU	EU	EU
Sodium Chloride 5%	EU	EU	EU	*	*	EU	EU	EU	EU	EU	LU	LU	EU	G	E
Steam 225° F	LU	LU	*	E	*	LU	*	*	*	*	EU	EU	EU	G	EU
Steam 300° F	LU	LU	*	E	*	LU	*	*	*	*	EU	EU	EU	G	EU
Water, Fresh 180° F	EU	EU	EU	EU	EU	EU	EU	*	EU	EU	E	EU	EU	E	EU

EU: EXCELLENT, UNRESTRICTED E: EXCELLENT G: GOOD LU: LIMITED, UNSATISFACTORY. * NO DATA AVAILABLE.

GENERAL NEMA STANDARDS

If detailed and specific data is needed on these standards, they can be obtained from the following source:
National Electric Manufacturer Assoc. (NEMA)

NEMA 1

General purpose enclosure:
Sheet metal construction, general protection, indoor use, not dust-tight or splash proof.

NEMA 2

Drip tight enclosure:
Used in high condensation areas.

NEMA 3R

Weather resistant:
Used outdoors for protection against snow, rain, sleet or other wind blown wet environments.

NEMA 4

Watertight:
Usually stainless steel construction with gasketed hinged covers and must pass hose test with enclosure remaining dry internally. Nema 4X to meet same requirements but constructed of fiberglass.

NEMA 5

Dust tight:
Gasketed enclosure used in dust areas of non-hazardous dust materials. Not to be used for combustible dusts.

NEMA 6

Submersible:
Must operate while submerged under water under prescribed pressures.

NEMA 7

Hazardous Locations (Class 1, Group C and D):
Explosion proof designed to contain explosions of flammable mixtures of specified gases or vapors so that gas-air mixtures existing in the atmosphere surrounding the enclosure will not be ignited.

NEMA 8

Hazardous Locations (Class 1, oil immersed):
Oil circuit breakers.

NEMA 9

Hazardous Locations (Class 11, Group E, F, G):
Combustible dust locations. Enclosures designed to prevent ingress of explosive amounts of hazardous dust.

NEMA 10

Bureau of Mines (Explosion Proof):
(special considerations).

NEMA 11

Acid and Fume Resistant, oil immersed, used indoors.

NEMA 12

Industrial Use:
To protect against fibers, filings, lint, dust, dirt and light splashing, seepage, chippings and condensation of non-corrosive liquids. Oil tight and dust tight.

NEMA 13

Special Design:
Oil tight and dust tight for indoor use.

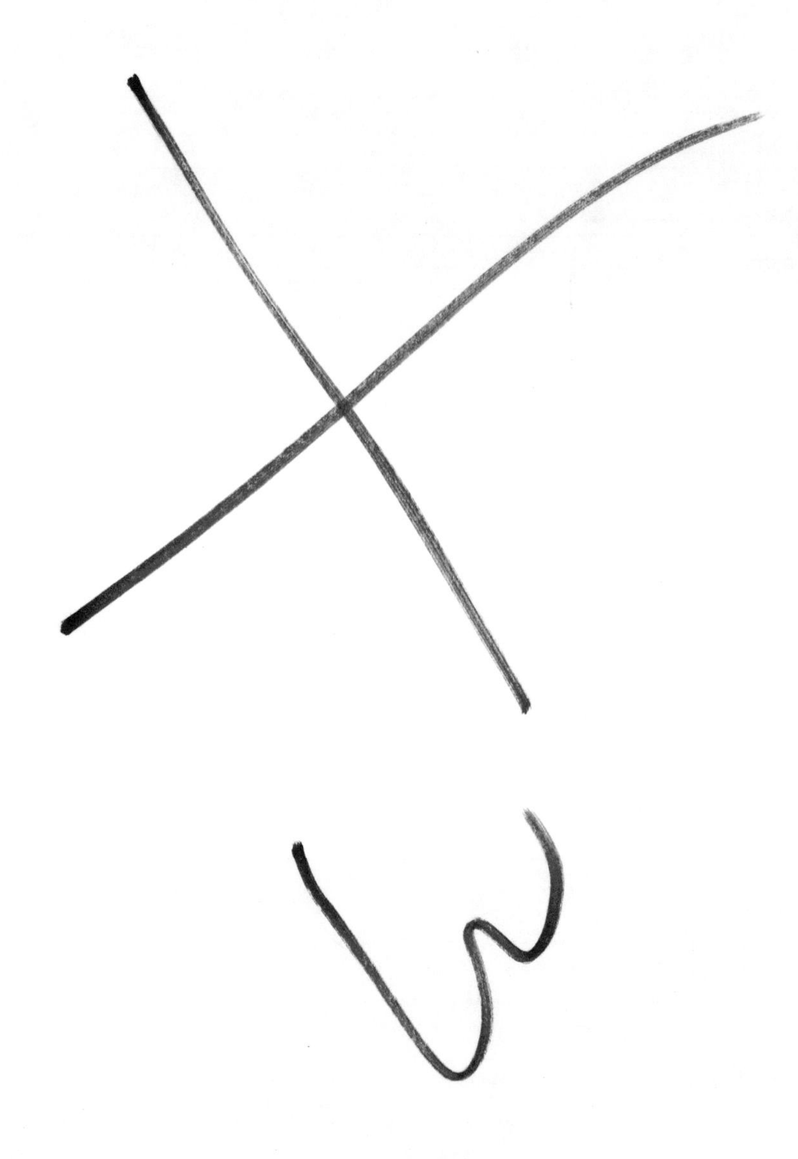